Recent Advances in Diffeologies and Their Applications

CONTEMPORARY MATHEMATICS

794

Recent Advances in Diffeologies and Their Applications

AMS-EMS-SMF Special Session
Recent Advances in Diffeologies and Their Applications
July 18–20, 2022
Université de Grenoble-Alpes, Grenoble, France

Jean-Pierre Magnot
Editor

AMERICAN MATHEMATICAL SOCIETY

2020 *Mathematics Subject Classification.* Primary 14D23, 14M25, 37K10, 46T05, 49Q10, 53D20, 57R30, 57R55, 58B10, 81R50.

Library of Congress Cataloging-in-Publication Data

Names: AMS-EMS-SMF Special Session on Recent Advances in Diffeologies and Their Applications (2022 : Université de Grenoble-Alpes, Grenoble, France), author | Magnot, Jean-Pierre, 1974– editor.

Title: Recent advances in diffeologies and their applications : AMS-EMS-SMF Special Session on Recent Advances in Diffeologies and Their Applications, July 18–20, 2022, Université de Grenoble-Alpes, Grenoble, France / Jean-Pierre Magnot, editor.

Description: Providence, Rhode Island : American Mathematical Society, [2024] | Series: Contemporary mathematics, 0271-4132 ; volume 794 | Includes bibliographical references.

Identifiers: LCCN 2023049999 | ISBN 9781470472542 (paperback) | 9781470476076 (ebook)

Subjects: LCSH: Differential topology–Congresses. | Differentiable manifolds–Congresses. | Geometry, Algebraic–Congresses. | Symplectic geometry–Congresses. | AMS: Algebraic geometry – Families, fibrations – Stacks and moduli problems. | Algebraic geometry – Special varieties – Toric varieties, Newton polyhedra. | Dynamical systems and ergodic theory – Infinite-dimensional Hamiltonian systems – Completely integrable systems, integrability tests, bi-Hamiltonian structures, hierarchies (KdV, KP, Toda, etc.). | Functional analysis – Nonlinear functional analysis – Infinite-dimensional manifolds. | Calculus of variations and optimal control; optimization – Manifolds – Optimization of shapes other than minimal surfaces. | Differential geometry – Symplectic geometry, contact geometry – Momentum maps; symplectic reduction. | Manifolds and cell complexes – Differential topology – Foliations; geometric theory. | Manifolds and cell complexes – Differential topology – Differentiable structures. | Global analysis, analysis on manifolds – Infinite-dimensional manifolds – Differentiability questions. | Quantum theory – Groups and algebras in quantum theory – Quantum groups and related algebraic methods.

Classification: LCC QA613.6 .R44 2024 | DDC 514/.72–dc23/eng/20231130

LC record available at https://lccn.loc.gov/2023049999

∞ The paper used in this book is acid-free and falls within the guidelines
established to ensure permanence and durability.
Visit the AMS home page at `https://www.ams.org/`

10 9 8 7 6 5 4 3 2 1 29 28 27 26 25 24

In memory of Jean-Marie Souriau

Contents

Preface

In the current state of knowledge in geometry, there are certain cases where the classical framework on differential manifolds does not enable one to work rigorously. It is then necessary to place oneself in a more or less generalized theoretical framework, adapted to the examples studied in order to rigorously prove natural results. One can also (and this happens even in mathematics) work with a formal geometry and invoke "natural properties" in order to overcome difficulties linked to the working formalism. In problems of these kinds, the category of diffeologies offers a coherent framework, easy to use and applicable to most (if not all) cases encountered. This is why we are currently seeing very recent works where diffeologies play a central role, not only in fields related to different geometries (algebraic, topological, differential) but in related fields such as deep learning or optimisation.

It is in this context of strong actual development of research related to diffeologies, both in terms of application and more theoretical aspects, that the special session "Recent advances on diffeologies and their applications" of the AMS-EMS-SMF congress took place in Grenoble, France, from 18 to 20 July 2022. Despite competition from many other leading conferences that were planned at the same time, and a world situation still marked by the pandemic and the sound of cannon fire not far from Western Europe, a rich and varied panel of presentations, presenting different aspects of the subject, stimulated exchanges and future collaborations between participants. Most of them gladly agreed to participate in this volume, to which I have been able to add the contributions of people who were unable to attend for various reasons.

I would therefore like to thank the conference organisers for giving Jordan Watts and me the opportunity to organise this session, and the editors of the Contemporary Mathematics series for this volume, which not only presents some of the most cutting-edge papers on diffeologies, but also an overview of some selected applications of diffeologies to a specific domain. Finally, I must say that I am very grateful to all the contributors for their enthusiastic participation in this volume.

Jean-Pierre Magnot
11 May 2023, Orcines, France

Contemporary Mathematics
Volume **794**, 2024
https://doi.org/10.1090/conm/794/15938

On diffeologies from infinite dimensional geometry to PDE constrained optimization

Nico Goldammer, Jean-Pierre Magnot, and Kathrin Welker

Abstract. We review how diffeologies complete the settings classically used from infinite dimensional geometry to partial differential equations, based on classical settings of functional analysis and with classical mapping spaces as key examples. As the classical examples of function spaces, we deal with manifolds of mappings in Sobolev classes (and describe the ILB setting), jet spaces and spaces of triangulations, that are key frameworks for the two fields of applications of diffeologies that we choose to highlight: evolution equations and integrable systems, and optimization problems constrained by partial differential equations.

Contents

Introduction

Diffeological spaces and Frölicher spaces are two frameworks, developed independently in the 20th century, in order to define smoothness on objects where basic differential properties cannot be stated in a standard way. Motivated by technical uncapacities to formulate properly applied problems, many such settings developed independently, that are generalizations of the classical finite dimensional differential geometry, exist in the literature. A non-exhaustive list of such settings can be found in e.g. [**87**]. One global underlying problem is the following:

How to express variational problems in a setting where differentiation is not easy?

One has immediately in mind spaces with singularities as elementary examples, such as the irrational torus or orbifolds where a rich differential geometry can be described by the use of diffeologies [**34**], but initialy the motivating (historical)

2020 *Mathematics Subject Classification.* Primary 46T05, 58J60, 58Z05.
Key words and phrases. Diffeology, Frölicher space, infinite dimensional geometry, partial differential equations, shape analysis.

examples for the development of diffeologies are groups of diffeomorphisms of non-compact manifolds and coadjoint orbits, in the works of French mathematician Jean-Marie Souriau (integrally written in French), reviewed and deeply expanded in [**33**]. The same way, the motivations of the Swiss mathematician Alfred Frölicher were certainly partly based on infinite dimensional considerations, till his book with Andreas Kriegl [**21**]. Therefore, motivated by historical considerations and by our own concerns, while most recent works in these two settings are developed in a background of algebraic or symplectic geometry (understood in a wide sense), we feel the need of an actualized presentation of diffeological spaces, and of Frölicher spaces as a subcategory, based on infinite dimensional examples arising in functional analysis, infinite dimensional geometry and numerical analysis. Indeed, since about ten years, under the impulsion of a new generation of motivated researchers, diffeologies have drastically increased their known technical capacities. Examples and counter-examples have successfully highlighted in specific constructions that point out singular problems, drawing attention by non-specialists to diffeologies as a potential framework for some of their technical difficulties. This is to this class of potential readers that our text is primarily addressed. We are ourselves primarily interested in two very different backgrounds, both linked with functional analysis, and the use of diffeologies have imposed its necessity to us. Indeed, after break-through in the theory of geometry and integrable systems [**51, 57**], diffeological spaces are also recently considered in applied mathematics (see, e.g., [**92**]). From the viewpoint of integrable systems, diffeologies enable to state smoothness, and hence well-posedness, on infinite dimensional (mostly algebraic) frameworks. From the viewpoint of shape optimization problems, diffeologies propose a more flexible framework than the one classically considered in the last decades, based on classical Riemannian infinite dimensional geometry.

One aim of this paper is to make a clear and understandable presentation of basic results on diffeologies and Frölicher spaces, based on the examples that we know better, but in a simple language, accessible to other researchers in the field of functional analysis and applied mathematics understood in a very wide sense. A further aim of this paper is to make a clear exposition of some basic concepts where diffeologies are interrelated with topics of functional analysis, infinite dimensional geometry and optimization for researchers whose basic knowledge remain e.g. in low dimensional topology or category theory, and who ignore such infinite dimensional settings.

The structure of the paper is as follows:

- A brief introduction in diffeological and Frölicher spaces is given in section 1. In particular, we present and summarize the various tangent space definitions of diffeological spaces, which appear in the literature. Other geometric objects generalized to diffeologies are also considered such as diffeological (Lie) groups, vector or principal (pseudo-) bundles, Riemannian metrics, groups of diffeomorphisms and automorphisms of various diffeological algebraic or set-theoric structures.
- Section 2 is concerned by (partial) differential equations in a two-fold way. We take profit of this exposition, in the two sections, to highlight or review on some tangent and cotangent structures that are well-known on (finite dimensional) manifolds and recovered in diffeological spaces with specific technical details. In section 2.1, we consider the problem of the

existence of an exponential map in a diffeological Lie group, solving a first order differential equation, and as a by-product holonomy-type equations by stating an infinite diemensional Ambrose-Singer theorem. In section 2.2 we present first order optimization techniques on diffeological spaces. Here, we concentrate on shape optimization, which has a wide range of application.
- The paper ends with section 3 where basic functional spaces, that is mapping spaces, are reviewed from the viewpoint of infinite dimensional geometry. Depending on the regularity required for mappings and on the source manifolds, one either observes Hilbert, Banach or ILB manifolds along the lines of [**72**], or topological spaces which can be unserstood as diffeological spaces [**52**]. Implicit functions, jet spaces, and spaces of triangulations are related examples where diffeologies apply in settings related to mapping spaces.

1. A presentation of diffeological and Frölicher spaces, with examples

1.1. Diffeological and Frölicher spaces. In this section we follow and complete the expositions appearing in [**51, 57, 58**]. The main reference for a comprehensive exposition on diffeologies is [**33**]. This reference will be completed all along the text for specific concerns.

DEFINITION 1.1. Let X be a set.

- A *p-parametrization* of dimension p on X is a map from an open subset O of $\mathbb{R}^p$ to X.
- A *diffeology* on X is a set $\mathcal{P}$ of parametrizations on X such that:
  - For each $p \in \mathbb{N}$, any constant map $\mathbb{R}^p \to X$ is in $\mathcal{P}$;
  - For each arbitrary set of indexes I and family $\{f_i : O_i \to X\}_{i\in I}$ of compatible maps that extend to a map $f : \bigcup_{i\in I} O_i \to X$, if $\{f_i : O_i \to X\}_{i\in I} \subset \mathcal{P}$, then $f \in \mathcal{P}$.
  - For each $f \in \mathcal{P}$, $f : O \subset \mathbb{R}^p \to X$, and $g : O' \subset \mathbb{R}^q \to O$, in which g is a smooth map (in the usual sense) from an open set $O' \subset \mathbb{R}^q$ to O, we have $f \circ g \in \mathcal{P}$.

If $\mathcal{P}$ is a diffeology on X, then $(X, \mathcal{P})$ is called a *diffeological space* and, if $(X, \mathcal{P})$ and $(X', \mathcal{P}')$ are two diffeological spaces, a map $f : X \to X'$ is *smooth* if and only if $f \circ \mathcal{P} \subset \mathcal{P}'$.

The notion of a diffeological space is due to J.M. Souriau, see [**86**], who was inspired by the remarks and constructions from Chen in [**9**]. A comprehensive exposition of basic concepts can be found in [**33**].

DEFINITION 1.2. Let $(X, \mathcal{P})$ be a diffeological space. Let $\mathcal{P}'$ be another diffeology on X. Then $\mathcal{P}'$ is a *subdiffeology* of $\mathcal{P}$ if and only if $\mathcal{P}' \subset \mathcal{P}$ or, in other words, if the identity mapping $(X, \mathcal{P}') \to (X, \mathcal{P})$ is smooth.

EXAMPLE 1.3. There exists on any nonempty set X a diffeology, minimal for inclusion, made only of constant parametrizations. It is called the *discrete diffeology.* This is a subdiffeology of any diffeology on X.

EXAMPLE 1.4. Let X be a set. Given a non-empty family A of functions

$$f : O_f \to X$$

where O_f is the domain of the function f, which is an open subset of an Euclidian space, a direct application of Zorn's lemma proves that there exists a diffeology $\mathcal{P}(A)$, minimal for inclusion, that contains A, in other words, for which A is a family of smooth maps. This is the *diffeology generated by A*.

The category of diffeological spaces is very large, and carries many different pathological examples even if it enables a very easy-to-use framework on infinite dimensional objects. therefore, a restricted category can be useful, the category of Frölicher spaces. This is for example the framework chosen in [**8**].

DEFINITION 1.5. A *Frölicher* space is a triple $(X, \mathcal{F}, \mathcal{C})$ such that
- $\mathcal{C}$ is a set of paths $\mathbb{R} \to X$,
- $\mathcal{F}$ is the set of functions from X to $\mathbb{R}$, such that a function $f : X \to \mathbb{R}$ is in $\mathcal{F}$ if and only if for any $c \in \mathcal{C}$, $f \circ c \in C^\infty(\mathbb{R}, \mathbb{R})$;
- A path $c : \mathbb{R} \to X$ is in $\mathcal{C}$ (i.e. is a *contour*) if and only if for any $f \in \mathcal{F}$, $f \circ c \in C^\infty(\mathbb{R}, \mathbb{R})$.

If $(X, \mathcal{F}, \mathcal{C})$ and $(X', \mathcal{F}', \mathcal{C}')$ are two Frölicher spaces, a map $f : X \to X'$ is *smooth* if and only if $\mathcal{F}' \circ f \circ \mathcal{C} \subset C^\infty(\mathbb{R}, \mathbb{R})$.

This definition first appeared in [**20**], see e.g. [**21**], but the actual terminology was fixed to our knowledge in Andreas Kriegl and Peter Michor's book [**42**], and independently by Paul Cherenack in [**10, 11**].

The comparison of these two frameworks does need to be exposed for a complete review. Its first steps were published in [**50**]; the reader can also see [**51, 54, 57, 91**] for extended expositions. In particular, it is explained in [**57**] that *Diffeological, Frölicher and Gateaux smoothness are the same notion if we restrict ourselves to a Fréchet context,* in a way that we explain here.

For this, we first need to analyze how we generate a Frölicher or a diffeological space, that is, how we implement a Frölicher or a diffeological structure on a given set X.

Any family of maps $\mathcal{F}_g$ from X to $\mathbb{R}$ generates a Frölicher structure $(X, \mathcal{F}, \mathcal{C})$ by setting, after [**42**]:
- $\mathcal{C} = \{c : \mathbb{R} \to X \text{ such that } \mathcal{F}_g \circ c \subset C^\infty(\mathbb{R}, \mathbb{R})\}$
- $\mathcal{F} = \{f : X \to \mathbb{R} \text{ such that } f \circ \mathcal{C} \subset C^\infty(\mathbb{R}, \mathbb{R})\}$.

We call $\mathcal{F}_g$ a *generating set of functions* for the Frölicher structure $(X, \mathcal{F}, \mathcal{C})$. One easily see that $\mathcal{F}_g \subset \mathcal{F}$. A Frölicher space $(X, \mathcal{F}, \mathcal{C})$ carries a natural topology, the pull-back topology of $\mathbb{R}$ via $\mathcal{F}$.

The same way, one can start alternatively from a generating set of contours $\mathcal{C}_g$, and set:
- $\mathcal{F} = \{f : X \to \mathbb{R} \text{ such that } f \circ \mathcal{C}_g \subset C^\infty(\mathbb{R}, \mathbb{R})\}$
- $\mathcal{C} = \{c : \mathbb{R} \to X \text{ such that } \mathcal{F} \circ c \subset C^\infty(\mathbb{R}, \mathbb{R})\}$.

In the case of a finite dimensional differentiable manifold X we can take $\mathcal{F}$ as the set of all smooth maps from X to $\mathbb{R}$, and $\mathcal{C}$ the set of all smooth paths from $\mathbb{R}$ to X. Then, the underlying topology of the Frölicher structure is the same as the manifold topology [**42**].

We also remark that if $(X, \mathcal{F}, \mathcal{C})$ is a Frölicher space, we can define a natural diffeology on X by using the following family of maps f defined on open domains $D(f)$ of Euclidean spaces, see [**50**]:

$$\mathcal{P}_\infty(\mathcal{F}) = \coprod_{p \in \mathbb{N}^*} \{ f : D(f) \to X \mid D(f) \text{ is open in } \mathbb{R}^p \text{ and } \mathcal{F} \circ f \in C^\infty(D(f), \mathbb{R})\}.$$

If X is a finite-dimensional differentiable manifold, setting $\mathcal{F} = C^\infty(X, \mathbb{R})$, this diffeology is called the *nebulae diffeology*, see e.g. [**33**]. Now, we can easily show the following:

PROPOSITION 1.6. [**50**] *Let*$(X, \mathcal{F}, \mathcal{C})$ *and* $(X', \mathcal{F}', \mathcal{C}')$ *be two Frölicher spaces. A map* $f : X \to X'$ *is smooth in the sense of Frölicher if and only if it is smooth for the underlying diffeologies* $\mathcal{P}_\infty(\mathcal{F})$ *and* $\mathcal{P}_\infty(\mathcal{F}')$.

Thus, Proposition 1.6 and the foregoing remarks imply that the following implications hold:

smooth manifold $\Rightarrow$ Frölicher space $\Rightarrow$ diffeological space

These implications can be analyzed in a refined way. The reader is referred to the Ph.D. thesis [**91**] for a deeper analysis of them.

REMARK 1.7. The set of contours $\mathcal{C}$ of the Frölicher space $(X, \mathcal{F}, \mathcal{C})$ does not give us a diffeology, because a diffeology needs to be stable under restriction of domains. In the case of paths in $\mathcal{C}$ the domain is always $\mathbb{R}$ whereas the domain of 1-plots can (and has to) be any interval of $\mathbb{R}$. However, $\mathcal{C}$ defines a "minimal diffeology" $\mathcal{P}_1(\mathcal{F})$ whose plots are smooth parametrizations which are locally of the type $c \circ g$, in which $g \in \mathcal{P}_\infty(\mathbb{R})$ and $c \in \mathcal{C}$. Within this setting, we can replace $\mathcal{P}_\infty$ by $\mathcal{P}_1$ in Proposition 1.6. The main technical tool needed to discuss this issue is Boman's theorem [**42**, p.26]. Related discussions are in [**50, 91**].

After this remark, one can push further the "restriction" procedure, for any diffeology $\mathcal{P}$ on X.

THEOREM 1.8. *Let* $\mathcal{P}$ *be a diffeology on* X. *Let* $\mathcal{F}(\mathcal{P})$ *be the set of smooth functions from* $(X, \mathcal{P})$ *to* $(\mathbb{R}, \mathcal{P}_\infty(\mathbb{R}))$ *where* $\mathbb{R}$ *is understood as a smooth manifold. Then:*

- *there exists a nebulae diffeology* $\mathcal{P}_\infty$ *generated by* $\mathcal{F}(\mathcal{P})$,
- $\mathcal{P} \subset \mathcal{P}_\infty$
- $\forall k \in \mathbb{N}^*$, *there exists a* $n-$ dimensional diffeology $\mathcal{P}_k \subset \mathcal{P}$ *made of plots* $p \in \mathcal{P}$ *which factors tthrough*
$$D(\mathcal{P}) \to \mathbb{R}^k \to X$$
- $\forall k \in \mathbb{N}$, *the set of smooth maps from* $(X, \mathcal{P}_k)$ *to* $(\mathbb{R}, \mathcal{P}_\infty(R))$ *coincides with* $\mathcal{F}(\mathcal{P})$.

This construction for $\mathcal{P} = \mathcal{P}_\infty$ can be found in [**51**]. It extends trivially to any diffeology $\mathcal{P}$. By this theorem, one can easily see that the categories of Frölicher spaces and diffeological spaces do not coincide. We now have to precise:

DEFINITION 1.9. [**91**] If $\mathcal{P} = \mathcal{P}_\infty$, the diffeology $\mathcal{P}$ is called reflexive (and it is the nebulae diffeology of a Frölicher space).

REMARK 1.10. Given a diffeological space $(X, \mathcal{P})$ there is a natural topology on X called $D-$topology in [**13**] which is the final topology for the set of maps $p \in \mathcal{P}$. This topology can be rather complicated and actually not-so-studied. To our best knowledge, this problem of topology associated to a diffeology is better studied actually in the context of Frölicher spaces, see e.g. [**21, 42**] for a serie of technical results.

Let us now give a class of examples of diffeologies.

EXAMPLE 1.11. Let $k \in \mathbb{N}$. Let $(X, \mathcal{F}, \mathcal{C})$ be a Frölicher space. The C^k–diffeology of $(X, \mathcal{F}, \mathcal{C})$ is the diffeology $\mathcal{P}_{(k)}$ defined by:

$$\forall d \in \mathbb{N}^*, \ \forall O \text{ open subset of } \mathbb{R}^d \quad \mathcal{P}_{(k)_O} = \{p : O \to X \,|\, \mathcal{F} \circ p \subset C^k(O, \mathbb{R})\}$$

and

$$\mathcal{P}_{(k)} = \bigcup_{d \in \mathbb{N}^*} \bigcup_{O \text{ open in } \mathbb{R}^d} \mathcal{P}_{(k)_O}.$$

This example is given in [**33**] for a finite dimensional manifold. The extension to Frölicher structures is straightforward.

1.2. Diffeologies on spaces of mappings, products, quotients, subsets and algebraic structures.

PROPOSITION 1.12. [**33**, **86**] *Let $(X, \mathcal{P})$ and $(X', \mathcal{P}')$ be two diffeological spaces. There exists a diffeology $\mathcal{P} \times \mathcal{P}'$ on $X \times X'$ made of plots $g : O \to X \times X'$ that decompose as $g = f \times f'$, where $f : O \to X \in \mathcal{P}$ and $f' : O \to X' \in \mathcal{P}'$. We call it the* product diffeology, *and this construction extends to an infinite (maybe not countable) product.*

We apply this result to the case of Frölicher spaces and we derive (compare with [**42**]) the following:

PROPOSITION 1.13. *Let $(X, \mathcal{F}, \mathcal{C})$ and $(X', \mathcal{F}', \mathcal{C}')$ be two Frölicher spaces equipped with their natural diffeologies $\mathcal{P}$ and $\mathcal{P}'$. There is a natural structure of Frölicher space on $X \times X'$ which contours $\mathcal{C} \times \mathcal{C}'$ are the 1-plots of $\mathcal{P} \times \mathcal{P}'$.*

We can also state the above result for infinite products; we simply take Cartesian products of the plots, or of the contours.

Now we consider quotients after [**86**] and [**33**, p. 27]: Let $(X, \mathcal{P})$ be a diffeological space, and let X' be a set. Let $f : X \to X'$ be a map. We define the *push-forward diffeology*, noted by $f_*(\mathcal{P})$, as the coarsest (i.e. minimal for inclusion) among the diffologies on X', which contains $f \circ \mathcal{P}$, that is, for which f is smooth. This construction is a specification of example 1.4, where the generating family A is set as $A = f \circ \mathcal{P}$. A smooth map $f : (X, \mathcal{P}) \to (X', \mathcal{P}')$ between two diffeoogical spaces is called a *subduction* if it is surjective anf if $f_*(\mathcal{P}) = \mathcal{P}'$. Conversly, if $(X', \mathcal{P}')$ is a diffeological space and if X is a set, if $f : X \to X'$, the *pull-back diffeology* noted by $f^*(\mathcal{P}')$ is the diffeology on X, maximal for inclusion, for which f is smooth.

PROPOSITION 1.14. [**33**, **86**] *Let $(X, \mathcal{P})$ b a diffeological space and $\mathcal{R}$ an equivalence relation on X. Then, there is a natural diffeology on $X/\mathcal{R}$, noted by $\mathcal{P}/\mathcal{R}$, defined as the push-forward diffeology on $X/\mathcal{R}$ by the quotient projection $X \to X/\mathcal{R}$.*

Given a subset $X_0 \subset X$, where X is a Frölicher space or a diffeological space, we equip X_0 with structures induced by X as follows:

(1) If X is equipped with a diffeology $\mathcal{P}$, we define a diffeology $\mathcal{P}_0$ on X_0 called the *subset or trace diffeology*, see [**33**, **86**], by setting

$$\mathcal{P}_0 = \{p \in \mathcal{P} \text{ such that the image of } p \text{ is a subset of } X_0\}.$$

(2) If $(X, \mathcal{F}, \mathcal{C})$ is a Frölicher space, we take as a generating set of maps $\mathcal{F}_g$ on X_0 the restrictions of the maps $f \in \mathcal{F}$. In this case, the contours (resp. the induced diffeology) on X_0 are the contours (resp. the plots) on X whose images are a subset of X_0.

EXAMPLE 1.15. Let us consider the infinite product diffeology on Y^X. This is the largest diffeology for which the evaluation maps ev_x are smooth for each $x \in X$. Therefore, any other diffeology on any subset of Y^X for which evaluation maps are smooth is a subdiffeology of the subset diffeology inherited from the this product diffeology.

EXAMPLE 1.16. Let M be a smooth (Riemannian) manifold. Then, the set $M^{\mathbb{N}}$ of sequences on M is a topological space with open sets of the form

$$O_0 \times O_1 \times \cdots \times O_k \times \dots$$

where the family $\{O_k, k \in \mathbb{N}\}$ is a family of open subsets of M such that $O_k = M$ except for a finite set of indexes k. With this topology, it seems difficult to define an atlas on $M^{\mathbb{N}}$ while an intuitive "natural" differentiation holds by considering differentiation on each component of the infinite product. By considering a diffeology $\mathcal{P}$ on M such that $\mathcal{F}(\mathcal{P}) = C^\infty(M, \mathbb{R})$ in the usual sense, the diffeology $\mathcal{P}$ also defines a diffeology on $M^{\mathbb{N}}$ by (countable) infinite dimensional product diffeology along the lines of Example 1.15 that encodes this "natural differentiation". An example of such a diffeology is obtained by setting $\mathcal{P} = \mathcal{P}_\infty(M)$. Following Proposition 1.13, it is an easy exercise to check that the n-dimensional diffeology $\mathcal{P}_n$ related to $\mathcal{P}$ generates also a n-dimensional diffeology on the infinite product.

Moreover, one can also consider subsets of $M^{\mathbb{N}}$ which inherit a subset diffeology, such as the so called *marked (infinite) configuration spaces* (see e.g. [**2**]) that we note here by

$$O\Gamma_M = \{(x_n) \in M^{\mathbb{N}} \,|\, \forall (m,n) \in \mathbb{N}^2, m \neq n \Leftrightarrow x_n \neq x_m$$
$$\text{and for all bounded set } B \subset M, |\{x_n \,|\, n \in \mathbb{N}\} \cap B| < +\infty\}.$$

It is an easy exercise to prove that he differentiable structure on $O\Gamma_M$, defined by another generalized notion of differentiability called Sikorski differentiability (see e.g. [**87**] for an overview), fits here with our diffeology, once we have remarked that each point $x_n \in M$ can be identified with its Dirac measure δ_{x_n} acting on $C^\infty(M, \mathbb{R})$.

Associated with marked infinite configuration spaces, the (unmarked) *infinite configuration space* Γ_M was studied by the same research group, see e.g. [**1**], can be understood as the space of marked configurations up to reindexation. More precisely, the infinite symmetric group $\mathfrak{S}_\infty$ made of bijections of $\mathbb{N}$ is acting on the indexes of the sequences of $O\Gamma_M$, and

$$\Gamma_M = O\Gamma_M / \mathfrak{S}_\infty.$$

Equipped with its quotient diffeology, this space inherits the same structure for generalized differentiability as the one described in the original works. Other similar examples can be found in [**35**].

EXAMPLE 1.17. [**55**] Let Y be a smooth complete manifold (not necessarily finite dimensional) and let $(X, \mathcal{P})$ be a diffeological space such that $X \subset Y$ as a set, and $\mathcal{P} \subset \mathcal{P}_\infty(Y)$ (in other words, the canonical inclusion map is smooth for the nebulae diffeology on Y). Then we define $\mathcal{C}(X, Y)$ as the set of sequences in X that converge in Y, in other words, the subset of $X^{\mathbb{N}}$ made of sequences that are Cauchy for the topology of Y. Then,

- the limit

$$\lim : \mathcal{C}(X, Y) \to Y$$

is not smooth when $\mathcal{C}(X,Y)$ is equipped with the subset topology inherited from the infinite product diffeology of $X^{\mathbb{N}}$, and when Y is equipped with $\mathcal{P}_\infty(Y)$, as soon as there exists a non constant smooth path in X.
- $\lim^*(\mathcal{P}_\infty(Y))$ may make the canonical projection (evaluation) maps
$$\mathcal{C}(X,Y) \to X$$
nonsmooth, consider for example $Y = \mathbb{R}$ and $X = \mathbb{Q}$ equipped with its subset diffeology.

Therefore, there is a true necessity to define the *Cauchy diffeology* on $\mathcal{C}(X,Y)$ for which the limit map and the evaluation maps are smooth in $\mathcal{C}(X,Y)$.

Our last general construction is the so-called functional diffeology. Its existence implies the following crucial fact: the category of diffeological spaces is Cartesian closed. Our discussion follows [**33**]. Let $(X,\mathcal{P})$ and $(X',\mathcal{P}')$ be diffeological spaces. Let $M \subset C^\infty(X,X')$ be a set of smooth maps. The *functional diffeology* on S is the diffeology $\mathcal{P}_S$ made of plots
$$\rho : D(\rho) \subset \mathbb{R}^k \to S$$
such that, for each $p \in \mathcal{P}$, the maps $\Phi_{\rho,p} : (x,y) \in D(p) \times D(\rho) \mapsto \rho(y)(x) \in X'$ are plots of $\mathcal{P}'$. We have, see [**33**, Paragraph 1.60]:

PROPOSITION 1.18. *Let X,Y,Z be diffeological spaces. Then,*
$$C^\infty(X \times Y, Z) = C^\infty(X, C^\infty(Y,Z)) = C^\infty(Y, C^\infty(X,Z))$$
as diffeological spaces equipped with functional diffeologies.

The next property will be useful for us:

PROPOSITION 1.19. *The composition of maps is smooth for functional diffeologies.*

Now, given an algebraic structure, we can define a corresponding compatible diffeological (resp. Frölicher) structure, see for instance [**48**]. For example, see [**33**, pp. 66-68], if $\mathbb{R}$ is equipped with its canonical diffeology (resp. Frölicher structure), we say that an $\mathbb{R}-$vector space equipped with a diffeology (resp. Frölicher structure) is a diffeological (resp. Frölicher) vector space if addition and scalar multiplication are smooth. We state:

DEFINITION 1.20. Let G be a group equipped with a diffeology (Frölicher structure). We call it a *diffeological (Frölicher) group* if both multiplication and inversion are smooth.

The same way, one can define diffeological fields, rings, algebras, groupoids, actions and so on.

1.3. The principle of pull-back on parametrizations. We explain here, with non-categorical vocabulary, some of the constructions that are necessary for the difinition of a *colimit,* which is often used to define geometric objects on a diffeology. We perform this contruction under three examples: the internal or kinematic tangent space, the Riemannian metric and the Haussdorf volume.

One has to remark that the plots of a diffeology are behaved like charts on a manifold through the property 1.1 in Definition 1.1. The notion of parametrization splits with the notion of chart at the point where a chart defines an open neighborhood of each point of its image set, which is not the case for a plot of a given diffeology. Moreover, for a given diffeological space $(X, \mathcal{P})$, the diffeology carries a notion of a given plot dimension in two ways:

- by the dimension of the domain of the plot p,
- and by the filtration of diffeologies defined in Theorem 1.8:

$$\mathcal{P}_1 \subset \mathcal{P}_2 \subset \cdots \subset \mathcal{P}.$$

Indeed, rephrasing the definition of $\mathcal{P}_n$, a plot $p \in \mathcal{P}$ lies in $\mathcal{P}_n$ if there exists $g \in \mathcal{P}_\infty(\mathbb{R}^n)$ and h a smooth map from $(\mathbb{R}^n, \mathcal{P}_\infty(\mathbb{R}^n))$ to $(X, \mathcal{P})$ such that $p = h \circ g$.

Now, let us consider a smooth manifold M equipped with its nebulae diffeology. Some geometric constructions (differential forms, Riemannian metrics, etc.) can be pulled back from M to the domain of any plot p. Therefore, there is a *pull-back principle* that we can apply to define the same contructions on the domain of any plot of a diffeology $\mathcal{P}$ provided that they are stable under pull-back via composition of smooth maps. Let us develop in few first examples how basic constructions on finite dimnsional manifolds extend to diffeological spaces through this principle.

1.3.1. *The kinematic (or internal) tangent cone.* We describe here the construction first established in [**48**] for diffeological groups, in [**18**] for Frölicher spaces, and extended in [**12**] for diffeologies, in a categorical vocabulary. Our aim is to express the same construction without the assumption that the reader is familiar with the categorical framework.

Let $(X, \mathcal{P})$ be a diffeological space. The domain $D(p)$ of each plot p of $\mathcal{P}$ can be considered as a smooth manifold, and first objects of interest are tangent vectors in the tangent space $TD(p)$. They are understood as germs of smooth path $\frac{d\gamma}{dt}|_{t=0}$ where $\gamma \in C^\infty(\mathbb{R}, D(p))$. Let $x \in X$ and let us consider

$$\mathcal{C}_x = \{\gamma \in C^\infty(\mathbb{R}, X) \,|\, \gamma(0) = x\}.$$

For each $p \in \mathcal{P}$, we also define

$$\mathcal{C}_{x,p} = \{\gamma \in C^\infty(\mathbb{R}, D(p)) \,|\, p \circ \gamma(0) = x\}.$$

This set of smooth paths passing at x enables to define the kinematic set

$$\mathcal{K}_x = \coprod_{p \in \mathcal{P}} \left\{ \frac{d\gamma}{dt}|_{t=0} \,|\, \gamma \in \mathcal{C}_{x,p} \right\} = \coprod_{p \in \mathcal{P}} \coprod_{x_0 \in p^{-1}(x)} T_{x_0} D(p).$$

Therefore, we identify $(X_1, X_2) \in \mathcal{K}_x^2$, where $X_1 = \frac{d\gamma_1}{dt}|_{t=0} \in T_{x_1} D(p_1)$ and $X_2 = \frac{d\gamma_2}{dt}|_{t=0} \in T_{x_2} D(p_2)$ if there exists a $p_3 \in \mathcal{P}$ and $(\gamma_{3,1}, \gamma_{3,2}) \in C_{x,p_3}$ such that

$$\left\{ \begin{array}{l} \forall i \in \{1,2\}, p_i \circ \gamma_i = p_{3,i} \circ \gamma_{3,i} \\ \frac{d\gamma_{3,1}}{dt}|_{t=0} = \frac{d\gamma_{3,2}}{dt}|_{t=0}. \end{array} \right. \tag{1.1}$$

This identification is reflexive, symmetric, but not transitive, as is shown in the following counter-example:

EXAMPLE 1.21. Let

$$X = \{(x, y, z) \in \mathbb{R}^3 \,|\, yz = 0\}.$$

We equip X with its subset diffeology, inherited from the nebulae diffeology of $\mathbb{R}^3$. Let us conside the paths

$$\gamma_1(t) = (t, t^2, 0)$$

and

$$\gamma_2(t) = (t, 0, t^2).$$

The natural intuition (which will be shown to have a defect later in the exposition, for another diffeology on X) says that

$$\frac{d\gamma_1}{dt}|_{t=0} = \frac{d\gamma_2}{dt}|_{t=0} = (1,0,0),$$

but there is no parametrization p_3 *at* $x = (0,0,0)$ *that fulfills* (1.1). Indeed, the diffeology of X is generated by the push-forwards of the plots of $\mathbb{R}^2$ to X by the maps $(x,y) \mapsto (x,y,0)$ and $(x,y \mapsto (x,0,y)$. In order to identify the two germs, one has to consider an intermediate path $\gamma_{1.5}(t) = (t,0,0)$.

Therefore, we define

DEFINITION 1.22. We define the equivalence relation $\sim$ on $\mathcal{K}_x$ as follows: $\forall (X_1, X_2) \in \mathcal{K}_x, u_1 \sim u_2$ if and only if one of the two following conditions is fulfilled:

(1)

$$\exists (\gamma_1,\gamma_2) \in \left(\coprod_{p\in\mathcal{P}} \mathcal{C}_{x,p}\right)^2, \frac{d\gamma_1}{dt}|_{t=0} = u_1 \in TD(p_1), \frac{d\gamma_2}{dt}|_{t=0} = u_2 \in TD(p_2)$$

and also $p_3 \in \mathcal{P}$ and $(\gamma_{3,1},\gamma_{3,2}) \in C_{x,p_3}$ such that condition (1.1) applies.

(2) there exists a finite sequence $(v_1,\ldots,v_k) \in \mathcal{K}_x^k$ such that $v_1 = u_1$, $v_k = u_2$, and such that condition (1) applies to v_i and v_{i+1} for each index $i \in \mathbb{N}_{k-1}$.

DEFINITION 1.23. The *internal* or *kinetic tangent cone* of X at $x \in X$ is defined by

$${}^iT_xX = \mathcal{K}_x/\sim,$$

The space ${}^iT_xX$ is endowed by the push-forward of the functional diffeology on $\mathcal{C}_x$.

The elements of ${}^iT_xX$ are called *germs of paths* on X at x. In fact, we will produce later in our exposition other tangent spaces, so that we avoid the terminology "tangent vector" which may be misleading.

1.3.2. *Differential forms and de Rham complex.*

DEFINITION 1.24. [**86**] Let $(X,\mathcal{P})$ be a diffeological space and let V be a vector space equipped with a differentiable structure. A $V-$valued $n-$differential form α on X (noted $\alpha \in \Omega^n(X,V)$) is a map

$$\alpha : \{p : O_p \to X\} \in \mathcal{P} \mapsto \alpha_p \in \Omega^n(p;V)$$

such that

• Let $x \in X$. $\forall p,p' \in \mathcal{P}$ such that $x \in Im(p) \cap Im(p')$, the forms α_p and $\alpha_{p'}$ are of the same order n.

• Moreover, let $y \in O_p$ and $y' \in O_{p'}$. If $(X_1,\ldots,X_n)$ are n germs of paths in $Im(p) \cap Im(p')$, if there exists two systems of $n-$vectors $(Y_1,\ldots,Y_n) \in (T_yO_p)^n$ and $(Y_1',\ldots,Y_n') \in (T_{y'}O_{p'})^n$, if $p_*(Y_1,\ldots,Y_n) = p'_*(Y_1',\ldots,Y_n') = (X_1,\ldots,X_n)$,

$$\alpha_p(Y_1,\ldots,Y_n) = \alpha_{p'}(Y_1',\ldots,Y_n').$$

We note by

$$\Omega(X;V) = \oplus_{n\in\mathbb{N}}\Omega^n(X,V)$$

the set of $V-$valued differential forms.

With such a definition, we feel the need to make two remarks for the reader:

• If there does not exist n linearly independent vectors $(Y_1, \ldots, Y_n)$ defined as in the last point of the definition, $\alpha_p = 0$ at y.

• Let $(\alpha, p, p') \in \Omega(X, V) \times \mathcal{P}^2$. If there exists $g \in C^\infty(D(p); D(p'))$ (in the usual sense) such that $p' \circ g = p$, then $\alpha_p = g^* \alpha_{p'}$.

PROPOSITION 1.25. *The set $\mathcal{P}(\Omega^n(X, V))$ made of maps $q : x \mapsto \alpha(x)$ from an open subset O_q of a finite dimensional vector space to $\Omega^n(X, V)$ such that for each $p \in \mathcal{P}$,*

$$\{x \mapsto \alpha_p(x)\} \in C^\infty(O_q, \Omega^n(O_p, V)),$$

is a diffeology on $\Omega^n(X, V)$.

Working on plots of the diffeology, one can define the product and the differential of differential forms, which have the same properties as the product and the differential of differential forms.

DEFINITION 1.26. Let $(X, \mathcal{P})$ be a diffeological space.

• $(X, \mathcal{P})$ is *finite-dimensional* if and only if

$$\exists n_0 \in \mathbb{N}, \quad \forall n \in \mathbb{N}, \quad n \geq n_0 \Rightarrow \dim(\Omega^n(X, \mathbb{R})) = 0.$$

Then, we set

$$\dim(X, \mathcal{P}) = \max\{n \in \mathbb{N} | \dim(\Omega^n(X, \mathbb{R})) > 0\}.$$

• If not, $(X, \mathcal{P})$ is called *infinite dimensional.*

Let us make a few remarks on this definition. If X is a manifold with $\dim(X) = n$ (in the classical sense) the nebulae diffeology $\mathcal{P}_\infty(X)$ is such that

$$\dim(X, \mathcal{P}_\infty(X)) = n.$$

Now, if $(X, \mathcal{F}, \mathcal{C})$ is the natural Frölicher structure on X, take $\mathcal{P}_1$ generated by the maps of the type $g \circ c$, where $c \in \mathcal{C}$ and g is a smooth map from an open subset of a finite dimensional space to $\mathbb{R}$. This is an easy exercise to show that

$$\dim(X, \mathcal{P}_1) = 1.$$

This first point shows that the dimension depends on the diffeology considered. Now, we remark that $\mathcal{F}$ is the set of smooth maps $(X, \mathcal{P}_1) \to \mathbb{R}$,

This leads to the following definition, since $\mathcal{P}(\mathcal{F})$ is clearly the diffeology with the biggest dimension associated to $(X, \mathcal{F}, \mathcal{C})$:

DEFINITION 1.27. The *dimension* of a Frölicher space $(X, \mathcal{F}, \mathcal{C})$ is the dimension of the diffeologial space $(X, \mathcal{P}(\mathcal{F}))$.

This definition totally fits with the following:

DEFINITION 1.28. Let $(X, \mathcal{P})$ be a diffeological space which is not totally disconnected, that is, for which $\mathcal{P}$ has non-constant plots. We define the dimension of $(X, \mathcal{P})$ as

$$\dim(X, \mathcal{P}) = min\left\{n \in \mathbb{N}^* \,|\, \mathcal{P}_n = \mathcal{P}\right\}.$$

1.3.3. *(Pseudo-)Riemannian metrics on diffeological spaces.* We now go to the extension of the basic structures of Riemannian manifolds to diffeological spaces.

DEFINITION 1.29. Let $(X, \mathcal{P})$ be a diffeological space. An *internal Riemannian metric* g on X (noted $g \in Met(X)$) is a map

$$g : \{p : O_p \to X\} \in \mathcal{P} \mapsto g_p$$

such that

- $x \in O_p \mapsto g_p(x)$ is a (smooth) metric on TO_p
- Moreover, let $y \in O_p$ and $y' \in O_{p'}$ such that $p(y) = p'(y')$. If (X_1, X_2) is a couple of germs of paths in $Im(p) \cap Im(p')$, if there exists two systems of 2−vectors $(Y_1, Y_2) \in (T_yO_p)^2$ and $(Y_1', Y_2') \in (T_{y'}O_{p'})^2$, if $p_*(Y_1, Y_2) = p'_*(Y_1', Y_2') = (X_1, X_2)$,

$$g_p(Y_1, Y_2) = g_{p'}(Y_1', Y_2').$$

$(X, \mathcal{P}, g)$ is a *internal Riemannian diffeological space* if g is a metric on $(X, \mathcal{P})$.

For any germ of path X we note $\|X\| = \sqrt{g(X, X)}$ This notation is independent on the chosen plot of the diffeology.

DEFINITION 1.30. We call *arc length* the map $L : C^\infty([0; 1], X) \to \mathbb{R}_+$ defined by

$$L(\gamma) = \int_0^1 \|\dot{\gamma}(t)\| dt.$$

Let $(x, y) \in X^2$. We define

$$d(x, y) = \inf \{L(\gamma) | \gamma(0) = x \wedge \gamma(1) = y\}$$

and we call *Riemannian pseudo-distance* the map $d : X \times X \to \mathbb{R}_+$ that we have just described.

The following proposition justifies the terms "pseudo-distance":

PROPOSITION 1.31. [**52**]

(1) $\forall x \in X, d(x, x) = 0.$
(2) $\forall (x, y) \in X^2, d(x, y) = d(y, x)$
(3) $\forall (x, y, z \in X^3, d(x, z) \leq d(x, y) + d(y, z).$

One could wonder whether d is a distance or not, i.e. if we have the stronger property:

$$\forall (x, y) \in X^2, d(x, y) = 0 \Leftrightarrow x = y.$$

Unfortunately, it seems to appear in examples arising from infinite dimensional geometry that there can have a distance which equals to 0 for $x \neq y$. This is what is described on a weak Riemannian metric of a space of proper immersions in the work of Michor and Mumford [**63**]. Moreover, the D-topology is not the topology defined by the pseudo-metric d. All these facts, which show that the situation on Riemannian Frölicher spaces is very different from the one known on manifolds, are checked in the following (toy) example.

REMARK 1.32. Let $Y = \coprod_{i \in \mathbb{N}^*} \mathbb{R}_i$ where $\mathbb{R}_i$ is the $i-$th copy of $\mathbb{R}$, equipped with its natural scalar product. Let $\mathcal{R}$ be the equivalence relation

$$x_i \mathcal{R} x_j \Leftrightarrow \begin{cases} (x_i \notin]0; \frac{1}{i}[\wedge x_j \notin]0; \frac{1}{j}[) \Rightarrow \begin{cases} x_i = x_j & \text{if } x_i \leq 0 \\ x_i + 1 - \frac{1}{i} = y_j + 1 - \frac{1}{j} & \text{if } x_i \geq \frac{1}{i} \end{cases} \\ (x_i \in]0; \frac{1}{i}[\vee x_j \in]0; \frac{1}{j}[) \Rightarrow i = j \wedge x_i = x_j \end{cases}$$

Let $X = Y/\mathcal{R}$. This is obviously a 1-dimensional Riemannian diffeological space. Let $\bar{0}$ be the class of $0 \in \mathbb{R}_1$, and let $\bar{1}$ be the class of $1 \in \mathbb{R}_1$. Then $d(\bar{0}, \bar{1}) = 0$. This shows that d is not a distance on X. In the $D-$topology, $\bar{0}$ and $\bar{1}$ respectively have the following disconnected neighborhoods:

$$U_{\bar{0}} = \left\{ \bar{x_i} | x_i < \frac{1}{2i} \right\}$$

and

$$U_{\bar{1}} = \left\{ \bar{x_i} | x_i > \frac{1}{2i} \right\}.$$

This shows that d does not define the $D-$topology.

1.4. Group of diffeomorphisms and linear group: two examples of diffeological groups. Let us start with the generalization of the notion of diffeomorphism [**86**].

DEFINITION 1.33. Let $(X, \mathcal{P})$ be a diffeological space. A (set theoric) bijection on X is a *diffeomorphism* if both f and f^{-1} are smooth. We note by $\mathrm{Diff}(X)$ the group of diffeomorphisms of X.

PROPOSITION 1.34. *Let us equip $C^\infty(X, X)$ with its functional diffeology $\mathcal{P}$. We define the maps*

$$i : f \in \mathrm{Diff}(X) \mapsto f \in C^\infty(X, X)$$

and

$$(.)^{-1} : f \in \mathrm{Diff}(X) \mapsto f^{-1} \in C^\infty(X, X).$$

Then $(\mathrm{Diff}(X), i^*(\mathcal{P}) \cap ((.)^{-1})^*(\mathcal{P}))$ *is a diffeological group.*

REMARK 1.35. When X is a compact boundaryless manifold, the group $\mathrm{Diff}(X)$ is an open submanifold of $C^\infty(X, X)$ in the sense of e.g. [**72**], that we will precise later. We anticipate this construction in order to point out here that in geenral, $\mathrm{Diff}(X)$ does *not* have the subset diffeology from $C^\infty(X, X)$, while it is the case when X is a compact boundaryless manifold, which is the most widely studied setting in the actual literature.

Let us now assume that X is a diffeological vector space with field of scalars a diffeological field $\mathbb{K}$.

PROPOSITION 1.36.

(1) *The space $\mathcal{L}(X)$ of smooth linear maps from X to X is an algebra with smooth addition, smooth multiplication and smooth scalar multiplication.*
(2) *The group of invertible elements of $\mathcal{L}(X)$, that we note by $GL(X)$, is a diffeological subgroup of* $\mathrm{Diff}(X)$.

REMARK 1.37. Following the recent review [**24**], to which we refer for a complete exposition, on topological vector spaces that are not normed spaces, multilinear mappings are merely hypocontinuous, and not continuous. This is the case, when E and F are locally convex complete topological vector spaces, for the evaluation map

$$\mathcal{L}(E, F) \times E \to F.$$

Switching to the (subset) functional diffeology of $\mathcal{L}(E, F)$, where E and F are understood as diffeological vector spaces, the same evaluation map remains smooth.

Following Iglesias-Zemmour, see [**33**], we do not assert that arbitrary diffeological groups have associated Lie algebras; however, the following holds, see [**57**, Proposition 2.20].

PROPOSITION 1.38. *Let G be a diffeological group. Then the internal tangent cone at the neutral element ${}^iT_eG$ is a diffeological vector space.*

The proof of Proposition **??** appearing in [**57**] uses explicitly the diffeologies $\mathcal{P}_1$ and $\mathcal{P}_\infty$ which appear in Proposition 3 and Remark 4 of this work.

DEFINITION 1.39. The diffeological group G is a *diffeological Lie group* if and only if the Adjoint action of G on the diffeological vector space ${}^iT_eG$ defines a smooth Lie bracket, that is, noting by c a path in $C^\infty(\mathbb{R}, G)$ such that $c(0) = e$ and $\partial_t c(0) = X$, and noting by Y an element of the internal tangent space ${}^iT_eG$,

- $\partial_t \left(\mathrm{Ad}_{c(t)} Y\right)|_{t=0} = [X, Y] \in {}^iT_eG$
- $(X, Y) \mapsto [X, Y]$ is smooth.

. In this case, we call ${}^iT_eG$ the Lie algebra of G and we denote it by $\mathfrak{g}$.

The question of criteria for a diffeological group to be a diffeological Lie group has been addressed in [**48**] for diffeological groups, and in [**47**] for Frölicher groups.

REMARK 1.40. Actually there is no proven example of diffeological group which is not a diffeological Lie group.

The basic properties of adjoint, coadjoint actions, and of Lie brackets, remain on diffeological Lie groups globally the same as in the case of finite-dimensional Lie groups, and the proofs are similar: see [**48**] and [**18**] for details.

1.5. On various tangent spaces. After our exposition of the construction of the internal tangent cone, we describe other existing generalizations of the notion of tangent space to a given diffeological space X. Indeed, for a finite dimensional manifold, tangent spaces carry so many technical aspects of interest that the notion of tangent space can be generalized by many ways. Some of them appear as non equivalent, while there is actually a shared notion of differential form and of cotangent space of a diffeological space. Let us describe briefly various notions of tangent space for a diffeological space, and let us compare them.

1.5.1. *The external tangent space.* Let us first adapt a classical algebraic construction to the diffeological category. The set of (diffeological) derivations on a diffeological space X, noted by $\mathfrak{Der}(X)$, is the space of smooth maps from $C^\infty(X, \mathbb{R})$ to itself which satisfies the Leibnitz rule. The same way, the (diffeological) pointwise derivations on X are the smooth maps d from $C^\infty(X, \mathbb{R})$ to $\mathbb{R}$ such that

$$\exists x \in X, \forall (f, g) \in (C^\infty(X, \mathbb{R}))^2, d(fg) = f(x)d(g) + g(x)d(f).$$

DEFINITION 1.41. The external tangent space ${}^eTX$ is defined as the set of diffeological pointwise derivations on X.

As a set of smooth maps, it can be endowed with the functional diffeology, and fixing the point $x \in X$, one defines ${}^eT_xX \subset {}^eTX$.

1.5.2. *At the intersection of the internal and the external tangent spaces.* This approach is the one chosen in [**51, 54, 57**], and refined in [**25**]. This leads to two objects of different type: a cone [**51, 54, 57**] which can be completed into a vector space [**25**] following exactly the same procedure as the one applied in [**12**]. This last construction is not described here: we do not want exhaustivity in our presentation.

The main idea behind the definition of this tangent space in [**25, 51, 54, 57**] is based on the definition of a tangent space on smooth manifolds through paths. More precisely, we consider the set of all paths that maps 0 to x and identify two of these paths with each other if their derivative coincide in 0. Therefore, an element in T_xX is an equivalence class $[c]$ for $c\colon \mathbb{R} \to X$ such that $c(0) = x$. The directional derivative of a function $f\colon X \to \mathbb{R}$ in direction v is given by the derivative $\frac{\partial(f\circ c)(t)}{\partial t}_{|t=0}$ for $c\colon \mathbb{R} \to X$ such that $c(0) = x$ and $c'(0) = v$. This approach is then applied to an arbitrary diffeological space to produce a tangent cone the following way:

DEFINITION 1.42. For each $x \in X$, we consider

$$C_x = \{c \in C^\infty(\mathbb{R}, X) | c(0) = x\}$$

and we take the equivalence relation $\mathcal{R}$ given by

$$c\mathcal{R}c' \Leftrightarrow \forall f \in C^\infty(X, \mathbb{R}), \partial_t(f \circ c)|_{t=0} = \partial_t(f \circ c')|_{t=0}.$$

Equivalence classes of $\mathcal{R}$ are called **germs** in [**51, 54, 57**] and are denoted by $\partial_t c(0)$ or $\partial_t c(t)|_{t=0}$. In the same references, the **internal tangent cone** at x is the quotient $C_x/\mathcal{R}$. If $X = \partial_t c(t)|_{t=0} \in {}^iT_X$, we define the pointwise derivation $f \mapsto Df(X) = \partial_t(f \circ c)|_{t=0}$. The reader may compare this definition to the one appearing in [**42**] for manifolds in the "convenient" $c^\infty-$setting. The internal tangent cone defined in section 1.3.1 trivially differs from this one, which is also a subset of ${}^eT_xX$. We note this "internal-external" tangent cone by $C^{ie}T_xX$.

EXAMPLE 1.43. Let

$$X = \left\{(x, y) \in \mathbb{R}^2 \,|\, xy = 0\right\},$$

equipped with its subset diffeology. Then, following [**55**],

$$C^{ie}T_0X = X$$

which shows that the internal-external tangent space is not in general a vector space.

EXAMPLE 1.44. Let us consider the example given in [**15**] of $\mathbb{R}^2$ equipped with the 1-dimensional diffology $\mathcal{P}_1(\mathbb{R}^2)$, called the spaghetti diffeology in the above reference. For reasons similar to the ones given in example 1.21, the internal tangent space of $\mathbb{R}^2$ equipped with the diffeology $\mathcal{P}_1(\mathbb{R}^2)$ is infinite dimensional, with uncountable dimension, while as a consequence of Bohman's theorem, $C^{ie}T_x\mathbb{R}^2$ is 2-dimensional vector space)

Following [**25**] we introduce the following terminology:

DEFINITION 1.45. Let X be a diffeological space and $c \in C_x$. The path derivative in direction c is defined by

$$\mathrm{d}_c\colon C^\infty(X, \mathbb{R}) \to \mathbb{R},\ f \mapsto \mathrm{d}_c(f) := \frac{\partial}{\partial t}\left(f(c(t))\right)_{|t=0} \in \mathbb{R}.$$

One calls $\mathrm{d}_c(f)$ the *path derivative* of the function $f \in C^\infty(X, \mathbb{R})$ in direction c.

As explained in Example 1.43, see e.g. [**25**], $C^{ie}T_xX$ does not provide a natural diffeological vector space structure. This motivates the next tangent space definition.

DEFINITION 1.46. Let X be a diffeological space and $x \in X$. The internal-external tangent space ${}^{ie}T_xX$ is given by

$$ {}^{ie}T_xX := \operatorname{span} C^{ie}T_xX. $$

REMARK 1.47. The set $C^{ie}T_xX$ is noted by C_xX and is called the tangent cone in [**25**]. Due to the multiple definitions of tangent objects in this work, we feel necessary to use the notation $C^{ie}T_xX$ because simplified notations woul be surely misleading.

Regarding the diffeologies on $C^{ie}T_xX$ and ${}^{ie}T_xX$ the following point of view is considered in [**25**] in order to obtain a *diffeological tangent space* from the set $C^{ie}T_xX$:

- An element in ${}^{ie}T_xX$ is a finite sum of elements in $C^{ie}T_xX$ that are multiplied by a scalar.
- The diffeology of $C^{ie}T_xX$ is the push-forward diffeology through the map
$$ c \in C_x \mapsto d_c \in C^{ie}T_xX $$
and a plot in ${}^{ie}T_xX$ is a mapping
$$ U \to {}^{ie}T_xX, \, u \mapsto \sum_{i=1}^{+\infty} \lambda_i(u)p_i(u) $$
where $U \subset \mathbb{R}^n$ denotes an open subset with $n \in \mathbb{N}$, $\lambda_i \colon U \to \mathbb{R}$ is smooth such that the sequence
$$ \lambda = (\lambda_1, \cdots, \lambda_i, \cdots) $$
is a plot of the subset diffeology of
$$ \mathbb{R}^\infty = \bigcup_{k \in \mathbb{N}^*} \mathbb{R}^k \subset \mathbb{R}^{\mathbb{N}} $$
and each p_i is a plot in $C^{ie}T_xX$ with domain U.

The diffeology introduced on T_xX is the so-called *weak-diffeology* defined in [**90**]. Equipped with the weak diffeology, T_xX becomes the finest possible diffeological vector space such that, with slight abuse of intuitionistic notations,

$$ \mathcal{P}_{C^{ie}T_xX} \subset \mathcal{P}_{T_xX}. $$

1.5.3. *The diff-tangent space.* We now extend the construction of [**55**] on Frölicher spaces to (general) diffeological spaces.

DEFINITION 1.48. We use here the notations that we used before for the definition of the internal tangent cone of section 1.3.1. Let ${}^dT_xX$ be the subset of ${}^iT_xX$ defined by

$$ {}^dT_xX = {}^dC_x/\mathcal{R} $$

with

$$ {}^dC_x = \{c \in C_x | \exists \gamma \in C^\infty(\mathbb{R}, \operatorname{Diff}(X)), c(.) = \gamma(.)(x) \text{ and } \gamma(0) = Id_X\}. $$

Through this definition, ${}^dT_xX$ is intrinsically linked with the tangent space at the identity ${}^iT_{Id_X}\operatorname{Diff}(X)$ described in [**48**] for any diffeological group (i.e. group equipped with a diffeology which makes composition and inversion smooth).

REMARK 1.49. Let $\gamma \in C^\infty(\mathbb{R}, \mathrm{Diff}(X))$ such that $\gamma(0)(x) = x$. Then $\lambda(x) = (\gamma(0))^{-1} \circ \gamma(.)(x)$ defines a smooth path $\lambda \in {}^dC_x$. Consequently,

$$^dC_x = \{c \in C_x | \exists \gamma \in C^\infty(\mathbb{R}, \mathrm{Diff}(X)), c(.) = \gamma(.)(x) \text{ and } \gamma(0) = Id_X\}$$

1.6. Fiber bundles, principal bundles and vector bundles in diffeologies. Let us now have a precise look at the notion of fiber bundle in classical (finite dimensional) fiber bundles. Fiber bundles, in the context of smooth finite dimensional manifolds, are defined by

- a smooth manifold E called total space
- a smooth manifold X called base space
- a smooth submersion $\pi : E \to X$ called fiber bundle projection
- a smooth manifold F called typical fiber, because $\forall x \in X, \pi^{-1}(x)$ is a smooth submanifold of E diffeomorphic to F.
- a smooth atlas on X, with domains $U \subset X$ such that $\pi^{-1}(U)$ is an open submanifold of E diffeomorphic to $U \times F$. We the get a system of local trivializations of the fiber bundle.

By the way, in order to be complete, a smooth fiber bundle should be the quadruple data (E, X, F, π) (because the definition of π and of X enables to find systems of local trivializations). For short, this quadruple setting is often denoted by the projection map $\pi : E \to X$.

There exists some diffeological spaces which carry no atlas, so that, the condition of having a system of smooth trivializations in a generalization of the notion of fiber bundles is not a priori necessary, even if this condition, which is additional, enables interesting technical aspects [**61**, pages 194-195]. So that, in a general setting, we do not need to assume the existence of local trivializations. Now, following [**75**], in which the ideas from [**86**, last section] have been devoloped to vector spaces, the notion of quantum structure has been introduced in [**86**] as a generalization of principal bundles, and the notion of vector pseudo-bundle in [**75**].The common idea consist in the description of fibered objects made of a total (diffeological) space E, over a diffeological space X and with a canonical smooth bundle projection $\pi : E \to X$ such as, $\forall x \in X$, $\pi^{-1}(x)$ is endowed with a (smooth) algebraic structure, but for which we do not assume the existence of a system of local trivialization.

(1) For a diffeological vector pseudo-bundle, the fibers $\pi^{-1}(x)$ are assumed diffeological vector spaces, i.e. vector spaces where addition and multiplication over a diffeological field of scalars (e.g. $\mathbb{R}$ or $\mathbb{C}$) is smooth. We notice that [**75**] only deals with finite dimensional vector spaces.
(2) For a so-called "structure quantique" (i.e. "quantum structure") following the terminology of [**86**], a diffeological group G is acting on the right, smoothly and freely on a diffeological space E. The space of orbits $X = E/G$ defines the base of the quantum structure $\pi : E \to X$, which generalize the notion of principal bundle by not assuming the existence of local trivialization. In this picture, each fiber $\pi^{-1}(x)$ is isomorphic to G.

From these two examples, we can generalize the picture.

DEFINITION 1.50. Let E and X be two diffeological spaces and let $\pi : E \to X$ be a smooth surjective map. Then (E, π, X) is a *diffeological fiber pseudo-bundle* if and only if π is a subduction.

Let us precise that we do not assume that there exists a typical fiber, in coherence with Pervova's diffeological vector pseudo-bundles. We can give the following definitions:

DEFINITION 1.51. Let $\pi : E \to X$ be a diffeological fiber pseudo-bundle. Then:

(1) Let $\mathbb{K}$ be a diffeological field. $\pi : E \to X$ is a diffeological $\mathbb{K}-$vector pseudo-bundle if there exists:
- a smooth fiberwise map $. : \mathbb{K} \times E \to E$,
- a smooth fiberwise map $+ : E^{(2)} \to E$ where

$$E^{(2)} = \coprod_{x \in X} \{(u,v) \in E^2 \,|\, (u,v) \in \pi^{-1}(x)\}$$

equipped by the pull-back diffeology of the canonical map $E^{(2)} \to E^2$, such that $\forall x \in X$, $(\pi^{-1}(x), +, .)$ is a diffeological $\mathbb{K}-$vector bundle.

(2) $\pi : E \to X$ is a *diffeological gauge pseudo-bundle* if there exists
- a smooth fiberwise involutive map $(.)^{-1} E \to E$,
- a smooth fiberwise map $. : E^{(2)} \to E$

such that $\forall x \in X$, $(\pi^{-1}(x), .)$ is a diffeological group with inverse map $(.)^{-1}$.

(3) $\pi : E \to X$ is a *diffeological principal pseudo-bundle* if there exists a diffeological gauge pseudo-bundle $\pi' : E' \to X$ such that, considering

$$E \times_X E' = \coprod_{x \in X} \{(u,v) \,|\, (u,v) \in \pi^{-1}(x) \times \pi'^{-1}(x)\}$$

equipped by the pull-back diffeology of the canonical map $E \times_X E' \to E \times E'$, there exists a smooth map $E \times_X E' \to E$ which restricts fiberwise to a smooth free and transitive right-action

$$\pi^{-1}(x) \times \pi'^{-1}(x) \to \pi^{-1}(x).$$

(4) $\pi : E \to X$ is a *Souriau quantum structure* if it is a diffeological principal pseudo-bundle with diffeological gauge (pseudo-)bundle $X \times G \to X$.

We now specialize to elementary classes of fiber bundles.

1.6.1. *Tangent bundles.* The first motivating examples for these constructions were, of course, the tangent cones and the tangent spaces which are all fiber pseudo-bundles and sometimes vector pseudo-bundles. Indeed, analyzing the constructions in sections 1.3.1, 1.5.2 and 1.5.3, everything starts with a set of paths that we note here generically by $C_x(X)$ which generates the tangent cone at x, that we note here generically by CT_xX, and its diffeology through the quotient by an equivalence relation $\mathcal{R}_x$. Let us now consider the *total* tangent space (in fact, tangent cone in general)

$$CTX = \coprod_{x \in X} CT_xX,$$

for which basic intuition asserts that there exists a much better diffeology than the coproduct diffeology where fibers are arcwise disconnected one to each other. For the construction of this desired diffeology in the context of the internal tangent cone (section 1.3.1) and of the internal-external tangent cone (section 1.5.2), one has only to remark that the evaluation maps at $t = 0$ define a fiber pseudo-bundle projection

$$\mathrm{ev}_0 : C^\infty(\mathbb{R}, X) \to X$$

which is a subduction, and such that

$$\forall x \in X, \mathrm{ev}_0^{-1}(x) = C_x(X).$$

The fiberwise equivalence relations $\mathcal{R}_x$ define a global equivalence relation $\mathcal{R}$ on $C^\infty(\mathbb{R}, X)$ for which

$$CTX = C^\infty(\mathbb{R}, X)/\mathcal{R},$$

which enables to defined a global quotient diffeology. Therefore, one can define from the kinetic tangent cone at $x \in X$ the global internal tangent cone of X and their diffeologies the following way:

DEFINITION 1.52. The internal or kinetic tangent cone of X as

$${}^iTX = \coprod_{x \in X} {}^iT_xX.$$

The space ${}^iTX$ is endowed by the push-forward of the functional diffeology on $C^\infty(\mathbb{R}, X)$.

The same definition holds for the definition of the total internal-external tangent cone $C^{ie}TX$, by replacing the equivalence relations. Considering ${}^dTX$, one has also to change in the definitions $C^\infty(\mathbb{R}, X)$ for its subset

$$\coprod_{x \in X} {}^dC_x$$

which is endowed with its subset diffeology.

DEFINITION 1.53. Let X be a Frölicher space. we define, by

$${}^dTX = \coprod_{x \in X} {}^dT_xX$$

the diff-tangent bundle of X.

By the way, we can get easily the following observations:

PROPOSITION 1.54. *Let $(X, \mathcal{P})$ be a reflexive diffeological space, and let $\mathcal{P}_{\mathrm{Diff}}$ be the functional diffeology on* $\mathrm{Diff}(X)$.

(1) *There exists a diffeology $\mathcal{P}(\mathrm{Diff}) \subset \mathcal{P}$ which is generated by the family of push-forward diffeologies :*

$$\{(\mathrm{ev}_x)_*(\mathcal{P}_{\mathrm{Diff}}) \,|\, x \in X\}.$$

(2) $\forall x \in X, {}^dT_xX$ *is the internal tangent cone of* $(X, \mathcal{P}(\mathrm{Diff}))$ *at* x.
(3) $\forall x \in X, {}^dT_xX$ *is a diffeological vector space*
(4) *The total diff-tangent space*

$${}^dTX = \coprod_{x \in X} {}^dTX \subset {}^iTX$$

is a vector pseudo-bundle for the subset diffeology inherited from ${}^iTX$ and also for the diffeology of internal tangent space of $(X, \mathcal{P}_{\mathrm{Diff}})$.

PROOF. (1) is a consequence of the definition of push-forward diffeologies the following way: the family

$$\{\mathcal{P} \text{ diffeology on } X \,|\, \forall x \in X,\ (\mathrm{ev}_x)_*(\mathcal{P}_{\mathrm{Diff}}) \subset \mathcal{P}\}$$

has a minimal element by Zorn Lemma.

(2) follows from remark 1.49.

(3): The diffeology $\mathcal{P}(\mathrm{Diff})$ coincides with the diffeology made of plots which are locally of the form $\mathrm{ev}_x \circ p$, where $x \in X$ and p is a plot of the diffeology of $\mathrm{Diff}(X)$. We have that ${}^iT_{Id}\,\mathrm{Diff}(X)$ is a diffeological vector space, following [**48**]. This relation follows from the differentiation of the multiplication of the group: given two paths γ_1, γ_2 in $C^\infty(\mathbb{R}, Diff(X))$, with $\gamma_1(0) = \gamma_2(0) = Id$, if $X_i = \partial_t \gamma_i(0)$ for $i \in \{1{,}2\}$, then

$$X_1 + X_2 = \partial_t(\gamma_1.\gamma_2)(0).$$

Reading locally plots in $\mathcal{P}(\mathrm{Diff})$, we can consider only plots of the for $\mathrm{ev}_x \circ p$, where p is a plot in $\mathrm{Diff}(X)$ such that $p(0) = Id_X$. By the way the vector space structure on ${}^dT_xX$ is inherited from ${}^iT_{Id}\,\mathrm{Diff}(X)$ via evaluation maps.

In order to finish to check (3), we prove directly (4) by describing its diffeology. For this, we consider

$$C_0^\infty(\mathbb{R}, \mathrm{Diff}(X)) = \{\gamma \in C_0^\infty(\mathbb{R}, \mathrm{Diff}(X)) \,|\, \gamma(0) = Id_X\}.$$

Let ${}^dC = \coprod_{x\in X}^d C_x$. The total evaluation map

$$\begin{array}{rccc} ev: & X \times C_0^\infty(\mathbb{R}, Diff(X)) & \to & {}^dC \\ & (x,\gamma) \mapsto & & ev_x \circ \gamma \end{array}$$

is fiberwise (over X), and onto. By the way we get a diffeology on ${}^dC$ which is the push-forward diffeology of $X \times C_0^\infty(\mathbb{R}, \mathrm{Diff}(X))$ by ev. Passing to the quotient, we get a diffeology on ${}^dTX$ which makes each fiber ${}^dT_xX$ a diffeological vector space trivially. □

Therefore, the diff-tangent space is a vector pseudo-bundle, while one supplementary effort is necessary to generate a diffeology on ${}^{ie}TX$. For this, let us consider the map

$$\begin{array}{cccc} l: & C^\infty(X, \mathbb{R}^\infty) \times C^{ie}TX^{\mathbb{N}} & \to & {}^{ie}TX \\ & (\lambda_1, \cdots, \lambda_k, \cdots) \times (d_{c_1}, \cdots, d_{c_k}, \cdots) & \mapsto & \sum_{k=1}^{+\infty} \lambda_k d_{c_k} \end{array}$$

which produces the desired diffeology on ${}^{ie}TX$ by push-forward.

1.6.2. *Riemannian metrics on vector pseudo-bundles.*

DEFINITION 1.55. Let $\pi : E \to X$ be a real diffeological vector pseudo-bundle. A *Riemannian metric* on E is a smooth map

$$g : E^{(2)} \to \mathbb{R}$$

that is fiberwise on each E_x a symmetric, definite and positive bilinear form.

We have here to point out a first group of difficulties which are not present in the context of (classical) finite dimensional manifolds:

- The metric g may not define an isomorphism between each fiber E_x and its diffeological dual $\mathcal{L}(E_x, \mathbb{R})$.
- The fibers of E may not be isomorphic.

As a first class of examples, one can mention the vector pseudo-bundle ${}^dTX$ when X is equipped with an internal Riemannian metric (see section 1.3.3), but one can wonder whether tangent cones can also carry Riemannian structures, whitout embedding in a larger Riemannian vector pseudo-bundle. To our actual state of knowledge, the only known such structure is described in section 1.3.3.

1.6.3. *Diffeological principal bundles and connections.* In [**33**, Article 8.32] Iglesias-Zemmour gives a definition of a connection on a principal G-bundle in terms of paths on the total space P, generalising the classical notion of path lifting for principal bundles with finite-dimensional Lie groups as structure groups.

DEFINITION 1.56. Let G be a diffeological group, and let $\pi\colon P \to X$ be a principal G-bundle. Denote by $\mathrm{Paths}_{\mathrm{loc}}(P)$ the diffeological space of *local paths* (see [**33**, Article 1.63]), and by $\mathrm{tpath}(P)$ the *tautological bundle of local paths*

$$\mathrm{tpath}(P) := \{(\gamma, t) \in \mathrm{Paths}_{\mathrm{loc}}(P) \times \mathbb{R} \mid t \in D(\gamma)\}.$$

A *diffeological connection* is a smooth map $H\colon \mathrm{tpath}(P) \to \mathrm{Paths}_{\mathrm{loc}}(P)$ satisfying the following properties for any $(\gamma, t_0) \in \mathrm{tpath}(P)$:

(1) the domain of γ equals the domain of $H(\gamma, t_0)$,
(2) $\pi \circ \gamma = \pi \circ H(\gamma, t_0)$,
(3) $H(\gamma, t_0)(t_0) = \gamma(t_0)$,
(4) $H(\gamma \cdot g, t_0) = H(\gamma, t_0) \cdot g$ for all $g \in G$,
(5) $H(\gamma \circ f, s) = H(\gamma, f(s)) \circ f$ for any smooth map f from an open subset of $\mathbb{R}$ into $D(\gamma)$,
(6) $H(H(\gamma, t_0), t_0) = H(\gamma, t_0)$.

Another formulation of this definition can be found in [**51**] under the terminology of *path-lifting.*

REMARK 1.57. Diffeological connections satisfy many of the usual properties that classical connections on a principal G-bundle (where G is a finite- dimensional Lie group) enjoy; in particular, they admit unique horizontal lifts of paths in $\mathrm{Paths}_{\mathrm{loc}}(M)$ [**33**, Article 8.32], and they pull back by smooth maps [**33**, Article 8.33].

PROPOSITION 1.58. *Let V be a vector space. Then, G acts smoothly from the right on the space $\Omega(P,V)$ of V-valued differential forms on P by setting*

$$\forall (g, \alpha) \in \Omega^n(P,V) \times G, \forall p \in \mathcal{P}(P), \quad (g_* \alpha)_{g.p} = \alpha_p \circ (dg^{-1})^n \,.$$

PROOF. G acts smoothly on P so that, if $p \in \mathcal{P}(P)$, $g.p \in \mathcal{P}(P)$. The right action is now well-defined, and smoothness is trivial. □

DEFINITION 1.59. Let $\alpha \in \Omega(P; \mathfrak{g})$. The differential form α is *right-invariant* if and only if, for each $p \in \mathcal{P}(P)$, and for each $g \in G$,

$$\alpha_{g.p} = Ad_{g^{-1}} \circ g_* \alpha_p \,.$$

Now, let us turn to connections and holonomy. Let $p \in P$ and let γ be a smooth path in P starting at p.

DEFINITION 1.60. A *connection* on P is a $\mathfrak{g}$−valued right-invariant 1-form θ, , such that, for each $v \in \mathfrak{g}$, for any path $c : \mathbb{R} \to G$ such that

$$\begin{cases} c(0) & = & e_G \\ \partial_t c(t)|_{t=0} & = & v \,, \end{cases}$$

and for each $p \in P$ we have:

$$\theta(\partial_t(p.c(t))_{t=0}) = v \,.$$

Now we assume that $dim(M) \geq 2$ and we fix a connection θ on P.

DEFINITION 1.61. Let $\alpha \in \Omega(P; \mathfrak{g})$ be a $G-$invariant 1-form. Let $\nabla\alpha = d\alpha - \frac{1}{2}[\theta, \alpha]$ be the horizontal derivative of α. The curvature 2-form induced by θ is

$$\Omega = \nabla\theta \,.$$

2. Diffeologies in functional equations

2.1. Principal bundles, fully regular Lie groups and holonomy.

2.1.1. *Motions on groups and (partial) differential equation.* A wide class of first order differential equations are usually solved globally by the use of the integration of smooth paths v in a Lie algebra into smooth paths g in a Lie group, satisfying a differential equation of the type

$$g(t)^{-1}\partial_t g(t) = v(t). \tag{2.1}$$

When v is constant, $g(t) = \exp(tv)$ produces, when it exists, the only solution g such that $g(0) = e$. This approach is the initial motivation for the development of the notion of Lie grops by Sophus Lie. For finite dimensional examples, this approach gives a geometric way to solve differential equations with value in Lie groups, but these techniques can be also of interest for infinite dimensional objects.

More precisely, the first examples of interest, where the exponential map is not easy to define, are linear differential equations of the form

$$\frac{df}{dt} = A(f) \tag{2.2}$$

where A is a differential operator of positive order acting in a space of smooth functions F.

EXAMPLE 2.1. When $F = C^\infty(S^1, \mathbb{R})$, consider $A = -\frac{d^2}{dx^2}$. Then there is an operator $\exp(tA)$ which is unbounded, that solves (2.2) with eigenvalues e^{tn^2}.

Modifying A into $\frac{d^2}{dx^2}$, the solution $\exp(tA)$ is a family of Fredholm operators of index 0 with eigenvalues $e^{-tn^2} \leq 1$.

From this elementary example, which can be expanded and complexified as far as wanted, one can easily understand that control of the regularity of the solutions for these equations, and the methods for approximating them numerically, highly depends on the nature of the operator A. The same way, equations of the form

$$\frac{dS}{dt} = [A(t), S(t)] \,, \tag{2.3}$$

where S is an operator-valued solution in a Lie algebra $\mathfrak{g}$ and A is a smooth path in $\mathfrak{g}$, have a formal solution

$$S(t) = \mathrm{Ad}_{G(t)}\, S(0)$$

where G is the solution of equation (2.1) .

These elementary facts can be generalized in two ways that we wish to mention here:

- Symmetries of a (partial) differential equation generate transformation groups which transform one solution to another. In practice, one often deals with *infintesimal symmetries* which are formally tangent vectors to the group of symmetries [**70**]. Integrating a Lie algebra of infinitesimal symmetries is less easy and requires technical abilities in the field of infinite dimensional Lie groups. One can find an example of such concerns for symmetries of the 3d-Euler equation in [**79**]. These indications help to solve the PDE by reduction, in general case by case and when it produces "enough" symmetries, but anyway the symmetries of PDEs are in general of interest for the global understanding of the space of solutions.

- Integrable systems are differential equations that are equivalent to a zero curvature condition, that is, they can be expressed as

$$dS \pm [S, S] = 0$$

where S is a well-chosen 1-form built from the initial PDE. The main clue for solving the problem is that this precise equation is a so-called *zero curvature equation*, that integrates globally when one deals with on a trivial finite dimensional principal bundle in which S is understood as a connexion 1-form.

In section 2, we expose what is the actual state-of-the-art of the application of diffeologies to the necessary technical properies on diffeological Lie groups, in view of potential applications to PDEs that can be expressed in an infinite dimensonal Lie group.

2.1.2. *On regular diffeological Lie groups.* Since we are interested in infinite-dimensional analogues of Lie groups, we need to consider tangent spaces of diffeological spaces, and we have to deal with Lie algebras and exponential maps. We state, after [**12, 18, 48**] the following definition:

DEFINITION 2.2. [**48**] A diffeological Lie group G with Lie algebra $\mathfrak{g}$ is called *regular* if and only if there is a smooth map

$$\mathrm{Exp} : C^\infty([0;1], \mathfrak{g}) \rightarrow C^\infty([0,1], G)$$

such that $g(t) = \mathrm{Exp}(v(t))$ is the unique solution of the differential equation

$$\begin{cases} g(0) = e \\ \frac{dg(t)}{dt} g(t)^{-1} = v(t) \end{cases} \tag{2.4}$$

We define the exponential function as follows:

$$\begin{aligned} exp : \mathfrak{g} &\rightarrow G \\ v &\mapsto exp(v) = g(1)\ , \end{aligned}$$

where g is the image by Exp of the constant path v.

REMARK 2.3. Equation (2.4) is called *right logarithmic* equation while Equation 2.1 is called *left logarithmic*. The correspondence between the two equations is actually well-exposed in [**42**], and the existence of the solution of the first one is equivalent to the existence of a solution to the second one.

When the Lie group G is a vector space V, the notion of regular Lie group specialize to what is called *regular vector space* in [**51**] and *integral vector space* in [**48**]; we follow the first terminology.

DEFINITION 2.4. [**48**] Let $(V, \mathcal{P})$ be a diffeological vector space. The space $(V, \mathcal{P})$ is *integral* or *regular* if there is a smooth map

$$\int_0^{(.)} : C^\infty([0;1]; V) \rightarrow C^\infty([0;1], V)$$

such that $\int_0^{(.)} v = u$ if and only if u is the unique solution of the differential equation

$$\begin{cases} u(0) = 0 \\ u'(t) = v(t) \end{cases} .$$

This definition applies, for instance, if V is a complete locally convex topological vector space equipped with its natural Frölicher structure given by the Frölicher completion of its nébuleuse diffeology, see [**33, 50, 51**]. We give now the corresponding notion for derivatives, after remarking that $\forall v \in V, V \subset {}^iT_vV$ through the identification of linear paths $t \mapsto tv$ with v :

DEFINITION 2.5. Let V be a diffeological vector space. Then V is **co-regular** if

$$\forall v \in V, {}^iT_vV = V.$$

This property, which seems natural, highly depends on the diffeology considered, see Example 1.44. This may explain why all the authors since [**48**] and till now, and even us, have considered this problem as minor or even negligible.

DEFINITION 2.6. Let G be a diffeological Lie group with Lie algebra $\mathfrak{g}$. Then, G is fully regular, i.e. regular with integral Lie algebra if $\mathfrak{g}$ is integral and G is regular in the sense of Definitions 2.2 and 2.4.

We finish this section with two structural results essentially proven in [**51**] in a restricted setting for the second one.

THEOREM 2.7. *Let*

$$1 \longrightarrow K \xrightarrow{i} G \xrightarrow{p} H \longrightarrow 1$$

be an exact sequence of Frölicher Lie groups, such that there is a smooth section $s : H \to G$*, and such that the trace diffeology on* $i(K) \subseteq G$ *coincides with the push-forward diffeology from* K *to* $i(K)$*. We consider also the corresponding sequence of Lie algebras*

$$0 \longrightarrow \mathfrak{k} \xrightarrow{i'} \mathfrak{g} \xrightarrow{p} \mathfrak{h} \longrightarrow 0 .$$

Then,

- *The Lie algebras* $\mathfrak{k}$ *and* $\mathfrak{h}$ *are integral if and only if the Lie algebra* $\mathfrak{g}$ *is integral;*
- *The Frölicher Lie groups* K *and* H *are regular if and only if the Frölicher Lie group* G *is regular.*

THEOREM 2.8. *Let* $(A_n)_{n\in\mathbb{N}}$ *be a sequence of coregular and integral Frölicher vector spaces equipped with a graded smooth multiplication operation on* $\bigoplus_{n\in\mathbb{N}^*} A_n$ *, i.e. a multiplication such that for each* $n, m \in \mathbb{N}^*$, $A_n.A_m \subset A_{n+m}$ *is smooth with respect to the corresponding Frölicher structures.*

- *Let us define the (non unital) algebra of formal series:*

$$\mathcal{A} = \left\{ \sum_{n\in\mathbb{N}^*} a_n | \forall n \in \mathbb{N}^*, a_n \in A_n \right\} ,$$

equipped with the Frölicher structure of the infinite product. Then, the space

$$1 + \mathcal{A} = \left\{ 1 + \sum_{n\in\mathbb{N}^*} a_n | \forall n \in \mathbb{N}^*, a_n \in A_n \right\}$$

is a regular Frölicher Lie group with integral Frölicher Lie algebra $\mathcal{A}$. *Moreover, the exponential map defines a smooth bijection* $\mathcal{A} \to 1 + \mathcal{A}$.

A result similar to Theorem 2.7 is also valid for Fréchet Lie groups, see [**42**].

2.1.3. *On the holonomy of a connection in a diffeological principal bundle.* Now, let P be a principal bundle. Let $p \in P$ and γ a smooth path in P starting at p, defined on $[0,1]$. Let $H_\theta\gamma(t) = \gamma(t)g(t)$, where $g(t) \in C^\infty([0,1];\mathfrak{g})$ is a path satisfying the differential equation:

$$\begin{cases} \theta\left(\partial_t H_\theta\gamma(t)\right) = 0 \\ H_\theta\gamma(0) = \gamma(0) \end{cases}$$

The first line of this equation is equivalent to the differential equation

$$g^{-1}(t)\partial_t g(t) = -\theta(\partial_t\gamma(t))$$

which is integrable, and the second line is equivalent to the initial condition $g(0) = e_G$. This shows that horizontal lifts are well-defined, as in the standard case of finite-dimensional manifolds. Moreover, the map $H_\theta(.)$ defines trivially a diffeological connection. This enables us to consider the holonomy group of the connection. Notice that a straightforward adaptation of the arguments of [**51**] shows that the holonomy group does not depend (up to conjugation and up to the choice of connected component of M) on the choice of the base point p. This definition allows us to consider reductions of the structure group. Following [**51**],

THEOREM 2.9. *We assume that* G_1 *and* G *are regular Frölicher groups with regular Lie algebras* $\mathfrak{g}_1$ *and* $\mathfrak{g}$. *Let* $\rho : G_1 \mapsto G$ *be an injective morphism of Lie groups. If there exists a connection* θ *on* P, *with curvature* Ω, *such that for any smooth* 1-*parameter family* $H_\theta c_t$ *of horizontal paths starting at* p, *and for any smooth vector fields* X, Y *in* M, *the map*

$$s,t \in [0,1]^2 \to \Omega_{H\theta_t(s)}(X,Y) \tag{2.5}$$

is a smooth $\mathfrak{g}_1$*-valued map (for the* $\mathfrak{g}_1-$ *diffeology), and if* M *is simply connected, then the structure group* G *of* P *reduces to* G_1, *and the connection* θ *also reduces.*

We can now state the announced Ambrose-Singer theorem, using the terminology of [**79**] for the classification of groups via properties of the exponential map [**51**]:

THEOREM 2.10. *Let* P *be a principal bundle whose structure group is a fully regular Frölicher Lie group* G. *Let* θ *be a connection on* P *and* H_θ *the associated diffeological connection.*

(1) *For each* $p \in P$, *the holonomy group* $\mathcal{H}_p^L$ *is a diffeological subgroup of* G, *which does not depend on the choice of* p *up to conjugation.*
(2) *There exists a second holonomy group* H^{red}, $\mathcal{H} \subset H^{red}$, *which is the smallest structure group for which there is a subbundle* P' *to which* θ *reduces. Its Lie algebra is spanned by the curvature elements, i.e. it is the smallest integrable Lie algebra which contains the curvature elements.*
(3) *If* G *is a Lie group (in the classical sense) of type I or II, there is a (minimal) closed Lie subgroup* $\bar{H}^{red}$ *(in the classical sense) such that* $H^{red} \subset \bar{H}^{red}$, *whose Lie algebra is the closure in* $\mathfrak{g}$ *of the Lie algebra of* H^{red}. $\bar{H}^{red}$ *is the smallest closed Lie subgroup of* G *among the structure groups of closed sub-bundles* $\bar{P}'$ *of* P *to which* θ *reduces.*

From [**51**] again, we have the following result:

PROPOSITION 2.11. *If the connection θ is flat and M is connected and simply connected, then for any path γ starting at $p \in P$, the map*

$$\gamma \mapsto H_\theta \gamma(1)$$

depends only on $\pi(\gamma(1)) \in M$, and it defines a global smooth section $M \to P$. Therefore, $P = M \times G$.

Let us precise a little bit more this result (see [**42**, section 40.2] for an analogous statement in the c^∞-setting):

THEOREM 2.12. *Let $(G, \mathfrak{g})$ be a regular Lie group with regular Lie algebra and let X be a simply connected Frölicher space. Let $\alpha \in \Omega^1(M, \mathfrak{g})$ such that*

$$d\alpha + [\alpha, \alpha] = 0 \,. \tag{2.6}$$

Then there exists a smooth map

$$f : X \to G$$

such that

$$df.f^{-1} = \alpha.$$

Moreover, we move from one solution f to another by applying the Adjoint action of G, pointwise in $x \in X$.

We remark that the theorem also holds if we consider the equation

$$d\alpha - [\alpha, \alpha] = 0$$

instead of (2.6); we only need to change left logarithmic derivatives for right logarithmic derivatives, and Adjoint action for Coadjoint action. The correspondence between solutions is given by the inverse map $f \mapsto f^{-1}$ on the group $C^\infty(X, G)$.

2.2. Optimization on diffeological spaces. In general, optimization methods aim at the minimization or maximization of an objective functional. Often, this functional depends on the solution of a partial differential equation (PDE). Examples of a PDE are the compliance of an elastic structure or a dissipated energy in a viscous flow. Since a maximization problem can be expressed as an minimization problem by considering the negative objective functional, often, only minimization problems are considered in the literature. In general, optimization methods are iterative methods that generate updates such that the objective functional is reduced (in the case of a minimization problem). We concentrate first on shape optimization to motivate the consideration of diffeological spaces in optimization techniques. Afterwards, we summarize the recent findings regarding optimization techniques on diffeological spaces.

2.2.1. *A motivation for considering diffeological spaces in (shape) optimization.* Shape optimization has many applications and a large variety of methods. Application examples are acoustic shape optimization [**81**], optimization of interfaces in transmission problems [**23, 73**], electrochemical machining [**30**], image restoration and segmentation [**31**] and inverse modelling of skin structures [**69**].

Shape optimization problems are usually solved by using iterative methods such that one starts with an initial shape and then gradually evolves it be morphing it into the optimal shape. In order to formulate optimization methods and enable their theoretical investigations, one needs to define first what we describe as a shape.

There are multiple options, e.g. the usage of landmark vectors [**16, 28, 37, 74, 85**], plane curves [**63–66**] or surfaces in higher dimensions [**5, 6, 40, 46, 62**], boundary contours of objects [**22, 49, 80**], multiphase objects [**93**], characteristic functions of measurable sets [**94**] and morphologies of images [**17**]. In general, a shape space does not have a vector space structure. Instead, the next-best option is to aim for a manifold structure with an associated Riemannian metric. In case a manifold structure cannot be established for the shape space in question, an alternative option is a diffeological space structure. In contrast to shape spaces as Riemannian manifolds, research for diffeological spaces as shape spaces has just begun, see e.g. [**38, 92**].

In the following, we summarize the main findings of [**92**], which suggests the use of diffeological spaces instead of smooth manifolds in the context of PDE constrained shape optimization.

In shape optimization, one investigates shape functionals. For a suitable shape space $\mathcal{U}$ a shape functional is a function $J\colon \mathcal{U} \to \mathbb{R}$. An unconstrained shape optimization problem is given by

$$\min_{\Omega\in\mathcal{U}} J(\Omega). \tag{2.7}$$

In applications, shape optimization problems are often constrained by equations like PDEs.

Regarding suitable shape spaces in optimization problems, [**92**] starts by concentrating on one-dimensional smooth shapes. A one-dimensional smooth shape is a C^∞-boundary of a simply connected and compact set Ω. These shapes can be interpreted as smooth single-closed curves represented by embeddings from S^1 into $\mathbb{R}^2$. Since we are only interested in the shape itself, i.e., the image of the curve, we are not interested in re-parametrisations. Therefore one can consider the shape space given in [**63**]:

$$B_e(S^1,\mathbb{R}^2) := \mathrm{Emb}(S^1,\mathbb{R}^2)/\mathrm{Diff}(S^1),$$

where $\mathrm{Emb}(S^1,\mathbb{R}^2)$ is the set of all embeddings from S^1 to $\mathbb{R}^2$ and $\mathrm{Diff}(S^1)$ the space of all diffeomorphisms from S^1 into itself. The space $B_e(S^1,\mathbb{R}^2)$ is a smooth manifold, see [**42**]. The tangent space is given by

$$T_cB_e(S^1,\mathbb{R}^2) \simeq \{h \mid h = \alpha n, \alpha \in C^\infty(S^1)\},$$

where n denotes the exterior unit normal field to the shape boundary $c : S^1 \to \mathbb{R}^2$. The space of one-dimensional smooth shapes is a Riemannian manifold for various Riemannian metrics like the almost local metrics and Sobolev metrics; see e.g. [**5, 6, 62, 63, 65**]. All these metrics arise from the standard L^2-metric by putting weights (almost local metrics), derivatives (Sobolev metrics) or both (weighted Sobolev metrics) in it.

In [**92**], for $\mathcal{U}$ the shape space $B_e(S^1,\mathbb{R}^2)$ combined with the first Sobolev metric and the Steklov-Poincaré metric (cf. [**83**]) is considered.

Remark 2.13. One can generalize the shape space $B_e(S^1,\mathbb{R}^2)$ and its results for a compact manifold M and a Riemannian manifold N with $\dim(M) < \dim(N)$ (cf., e.g., [**62**]).

In general, to formulate optimization methods on a Riemannian shape space $(\mathcal{U}, g)$, a Riemannian shape gradient with respect to g is needed. A connection of the shape space $B_e(S^1, \mathbb{R}^2)$ together with the first Sobolev metric and the Steklov-Poncaré metric to shape calculus is given in [**92**]. Moreover, both approaches are compared to each other (cf. [**82, 92**]). In particular, it should be mentioned that working with the Steklov-Poincaré metric has several computational advantages regarding the finite element mesh, which needs to be considered to solve PDE constraints. We refer the reader to [**82–84**].

REMARK 2.14. It is well known that shape optimization algorithms may invoke substantial transformations of the initial shape in the course of the optimization process, which often leads to a deterioration of the cell aspect ratios of the underlying computational meshes. Usually, optimization problems in function spaces can be solved using one of two different approaches: discretize–then–optimize (DO) and optimize–then–discretize (OD). Shape optimization problems are no exception. We refer the interested reader to [**7**] for the description of differences between the OD and DO approaches by a prototypical example of a PDE-constrained shape optimization problem.

Regardless the use and benefits of these OD-approaches (for which we refer to the literature), the shape space $B_e(S^1, \mathbb{R}^2)$ itself limits the application of the methods since it only contains smooth shapes. From a numerical point of view it is desirable to weak the smoothness assumption of the shapes. Numerical experiments show that also shapes with kinks can be handled with the Steklov-Poincaré metric; see [**82–84**]. Thus, another shape space definition is necessary. In [**92**], the space of so called $H^{1/2}$-shapes is defined:

DEFINITION 2.15. Let $\Gamma_0 \subset \mathbb{R}^d$ be a d-dimensional Lipschitz shape (the boundary of a non-trivial Lipschitz domain). The space of all d-dimensional $H^{1/2}$-shapes is given by

$$\mathcal{B}^{1/2}\left(\Gamma_0, \mathbb{R}^d\right) := \mathcal{H}^{1/2}\left(\Gamma_0, \mathbb{R}^d\right) / \sim,$$

where

$$\begin{aligned}&\mathcal{H}^{1/2}\left(\Gamma_0, \mathbb{R}^d\right)\\&:= \left\{w : w \in H^{1/2}\left(\Gamma_0, \mathbb{R}^d\right) \text{ injective, continuous; } w\left(\Gamma_0\right) \text{ Lipschitz shape}\right\}\end{aligned}$$

and the equivalence relation $\sim$ is given by

$$w_1 \sim w_2 \Leftrightarrow w_1\left(\Gamma_0\right) = w_2\left(\Gamma_0\right), \text{ where } w_1, w_2 \in \mathcal{H}^{1/2}\left(\Gamma_0, \mathbb{R}^d\right).$$

Since the space $\mathcal{B}^{1/2}$ is a challenging one, it is so far unclear if it is a Riemannian manifold or even a manifold. Therefore, [**92**] came up with the idea of using a different point of view and drop the restrictiveness of Riemannian manifolds and starts to consider the opportunities within diffeological spaces. Due to the wide variety of natural diffeological spaces, the $\mathcal{B}^{1/2}$ shape space is a diffeological space (cf. [**92**]). In addition, $\mathcal{B}^{1/2}$ is way less restrictive than B_e. However, diffeological spaces are yet to be established in the area of optimization. The first results regarding optimization techniques on diffeological spaces are summarized in the next subsection

2.2.2. *General concepts for optimization on diffeological spaces.* In the following, we give a brief summary of [**25**], in which optimization approaches on diffeological spaces are formulated. In order to formulate optimization methods on diffeological spaces, the steepest decent method on manifolds is considered extended to diffeological spaces. For convenience, we formulate the steepest descent method on a complete Riemannian manifold (M, g) in algorithm 1.

Require: Objective function f on M; Levi-Civita connection ∇ on (M, g);

step size strategy.

Goal: Find the solution of $\min_{x \in M} f(x)$.

Input: Initial data $x_0 \in M$.

for $k = 0, 1, \dots$ **do**

[1] Compute $\mathrm{grad} f(x_k)$ denoting the Riemannian shape gradient of f in x_k.

[2] Compute step size t_k.

[3] Set $x_{k+1} := \exp_{x_k}(-t_k \mathrm{grad} f(x_k))$, where $\exp\colon TM \to M$ denotes the exponential map

end for

Algorithm 1: Steepest descent method on a complete Riemannian manifold (M, g)

In [**25**], diffeological counterparts to the known and needed objects for the steepest decent method on smooth manifolds are considered, i.e., diffeological variants of a Riemannian manifold, a Riemannian gradient, a Levi-Civita connection, as well as a retraction. In the following, we introduce these objects from [**25**].

REMARK 2.16. The article [**25**] does not consider a diffeological version of the exponential map, since for optimization purpose the exponential map is usually replaced by a retraction that approximates the exponential map. Furthermore, the existence of an exponential map in a diffeological sense is totally unclear.

2.2.3. *Diffeological gradient.* For optimization purposes the existence of a gradient is more or less indispensable, since it provides a direction of decent under special assumptions. In order to define a diffeological version of the gradient we first define a diffeological Riemannian space, which results in the definition of a diffeological gradient. We copy here the definition given in [**25**], that we compare with the previous definitions.

DEFINITION 2.17 (Diffeological Riemannian space). Let X be a diffeological space. We say X is a diffeological Riemannian space if there exists a smooth map

$$\mathrm{Sym}({}^{ie}TX, \mathbb{R}),\ x \mapsto g_x$$

such that $g_x\colon {}^{ie}T_xX \times {}^{ie}T_xX \to \mathbb{R}$ is smooth, symmetric and positive definite. Then, we call the map g a diffeological Riemannian metric.

REMARK 2.18. This definition fits with the defninition of a Riemannian metric along the lines of Definition 1.55 on the diffeological vector bundle ${}^{ie}TX$ but differs from Definition 1.29 of an internal Riemannian metric.

DEFINITION 2.19 (Diffeological gradient). Let X be a diffeological Riemannian space. The diffeological gradient $\operatorname{grad} f$ of a function $f \in C^\infty(X)$ in $x \in X$ is defined as the solution of

$$g_x(\operatorname{grad} f, \mathrm{d}_c) = \mathrm{d}_c(f).$$

REMARK 2.20. The existence of a diffeological gradient for a diffeological Riemannian space is not guaranteed. Moreover, so far we do not have any conditions to guaranty a diffeological gradient which does not result in a trivial diffeological Riemannian space. This should not be surprising, since even in the more restrictive setting of infinite dimensional manifolds the existence of a gradient is not guaranteed.

2.2.4. *Towards updates of iterates: A diffeological retraction.* In order to update the iterates, one generally uses the exponential map. However, the exponential map is an expensive operation in optimization techniques. Thus, one uses so-called retractions to update the iterates.

DEFINITION 2.21 (Diffeological retraction). Let X be a diffeological space. A diffeological retraction of X is a map $\mathcal{R}\colon TX \to X$ such that the following conditions hold:

(i) $\mathcal{R}_{|T_x X}(0) = x$
(ii) Let $\xi \in T_x X$ and $\gamma_\xi\colon T_0\mathbb{R} \to T_x X,\ t \mapsto \mathcal{R}_{|T_x X}(t\xi)$. Then $T_0\gamma_\xi(0) = \xi$.

Since finding a diffeological retraction can be quite challenging, we also define the so-called weak diffeological retraction.

DEFINITION 2.22. Let X be a diffeological space and CX the tangent cone bundle of X. A *weak diffeological retraction* of X is a map $\mathcal{R}\colon CX \to X$ such that the following conditions hold:

(i) $\mathcal{R}_{|C_x X}(0) = x$
(ii) Let $\xi \in C_x X$ and $\gamma_\xi\colon C_0\mathbb{R} \to C_x X,\ t \mapsto \mathcal{R}_{|C_x X}(t\xi)$. Then, $C_0\gamma_\xi = \xi$.

In (ii), the map $C_0\gamma_\xi$ is a tangential cone map and given by $C_0\gamma_\xi\colon C_0\mathbb{R} \to C_0 X, \mathrm{d}_\alpha \mapsto \mathrm{d}_{\gamma_\xi \circ \alpha}$.

REMARK 2.23. In [**25**], there are not only retractions considered. The definition of a diffeological Levi-Civita connection as well as a proof of uniqueness of such a connection is also given. But similar to the Riemannian gradient a condition for the existence is not given.

2.2.5. *First order optimization techniques on diffeological spaces.* Considering algorithm 1, [**25**] came up with a diffeological version of the steepest descent algorithm using a retraction instead of an exponential map. For convenience, we repeat this algorithm below (cf. algorithm 2).

Require: Objective function f on a diffeological Riemannian space X;

diffeological retraction $\mathcal{R}$ on X.
Goal: Find the solution of $\min_{x\in X} f(x)$.
Input: Initial data $x_0 \in X$; constants $\hat{\alpha} > 0$ and $\sigma, \rho \in (0,1)$ for Armijo

backtracking strategy

for $k = 0, 1, \ldots$ **do**
[1] Compute $\text{grad}f(x_k)$ denoting the diffeological shape gradient of f in x_k.
[2] Compute Armijo backtracking step size:
Set $\alpha := \hat{\alpha}$.
while $f\big(\mathcal{R}_{|T_{x_k}X}(-t_k\text{grad}f(x_k))\big) > f(x^k) - \sigma\alpha \left\|\text{grad}f(x_k)\right\|^2_{T_{x_k}X}$
Set $\alpha := \rho\alpha$.
end while
Set $t_k := \alpha$.
[3] Set $x_{k+1} := \mathcal{R}_{|T_{x_k}X}(-t_k\text{grad}f(x_k))$.
end for

Algorithm 2: Steepest descent method on the diffeological space X with Armijo backtracking line search

REMARK 2.24. Algorithm 2 is applied to an example in [**25**]. For more details about the algorithm we refer the interested reader to the corresponding article.

So far, there is no convergence proof of the algorithm. A convergence proof would be of grate interest for the diffeological as well as for the optimization community. Research work on diffeological optimization theory needs to be done in the future to generalize shape optimization to a diffeological setting.

3. On mapping spaces

3.1. On the ILB setting. We consider in this subsection very particular class of Fréchet Lie groups, that are the projective limit of Banach Lie groups as infinite dimensional Lie groups, that is, where charts and atlases enjoy also the projective limit property. We rely heavily on [**72**] for terminology.

We must first note that infinite-differentiability in the Fréchet sense is equivalent to smoothness in the diffeological sense when we equip a Fréchet space/manifold with the diffeology comprising Fréchet infinitely-differentiable parametrisations. The framework that we develop here is more restrictive, but more simple and hence more accessible, than the one fully defined in [**72**]. In the sequel the abreviation ILB means "Inverse Limit of Banach".

DEFINITION 3.1 (ILB-Manifolds & Bundles).

(1) A Fréchet manifold M modeled on a Fréchet space F is an *ILB-manifold* if
 - there exists a sequence of Banach spaces (B_n) such that
   - $\forall n \in \mathbb{N}$, B_{n+1} is a subset of B_n, dense for the topology of B_n and such that the closed unit ball of B_{n+1} is a compact subset of the closed unit ball of B_n.
   - F is the projective limit of the sequence (B_n) as a topological vector space.

   With such assumptions $(F, (B_n)_{n\in\mathbb{N}})$ is called an *ILB vector space.*
 - There exists a family of Banach manifolds $\{M_n\}_{n\in\mathbb{N}}$ in which there is a smooth and dense inclusion $M_{n+1} \hookrightarrow M_n$ for each n, and such that there exists a Banach atlas on M_0 that restricts to an atlas on M_n for each n.

(2) An *ILB-map* f between two ILB-manifolds M and N is a smooth map $f\colon M \to N$ along with a family of smooth maps $\{f_n\colon M_n \to N_n\}$ such that f and all f_n commute with all inclusions maps $M \hookrightarrow M_{n+1} \hookrightarrow M_n$ and $N \hookrightarrow N_{n+1} \hookrightarrow N_n$.

(3) An *ILB-principal bundle* is an ILB-map between two ILB-manifolds $(\pi\colon P \to M,\ \pi_n\colon P_n \to M_n)$ such that for each n, the map $\pi_n\colon P_n \to M_n$ is a principal G_n-bundle where G_n is a Banach Lie group.

(4) An ILB-map F between two ILB-principal bundles $\pi\colon P \to M$ and $\pi'\colon P' \to M'$ is an *ILB-bundle map* if $F_n\colon P_n \to P'_n$ is a (G_n-equivariant) bundle map for each n. Note that F induces an ILB-map $M \to M'$. An ILB-bundle map is an *ILB-bundle isomorphism* if it has an inverse ILB-bundle map.

When one replaces the sequence of Banach spaces by a sequence of Hilbert spaces, we replace ILB by ILH (Inverse limit of Hilbert) in the definitions. Let us illustrate this setting with the two motivating examples of this setting. For this, we consider a smooth compact boundaryless manifold equipped with a fixed finite atlas

$$\{\phi_k : O_k \subset \mathbb{R}^m \to M\}_{k\in\mathbb{N}_N}.$$

EXAMPLE 3.2. The space $C^r(M,\mathbb{R}^n)$ of $r-$times continuously differentiable maps, equipped with the norm

$$\|f\|_{r,\{\phi_k\}_{k\in\mathbb{N}_N}} = \sup_{k\in\mathbb{N}_N}\sup_{i\in\mathbb{N}_r}\sup_{x\in M}\|D^i f\circ\phi_k\|_{L^i(\mathbb{R}^m,\mathbb{R}^n)}$$

is a Banach norm on $C^r(M,\mathbb{R}^n)$ which does not topologically depend on the choice of the atlas $\{\phi_k\}_{k\in\mathbb{N}_N}$. Then the sequence $(C^r(M,\mathbb{R}^n))_{r\in\mathbb{N}}$ has $C^\infty(M,\mathbb{R}^n)$ as a projective limit.

Let us now consider N a $n-$dimensional (locally compact and paracompact) Riemannian manifold. Following [**36**], the exponential map on E defines pointwise an atlas on $C^0(M,N)$, with domain in $C^0(M,\mathbb{R}^n)$, which by restriction of the domain, defines a structure of smooth Banach manifold on $C^r(M,N)$, modeled on $C^r(M,\mathbb{R}^n)$, for $r \in \mathbb{N}\cup\{\infty\}$. This defines an ILB manifold. If moreover we replace G by a Lie finite dimensional group, we get an ILB Lie group.

Let us keep the notations of the last example. There always exist a smooth embedding of N into $\mathbb{R}^{2n+1}$ following e.g. [**32**]. We now assume that this is a Riemannian embedding, that is, the Riemannian metric on N is the pull-back of the

Riemannian (Euclidian) metric on $\mathbb{R}^{2n+1}$. We also assume that M is Riemannian. Therefore, for the Riemannian volume dx on M, the space $L^2(M, \mathbb{R}^{2n+1})$, defined as the completion of $C^\infty(M, \mathbb{R}^{2n+1})$ for the norm

$$\|f\|_{L^2} = \int_M f^2 dx$$

is a Hilbert space. Moreover, if Δ is the (positive) Laplacian on M, one can define the Sobolev spaces $H^s(M, \mathbb{R}^{2n+1})$ for $s \in \mathbb{R}$ as the completion of $C^\infty(M, \mathbb{R}^{2n+1})$ for the norm

$$\|f\|_{H^s} = \int_M \left((1+\Delta)^{s/2} f\right)^2 dx,$$

where the real power $(1+\Delta)^s$ is defined as the operator (unbounded for $s > 0$) with eigenvalue $(1+\lambda_u)^s$ for the eigenvector u of Δ with respect to the eigenvalue $\lambda_u \geq 0$. For completeness, we have to recall that Δ extends to an unbounded, self-adjoint and positive operator on $L^2(M, \mathbb{R}^{2n+1})$, with smooth eigenvectors, which explains the deep ease for taking some real powers of $(1 + \Delta)$. With these definitions, a special case of the so-called Sobolev embedding theorems is:

THEOREM 3.3. *The inclusion map $H^{s+k}(M, \mathbb{R}^{2n+1}) \subset C^k(M, \mathbb{R}^{2n+1})$, defined $\forall s > m/2$, is a bounded map and $H^{s+k}(M, \mathbb{R}^{2n+1})$ is a dense subset of $C^k(M, \mathbb{R}^{2n+1})$ for the C^k-topology.*

We can now turn to our second example, following again [**36**] as a main reference, completed by [**72**].

EXAMPLE 3.4. Let $s > m/2$. The set

$$H^s(M, N) = H^s(M, \mathbb{R}^{2n+1}) \cap C^0(M, N)$$

is a (smooth) Hilbert manifold modeled on $H^s(M, \mathbb{R}^n)$, and this family defines ILH structures on $C^\infty(M, N)$. The atlas on $H^s(M, N)$ is the restriction of the atlas on $C^0(M, N)$ where charts are defined pointwise by the exponential map on N. As a consequence, $H^s(M, N)$ is also the closure of $C^\infty(M, N)$ in $H^s(M, \mathbb{R}^{2n+1})$.

After these two examples, we have here to insist on the fact that there exist examples where refined properties of Fréchet/Banach manifold require more attention. One can see for instance the manifold of paths defined in [**88**].

We now turn to a first point where diffeologies invite themselves in the ILB setting.For this, we consider a ILB space $(F, (B_n)_{n\in\mathbb{N}})$. The linear groups $GL(B_n)$ are Banach Lie groups, as an immediate consequence of the implicit functions theorem on the Banach algebra $L(B_n)$ of bounded linear maps on B_n, while example 1.37 explains that the use of a diffeology for $GL(F)$ seem to be more natural, even if it differs from the subset diffeology inherited from the functional diffeology of $L(F)$.

This problem has been highlighted in the first works concenring ILB spaces. Indeed, in [**72**], H. Omori considers the spaces

$$\mathcal{L}_n = \cap_{k=0}^n L(B_k)$$

and invokes "natural differentiation" on

$$\mathcal{L}_\infty = \cap_{n\in\mathbb{N}} \mathcal{L}_k$$

as well as on

$$G\mathcal{L}_\infty = \cap_{n\in\mathbb{N}} GL(B_k).$$

Obviously, $G\mathcal{L}_\infty$ cannot be equipped with any atlas modeled on $\mathcal{L}_\infty$ in the actual state of knowledge, that is, we cannot construct explicitely any such atlas with the actual known techniques. Therefore, the extension of the definition of e.g. connections to an ILB setting require many refinements that can be found in [**89**].

In the context of diffeologies, we have another way to consider smoothness of $G\mathcal{L}_\infty$ through the following result:

THEOREM 3.5. [**51**] *Let $(G_n)_{n\in\mathbb{N}}$ be a sequence Banach Lie groups, decreasing for $\subset$, for which inclusion maps are smooth. Then $G = \bigcap_{n\in\mathbb{N}} G_n$ is a regular Frölicher Lie group.*

This theorem obviously applies to $G\mathcal{L}_\infty$.

3.2. Mappings with non compact source. We first assume that E is a vector bundle with typical fiber V. Let us first consider a fixed local section s of E, with (open) domain D. For any open subset of M such that $\bar{O} \subset D$, the evaluation map is smooth. Since M is paracompact, one can define a smooth map $\phi : M \to \mathbb{R}_+$ such that

$$\phi(x) = \begin{cases} 1 & \text{if } x \in O \\ 0 & \text{if } x \notin D \end{cases}$$

thus, talking about smoothness with respect to local section, the same is to talk about smoothness with respect to global sections. For the same reasons, one can assume M to be Riemannian. These remarks combined with the existence of local trivialisations on E, we wish to characterize all the topologies on smooth sections of E, that we note by $C^\infty(M, E)$, such that the evaluation maps

$$\mathrm{ev} : (x, s) \in M \times C^\infty(M, E) \mapsto \mathrm{ev}_x(s) = s(x) \in E$$

is continuous and moreover, for a given local trivialization $E|_U \sim U \times V$ of E over the open subset U of M, and $\forall \alpha \in \mathbb{N}^*$, the map

$$\mathrm{ev}_U^{(\alpha)} : (x, s) \in M \times C^\infty(M, E) \mapsto \mathrm{ev}_{U,x}^{(\alpha)}(s) = D_x^\alpha s \in L^\alpha(\mathbb{R}^m, E)$$

is continuous. For this purpose, the weaker topology is the pull-back topology on $C^\infty(M, E)$ through the family of maps $\left\{\mathrm{ev}, \mathrm{ev}_U^{(\alpha)}\right\}$. This is a standard exercise to show that this topology coincides with the topology defined, for a locally finite family $\{U_i \,|\, i \in I\}$ of open subset of M over which E is trivial, by the semi-norms $\|.\|_{\alpha,K}$ indexed by $\alpha \in \mathbb{N}$ and by the compact subsets K of M defined, for $s \in C^\infty(M, E)$, by

$$\|s\|_{0,K} = \sup_{x\in K} \|s(x)\|$$

and, for $\alpha > 0$,

$$\|s\|_{\alpha,K} = \sup\left\{ \sup_{x\in K\cap U_i} \|D^\alpha s(x)\| \,|\, U_i \cap K \neq \emptyset \right\}.$$

In other words, this is the topology of uniform convergence of derivatives at any order of sections s, on any compact subset K of M. With this topology, no global estimate on S is assumed.

Let us now assume that E is a fiber bundle with typical fiber F which is a smooth (finite dimensional) manifold for dimension k. Let $\{V_j \,|\, j \in J\}$. be a locally finite open cover of F such that each open domain V_j is of compact closure in F. Each V_j is identified (by the chosen atlas on F) with an open subset $\tilde{V}_j$ of $\mathbb{R}^k$. Let

$U_i \times F$ be a local trivialization of E. then on $\mathrm{ev}^{-1}(V_j) \subset U_i \times C^\infty(U_i, F)$ one can consider the distances $d_{i,j,K,\alpha}$ defined by the semi-norms $\|.\|_{\alpha,K}$ on (local) smooth sections with domain in U_i with values in $V_j \sim \tilde{V}_j \subset \mathbb{R}^k$. By the way, *$C^\infty(M,E)$ is a metrizable topological space* for the weaker topology which makes $\left\{\mathrm{ev}, \mathrm{ev}_U^{(\alpha)}\right\}$ be a family of continuous maps. *However, if M is not compact, there is actually no atlas for which $C^\infty(M,N)$ is a C^0-manifold for this topology* in contrast with the case when M is compact described before. In order to circumvent this problem, there are two ways:

- consider an exhaustive sequence of compact subsets of M, that we note by $(K_n)_{n\in\mathbb{N}}$ and remark that the topology of $C^\infty(M,E)$ is the inductive limit of the Fréchet manifold topologies of the increasing family
$$\{C^\infty(K_n, E|_{K_n}) \,|\, n \in \mathbb{N}\}.$$
- consider the Frölicher structure on $C^\infty(M,E)$ that makes the jet maps
$$j^k : f \in C^\infty(M,E) \to J^k(M,E), \text{ for } k \geq 1$$
and
$$j^\infty : f \in C^\infty(M,E) \to J^\infty(M,E)$$
smooth (see [**26**] for basic definitions on jet spaces, or next section).

Other stronger topologies exist in $C^\infty(M,N)$. In order to be not so incomplete, we have to mention here the smooth Whitney topology [**26**], which is also based on Jet spaces, and for which $C^\infty(M,N)$ is a complete locally convex smooth manifold [**42**].

3.3. Jets with non-compact source and a digression on geometric theory of PDEs. Following [**59**, Appendix 2], the last structures on $C^\infty(M,E)$ extend to $J^k(M,E)$ and to $J^\infty(M,E)$. For this, we consider a finite dimensional fiber bundle E with typical fiber F (with $\dim(F) = n$) over a smooth manifold M of dimension m, which can be non compact. In the exposition, we also consider a generic trivialization $\varphi : U \times F \to E$ of E over an open subset U of M. The manifold M is the space of independent variables x_i, $1 \leq i \leq n$, and the typical fiber is the space of dependent variables u^α, $1 \leq \alpha \leq m$. The reader may assume that the bundle π is simply $\mathbb{R}^n \times \mathbb{R}^m \to \mathbb{R}^n$: even if it looks like we are "trivializing" everything because we are working locally, the fact that we are interested in properties of differential equations and their solutions makes geometry highly non-trivial, as we show below.

Let $s_1(x_i) = (x_i, s_1^\alpha(x_i))$ and $s_2(x_i) = (x_i, s_2^\alpha(x_i))$ be two local sections of the bundle $E \to M$ defined about a point $p = (x_i)$ in M. We say that s_1 and s_2 *agree to order k at $p \in M$* if s_1, s_2 and *all* the partial derivatives of the sections s_1 and s_2, up to order k, agree at p. This notion determines a coordinate-independent equivalence relation on local sections of E. We let $j^k(s)(p)$ represent the equivalence class of the section s at p, and we call this equivalence class *the k–jet of s at p*.

DEFINITION 3.6. The *k–order jet bundle of E* is the space
$$J^k E = \bigcup_{p \in M} J^k(p),$$
in which $J^k(p)$ denotes the set of all the k–jets $j^k(s)(p)$ of local sections s at p.

The jet bundle J^kE possesses a natural manifold structure; it fibers over J^lE ($l < k$) and also over M: local coordinates on J^kE are $(x_i, u_0^\alpha, u_{i_1}^\alpha, u_{i_1 i_2}^\alpha, \dots, u_{i_1\dots i_k}^\alpha)$, in which

$$u_0^\alpha(j^k(s)(p)) = s^\alpha(p)\,, \qquad u_i^\alpha(j^k(s)(p)) = \frac{\partial s^\alpha}{\partial x_i}(p)\,, \qquad u_{i_1 i_2}^\alpha(j^k(s)(p)) = \frac{\partial^2 s^\alpha}{\partial x_{i_1}\partial x_{i_2}}(p)\,,$$

and so forth, where $j^k(s)(p) \in J^kE$ and $(x^i) \mapsto (x^i, s^\alpha(x^i))$ is any local section in the equivalence class $j^k(s)(p)$. In these coordinates, the projection map $\alpha^k : J^kE \to M$ (source map) factors through a projection map $\pi^k : J^k(E) \to E$ for which, on a local generic trivialization φ as defined before, where $E \sim U \times F$ locally, $\pi^k = \alpha^k \times \beta^k$ where β^k is the local target map. Moreover, the projections $\pi_l^k : J^kE \to J^lE$, $l < k$, are defined in obvious ways. For instance, π_M^k is simply $(x_i, u_0^\alpha, u_{i_1}^\alpha, \dots, u_{i_1\dots i_k}^\alpha) \mapsto (x_i)$. Let us precise a little more: for $p \in E$, $(\pi_k)^{-1}(p)$ is a vector space modelled on

$$\bigoplus_{i=1}^k L^i(\mathbb{R}^m, \mathbb{R}^n)$$

where $L^i(\mathbb{R}^m, \mathbb{R}^n)$ is the space of $i-$linear symmetric mappings on $\mathbb{R}^m$ with values in $\mathbb{R}^n$. Therefore, $J^k(M,E)$ can be understood as:

- a finite dimensional vector bundle over E with typica fiber $\bigoplus_{i=1}^k L^i(\mathbb{R}^m, \mathbb{R}^n)$
- a finite dimensional fiber bundle over M with typical fiber
$$F \times \left(\bigoplus_{i=1}^k L^i(\mathbb{R}^m, \mathbb{R}^n) \right).$$

An extended exposition of the properties of $k-$jets can be found in [**26**].

Any local section $s : (x_i) \mapsto (x_i, s^\alpha(x_i))$ of E lifts to a unique local section $j^k(s)$ of J^kE called the *kth prolongation* of s. In coordinates, $j^k(s)$ is the section

$$j^k(s) = \left(x_i, s^\alpha(x_i), \frac{\partial s^\alpha}{\partial x_{i_1}}(x_i), \dots, \frac{\partial^k s^\alpha}{\partial x_{i_1}\dots\partial x_{i_k}}(x_i), \dots \right).$$

The *infinite jet bundle* $\pi_M^\infty : J^\infty E \to M$ is the inverse limit of the tower of jet bundles $M \leftarrow E \cdots \leftarrow J^kE \leftarrow J^{k+1}E \leftarrow \cdots$ under the standard projections $\pi_l^k : J^kE \to J^lE$, $k > l$. We have here to precise that we consider the inverse limit and not the inductive limit. That is, comparing with $J^k(M,E)$ which is a finite dimensional vector bundle over E, taking the inverse limit, we also get $J^\infty(M,E)$ as a vector bundle over E with typical fiber the *formal series*

$$\sum_{i=1}^{+\infty} L^i(\mathbb{R}^m, \mathbb{R}^n).$$

We describe the space $J^\infty E$ locally, by sequences

$$(x_i, u^\alpha, u_{i_1}^\alpha, \dots, u_{i_1 i_2 \dots i_k}^\alpha, \dots)\,, \quad 1 \le i_1 \le i_2 \le \dots \le i_k \le n\,, \tag{3.1}$$

obtained from the standard coordinates on the finite–order jet bundles J^kE.

The limit $J^\infty E$ is a topological space: a basis for the topology on $J^\infty E$ is the collection of all sets of the form $(\pi_k^\infty)^{-1}(W)$, in which W is an open subset of J^kE. We would like to consider $J^\infty E$ as a manifold which needs to be at least an infinite dimensional manifold/ we need to re-define all of the standard differential geometry notions in this new context, see [**3, 4**]:

Let $\pi_k^\infty : J^\infty E \to J^k E$ denote the canonical projection from $J^\infty E$ onto $J^k E$. A function $f : J^\infty E \to \mathbb{R}$ is *smooth* if it factors through a finite-order jet bundle, that is, if $f = f_k \circ \pi_k^\infty$ for some smooth function $f_k : J^k E \to \mathbb{R}$. Real-valued smooth functions on $J^\infty E$ are also called *differential functions*, see [**70**].

This notion of smoothness coincides with the notion of Frölicher space, where a generating set of functions $\mathcal{F}_0$ serves to define which set of mapping is smooth. Moreover, considering the typical fibers $\sum_{i=1}^{+\infty} L^i(\mathbb{R}^m, \mathbb{R}^n)$ of $J^\infty(M,E)$ as a vector bundle over E, since each $L^i(\mathbb{R}^m, \mathbb{R}^n)$ is finite dimensional, it us easy to prove that $\sum_{i=1}^{+\infty} L^i(\mathbb{R}^m, \mathbb{R}^n)$ is a Fréchet space, equipped with the semi-norms

$$\|.\|_i = \|.\|_{L^i(\mathbb{R}^m,\mathbb{R}^n)} \circ \pi_i$$

where $\pi_i : \sum_{i=1}^{+\infty} L^i(\mathbb{R}^m, \mathbb{R}^n) \to L^i(\mathbb{R}^m, \mathbb{R}^n)$ is the canonical projection.

When one tries to define the tangent space $TJ^\infty(M,E)$, it is natural to consider all derivations X on $\mathcal{F}_0$. In this very particular setting, by restriction of the derivation X to $C^\infty(J^k(M,E),\mathbb{R}) \subset \mathcal{F}_0$, since $J^k(M,E)$ is a finite dimensional manifold, then the restriction of X to $J^k(M,E)$ can be expressed as a (classical) vector field over $J^k(M,E)$ and, by inverse limit procedure, $TJ^\infty(M,E)$ is also the inverse limit of the sequence

$$TJ^1(M,E) \leftarrow \cdots \leftarrow TJ^k(M,E) \leftarrow \ldots$$

which shows that Olver's definition of $TJ^\infty(M,E)$ coincides with the standard definition of the tangent space to the Fréchet manifold $J^\infty(M,E)$.

These structures are useful to talk about partial differential equations understood as *partial differential relations* along the lines of e.g. [**27**]. Indeed, a basic partial differential equation of the form $A(u) = 0$, where A is a linear differential operator of order k, envolves a solution u and its partial derivatives till the order k, and hence can be encoded as $\Xi(j^k(u)) = 0$ where

$$\Xi : J^k(M,\mathbb{R}) \to \mathbb{R}$$

is a smooth map. The same holds for most non-linear partial differential equations of interest, but it can occur that the space $\Xi^{-1}(0)$ can be equipped with many "singularities" as shows this very non-linear example based on a non-linear differential equation:

EXAMPLE 3.7. When $\Xi(x,y,y') = x\cos(y)\sin(4y')$, the subset $\Xi^{-1}(0)$ has an infinity of singular points. Let us assume that we restrict $\mathcal{E}$ to a neighbourhood U of a regular point. The solutions $x \mapsto (x, y(x))$ we would find are simply constant solutions or affine solutions and we may lose information depending on our choice of U. Now, solutions extend to piecewise affine (and hence only piecewise smooth) solutions on $\Xi^{-1}(0)$.

The problem of the geometry of singularities needs the definition of a generalized way to define differential geometric objects.

Let us now consider a partial differential equation $\Xi = 0$ in $J^k(M,E) \subset J^\infty(M,E)$. Following Example 3.7, this is a bit too optimistic to hope that what we call here *the vanishing set* along the lines of [**59**, Appendix 2], which is the subset $\Xi^{-1}(0)$, has a structure of smooth manifold. Therefore, in the classical literature, one invokes the use of a *locus*, that is, a smooth submanifold of $J^k(M,E)$ with range in $\Xi^{-1}(0)$. Typically, this locus can be defined by an unique embedding of an open subset of an Euclidian space O to $J^k(M,E)$. This enables to define tangent space, differential forms and other objects on $\{\Xi = 0\}$, actually with a slightly non rigorous arguments concerning their global definition on the full space $\Xi^{-1}(0)$. until [**59**] in the context of diffeologies. Indeed, the subset diffeology defines in a natural way a diffeological structure on the vanishing set, which completes the basic setting for the geometry on the main basic example of *diffieties* given by $\Xi^{-1}(0)$, along the lines of [**60**].

REMARK 3.8. The spaces of jets are also useful for alternate definitions of connections and their generalizations, see [**41**], and an application of diffeologies to higher connections may be feasible. We do not have enough place here to develop this aspect of the theory.

3.4. Mappings with low regularity. In the sequel we consider an algebra $\mathcal{M}$ with complex coefficients, with the hermitian product of matrices $(A, B) \mapsto tr(AB^*)$. If there is no possible confusion, we note this matrix norm by $\|.\|$ or by $\|.\|_{\mathcal{M}}$ if necessary. We now extend the construction of Example 3.4 the following way, along the lines of [**52**].

DEFINITION 3.9. Let $s \in \mathbb{R}$, let M be a compact boundaryless Riemannian manifold and let N be a Riemannian manifold embedded in $\mathcal{M}$. We define $H^s(M, N)$, as the completion of $C^\infty(M, N)$ in $H^s(M, \mathcal{M})$.

We remark that the condition $s < \dim(M)/2$ is not assumed here. This implies a weaker differential structure on $H^s(M; \mathcal{M})$ when $s \leq dim(M)/2$ which can be viewed as follows, generalizing [**52**] in a straightforward way:

PROPOSITION 3.10. *$H^s(M, N)$ is a diffeological space, equipped with the subset diffeology inherited from $H^s(M, \mathcal{M})$.*

Let us now consider a compact connected Lie group G of matrices.

For $s > \dim(M)/2$, it is well-known, that $H^s(M, G)$ is a Hilbert Lie group. The case where $M = S^1$, that is the loop space or the loop group, is the most widely studied in the literature, see e.g. [**78**]. The biggest Sobolev order where $H^s(S^1, G)$ fails to be a Hilbert manifold, and also a group, is $s = 1/2$.

PROPOSITION 3.11. *Let $s \leq 1/2$.. Then $H^s(S^1, G)$ and $H^s_0(S^1, G)$ are Frölicher space.*

This proposition completes the classical setting, in which the study of the $H^{1/2}-$metric is a key point, from a geometric viewpoint, see e.g. [**19, 78**], as from the viewpoint of shape analysis, see e.g. [**92**].

3.5. Groups of diffeomorphisms are regular... or not. Let M be a locally compact, non-compact manifold, which is assumed to be Riemannian without restriction, equipped with its nébuleuse diffeology. We equip the group of diffeomorphisms $\mathrm{Diff}(M)$ with the topology of convergence of the derivatives an any order, uniformly on each compact subset of M, usually called $C^\infty-$compact-open topology or weak topology in [**32**]. Traditionnally, $\mathrm{Vect}(M)$ is given as the Lie algebra of $\mathrm{Diff}(M)$, but [**42**, section 43.1] shows that this strongly depends on the topology of $\mathrm{Diff}(M)$. Indeed, the Lie algebra of vector fields described in [**42**, section 43.1] is the Lie algebra of compactly supported vector fields, which is not the (full) Lie algebra $\mathrm{Vect}(M)$. In another context, when M is compact, $\mathrm{Vect}(M)$ is the Lie algebra of $\mathrm{Diff}(M)$, which can be obtained by Omori's regularity theorems [**71, 72**] and recovered in [**12**]. What is well known is that infinitesimal actions of Diff(M) on $C^\infty(M, \mathbb{R})$ generate vector fields, viewed as order 1 differential operators. The bracket on vector fields is given by

$$(X, Y) \in \text{Vect}(M) \mapsto [X, Y] = \nabla_X Y - \nabla_Y X,$$

where ∇ is the Levi-Civita connection on TM. This is a Lie bracket, stable under the Adjoint action of Diff(M). Moreover, the compact-open topology on Diff(M) generates a corresponding $C^\infty-$compact-open topology on Vect(M). This topology is itself the $D-$topology for the the functional diffeology on Diff(M). Following [**48**, Definition 1.13 and Theorem 1.14], Vect(M) equipped with the C^∞ compact-open topology is a Fréchet vector space, and the Lie bracket is smooth. Moreover, we feel the need to remark that the evaluation maps

$$T^*M \times Vect(M) \to \mathbb{R}$$

separate Vect(M). Thus Diff(M) is a diffeological Lie group matching with the criteria of [**48**, Definition 1.13 and Theorem 1.14].

Let F be the vector space of smooth maps $f \in C^\infty(]0;1[;\mathbb{R})$. We equip F with the following semi-norms:

For each $(n,k) \in \mathbb{N}^* \times \mathbb{N}$,

$$\|f\|_{n,k} = \sup_{\frac{1}{n+1} \leq x \leq \frac{n}{n+1}} |D_x^k f|.$$

This is a Fréchet space, and its topology is the smooth compact-open topology, which is the D topology of the compact-open diffeology. Let

$$\mathcal{A} = \{f \in C^\infty(]0;1[;]0;1[) | \lim_{x\to 1} f(x) = 1 \wedge \lim_{x \to 0} f(x) = 0\}.$$

Finally, we set

$$\mathcal{D} = \{f \in \mathcal{A} | \inf_{x \in]0;1[} f'(x) > 0\}.$$

$\mathcal{D}$ is a contractible set of diffeomorphisms of the open interval $]0;1[$ which is an (algebraic) group for composition of functions. Composition of maps, and inversion, is smooth for the functional diffeology. Unfortunately, $\mathcal{D}$ is not open in $\mathcal{A}$. As a consequence, we are unable to prove that it is a Fréchet Lie group. However, considering the smooth diffeology induced on $\mathcal{D}$ by $\mathcal{A}$, the inversion is smooth. As a consequence, $\mathcal{D}$ is (only) a diffeological Lie group.

Following [**54**], we can state:

THEOREM 3.12. *$\mathcal{D}$ is a non-regular Frölicher Lie group.*

We now give a consequence of this first theorem for the non-regularity of the group of diffeomorphisms of a non-compact boundaryless manifold, following [**54**].

THEOREM 3.13. *$Diff(M)$ is a non-regular Frölicher Lie group.*

REMARK 3.14. In this statement and as we stated before, Diff(M) is equipped with a topology which is called weak topology in some parts of the literature. It appears to be one of the weakest topologies, inherited from a Frölicher structure, that make evaluation maps continuous and smooth. This is in great contrast with the large class of examples of manifold structures on Diff(M) for which Diff(M) is regular [**42, 43**]. Therefore, as in the context of loop spaces described before, the geometry and the topology depend a lot on the chosen geometric structures.

3.6. Geometry of finite and infinite triangulations. The notion of triangulation is itself a bit confuse in the litterature. One can define:

- either parametrized smooth triangulations in a manifold M:

  DEFINITION 3.15. A *smooth triangulation* of M is a family $\tau = (\tau_i)_{i \in I}$ where $I \subset \mathbb{N}$ is a set of indexes, finite or infinite, each τ_i is a smooth map $\Delta_n \to M$, and such that:
  1. $\forall i \in I, \tau_i$ is a (smooth) embedding, i.e. a smooth injective map such that $(\tau_i)_* (\mathcal{P}(\Delta_n))$ is also the subset diffeology of $\tau_i(\Delta_n)$ as a subset of M.
  2. $\bigcup_{i \in I} \tau_i(\Delta_n) = M$. (covering)
  3. $\forall (i,j) \in I^2$, $\tau_i(\Delta_n) \cap \tau_j(\Delta_n) \subset \tau_i(\partial\Delta_n) \cap \tau_j(\partial\Delta_n)$. (intersection along the borders)
  4. $\forall (i,j) \in I^2$ such that $D_{i,j} = \tau_i(\Delta_n) \cap \tau_j(\Delta_n) \neq \emptyset$, for each $(n-1)$-face F of $D_{i,j}$, the "transition maps" "$\tau_j^{-1} \circ \tau_i$" $: \tau_i^{-1}(F) \to \tau_j^{-1}(F)$ are affine maps.

- or non-parametrized smooth triangulations, which consider only the images in M of the tarndard simplex through a smooth simplex of the triangulation.

In open domains $\Omega \subset \mathbb{R}^n$, one often restricts to **affine** triangulations, for which, for each $k \in \mathbb{N}_n$, $(n-k)-$ dimensional faces lie in $(n-k)-$dimensional affine subspaces of $\mathbb{R}^n$ and, in the parametrized case, embeddings are (restrictions of) affine maps.

3.6.1. *Diffeologies associated to a fixed triangulation.* There exists actually many diffeologies on a fixed triangulated manifold, and more generally, on a CW-complex and the first described were in [**67, 68**] many years before [**14**], see e.g. [**39, 44, 45**] while other approaches are still in progress. Each of them has its own technical interest. We give here a selected one adapted to our needs. we begin with a lemma from [**53**] which is adapted from so-called gluing results present in [**67, 76, 77**] to the context which is of interest for us.

LEMMA 3.16. *Let us assume that X is a topological space, and that there is a collection $\{(X_i, \mathcal{F}_i, \mathcal{C}_i)\}_{i \in I}$ of Frölicher spaces, together with continuous maps $\phi_i : X_i \to X$. Then we can define a Frölicher structures on X setting*

$$\mathcal{F}_{I,0} = \{f \in C^0(X, \mathbb{R}) | \forall i \in I, \quad f \circ \phi_i \circ \mathcal{C}_i \subset C^\infty(\mathbb{R}, \mathbb{R})\},$$

wa define $\mathcal{C}_I$ the contours generated by the family $\mathcal{F}_{I,0}$, and $\mathcal{F}_I = \mathcal{F}(\mathcal{C}_I)$.

Let M be a smooth manifold for dimension n. Let

$$\Delta_n = \{(x_0, \ldots, x_n) \in \mathbb{R}_+^{n+1} | x_0 + \cdots + x_n = 1\} \tag{3.2}$$

be the standard $n-$ simplex, equipped with its subset diffeology. It is easy to show that this diffeology is reflexive through Boman's theorem already mentionned, and hence we can call it Frölicher space $(\Delta_n, \mathcal{F}_{\Delta_n}, \mathcal{C}_{\Delta_n})$, and we denote its associated reflexive diffeology by $\mathcal{P}(\Delta_n)$. Under these conditions, we equip the triangulated manifold (M, τ) with a Frölicher structure $(\mathcal{F}_I, \mathcal{C}_I)$, generated by the smooth maps τ_i applying Lemma 3.16. The following result is obtained from the construction of $\mathcal{F}$ and $\mathcal{C}$:

THEOREM 3.17. *The inclusion $(M, \mathcal{F}, \mathcal{C}) \to M$ is smooth.*

3.6.2. *Geometry of the space of triangulations.* Let us now fully develop an approach based on the remarks given in [**53**]. For this, the space of triangulations of Ω is considered itself as a Frölicher space, and the mesh of triangulations which makes the finite element method converge will take place, as the function f, among the set of parameters Q. We describe here step by step the Frölicher structure on the space of triangulations. By the way,

Under these conditions, we equip the triangulated manifold (M, τ) with a Frölicher structure $(\mathcal{F}_I, \mathcal{C}_I)$, generated by the smooth maps τ_i applying Lemma 3.16. The following result from [**55**] is obtained from the construction of $\mathcal{F}$ and $\mathcal{C}$:

THEOREM 3.18. *The inclusion $(M, \mathcal{F}, \mathcal{C}) \to M$ is smooth.*

REMARK 3.19. Maps in $\mathcal{F}_I$ can be intuitively identified as some piecewise smooth maps $M \to \mathbb{R}$, which are of class C^0 along the 1-skeleton of the triangulation. We have proved also that $\mathcal{C}_I \subset \mathcal{P}_\infty(M)$. Some characteristic elements of $\mathcal{C}_I$ can be understood as paths which are smooth (in the classical sense) on the interiors of the domains of the simplexes of the triangulation, and that fulfill some more restrictive conditions while crossing the 1-skeleton of the triangulation. For example, paths that are (locally) stationnary at the 1-skeleton are in $\mathcal{C}_I$.

REMARK 3.20. While trying to define a Frölicher structure from a triangulation, one could also consider

$$\mathcal{C}_{I,0} = \left\{\gamma \in C^0(\mathbb{R}, M) \,|\, \forall i \in I, \forall f \in C^\infty_c(\phi_i(\Delta_n), \mathbb{R}), f \circ \gamma \in C^\infty(\mathbb{R}, \mathbb{R})\right\}$$

where $C^\infty_c(\phi_i(\Delta_n), \mathbb{R})$ stands for compactly supported smooth functions $M \to \mathbb{R}$ with support in $\phi_i(\Delta_n)$. Then define

$$\mathcal{F}'_I = \{f : M \to \mathbb{R} \,|\, f \circ \mathcal{C}_{I,0} \in C^\infty(\mathbb{R}, \mathbb{R})\}$$

and

$$\mathcal{C}'_I = \{C : \mathbb{R} \to M \,|\, \mathcal{F}'_I \circ c \in C^\infty(\mathbb{R}, \mathbb{R})\}.$$

We get here another construction, but which does not understand as smooth maps $M \to \mathbb{R}$ the maps δ_k already mentionned.

Now, let us fix the set of indexes I and fix a so-called *model triangulation* τ. This terminology is justified by two ideas:

- Anticipating next constructions, this model triangulation τ will serve at defining a sequence of refined trinagulations. This is our "starting triangulation" for the refinement procedure in the finite elements method.
- Changing τ into $g \circ \tau$, where g is a diffeomorphism, we get another model triangulation, which has merely the same properties as τ. But each "starting" trinagulation cannot be obtained by transforming a fixed triangulation by using a diffeomorphism. For example, on the 2-sphere, a tetrahedral triangulation τ_1 and an octahedral triangulation τ_2 separately generate two sequences of refined triangulations, and there is a topological obstruction for changing τ_1 into τ_2 by the action of a diffeomorphism of the sphere.

We denote by $\mathcal{T}_\tau$ the set of triangulations τ' of M such that the corresponding 1-skeletons are diffeomorphic to the 1-skeleton of τ (in the Frölicher category). The set $\mathcal{T}_\tau$ contains, but is not reduced to, the orbit of τ by the action of the group of diffeomorphisms. Indeed, one can reparametrize each simplex with adequate compatibility on the border. Intuitively speaking, reparametrizations need not to be smooth in the usual sense while "crossing the border of a simplex". This choice is motivated by the Frölicher structure that we identify as useful for the finie elements method, ddefined hereafter.

DEFINITION 3.21. Since $\mathcal{T}_\tau \subset C^\infty(\Delta_n, M)^I$, we can equip $\mathcal{T}_\tau$ with the subset Frölicher structure, in other words, the Frölicher structure on $\mathcal{T}_\tau$ whose generating family of contours $\mathcal{C}$ are the contours in $C^\infty(\Delta_n, M)^I$ which lie in $\mathcal{T}_\tau$.

We define the full space of triangulations $\mathcal{T}$ as the disjoint union of the spaces of the type $\mathcal{T}_\tau$, with disjoint union Frölicher structure. With this notation, in the sequel and when it carries no ambiguity, the triangulations in $\mathcal{T}_\tau$ is equipped with a fixed set of indexes I (which is impossible to fix for $\mathcal{T}$). We need now to describe the procedure which intends to refine the triangulation and define a sequence of triangulations $(\tau_n)_{n\in\mathbb{N}}$. We can now consider the refinement operator, which is the operator which divides a simplex Δ_n into a triangulation.

DEFINITION 3.22. Let $m \in \mathbb{N}$, with $m \geq 3$. Let

$$\mu = \{\mu_i : \Delta_n \to \Delta_n \,|\, i \in \mathbb{N}_m \}$$

be a smooth triangulation of Δ_n Let $\tau \in \mathcal{T}$. Then we define

$$\mu(\tau) = \{f_i \circ \mu_i \,|\, i \in \mathbb{N}_m \text{ and } \tau = (f_i)_{i\in I}\}.$$

We say that μ defines a *refinement map* if $\forall n \in \mathbb{N}^*, \mu^n(\tau)$ is a triangulation.

With this definition, $\mu(\tau)$ is trivially a triangulation of M if τ is a triangulation of M. The conditions imposed in the definition ensures that the refinement map maps a triangulation to another triangulation, that is, if τ is a triangulation, $\mu(\tau)$ is also a triangulation. The delicate needed condition is that the new 0-vertices added to *tau* in $\mu(\tau)$ are matching.

DEFINITION 3.23. Let $\tau \in \mathcal{T}$. We define the $\mu-$refined sequence of triangulations $\mu^{\mathbb{N}}(\tau) = (\tau_n)_{n\in\mathbb{N}}$ by

$$\begin{cases} \tau_0 & = & \tau \\ \tau_{n+1} & = & \mu(\tau_n) \end{cases}$$

PROPOSITION 3.24. *The map*

$$\mu^{\mathbb{N}} : \mathcal{T} \to \mathcal{T}^{\mathbb{N}}$$

is smooth (with $\mathcal{T}^{\mathbb{N}}$ equipped with the infinite product Frölicher structure).

Let Ω be a bounded connected open subset of $\mathbb{R}^n$, and assume that the border $\partial\Omega = \bar{\Omega} - \Omega$ is a polygonal curve. Since $\mathbb{R}^n$ is a vector space, we can consider the space of affine triangulations:

$$\text{Aff}\, \mathcal{T}_\tau = \{\tau' \in \mathcal{T}_\tau | \forall i, \tau'_i \text{ is (the restriction to } \Delta_n \text{ of) an affine map } \}.$$

We define Aff $\mathcal{T}$ from Aff $\mathcal{T}_\tau$ the same way we defined $\mathcal{T}$ from $\mathcal{T}_\tau$, via disjoint union. We equip $\text{Aff}(\mathcal{T}_\tau)$ and $\text{Aff}(\mathcal{T})$ with their subset diffeology. We use here the notations of last Lemma.

THEOREM 3.25. *Let*

$$c : \mathbb{R} \to \text{Aff}(\mathcal{T}_\tau)$$

be a path on $\text{Aff}(\mathcal{T}_\tau)$. *Then*

$$c \text{ is smooth} \Leftrightarrow \forall (i,j) \in I \times \mathbb{N}_{n+1}, t \mapsto x_j(c(t)_i) \text{ is smooth.}$$

PROPOSITION 3.26. *Let μ be a fixed affine triangulation of Δ_n. The map $\mu^{\mathbb{N}}$ restricts to a smooth map from the set of affine triangulations of Ω to se set of sequences of affine triangulations of Ω.*

3.7. Diffeologies and implicit functions in the ILB setting. We set the following notations, from the standard reference [**72**] and along the lines of [**56**]: Let $\mathbf{E} = (E_\infty, (E_i)_{i\in\mathbb{N}})$ and $\mathbf{F} = (F_\infty, (F_i)_{i\in\mathbb{N}})$ be two ILB vector spaces. Let O_0 be an open neighborhood of $(0,0)$ in $E_0 \times F_0$, let $\mathbf{O} = (O_i)_{i\in I}$ with $O_i = O_0 \cap (E_i \times F_i)$, for $i \in \mathbb{N} \cup \{\infty\}$.

Let us now propose, along the lines of [**55**], a diffeological approach to the main result of [**56**] that we recall here. For this, we consider a function f_0 of class C^∞such that

(1) $f_0(0;0) = 0$
(2) $D_2 f_0(0;0) = Id_{F_0}$.

Moreover, let us assume that f_0 restricts to $C^\infty-$maps $f_i : U_i \times V_i \to F_i$, and defines an ILB map

$$\mathbf{f} : \mathbf{E} \to \mathbf{F}.$$

let

$$U_\infty = E_\infty \cap U_0 \quad \text{and} \quad V_\infty = V_0 \cap F_\infty.$$

We do not assume here any other assumption, contrasting with e.g. the classical Nash-Moser theorem [**29**] where additional norm estimates are necessary. Under our weakened conditions, one can state [**56**, Theorem 2.2]:

THEOREM 3.27. *There exists a non-empty domain $D_\infty \subset U_\infty$, possibly non-open in U_∞, and a function*

$$u_\infty : D_\infty \to V_\infty$$

such that

$$\forall x \in D_\infty, \quad f_\infty(x; u_\infty(x)) = 0.$$

Moreover, there exists a sequence $(c_i)_{i\in\mathbb{N}} \in (\mathbb{R}_+^*)^{\mathbb{N}}$ and a Banach space B_{f_∞} such that

- *$B_{f_\infty} \subset E_\infty$ (as a subset)*
- *the canonical inclusion map $B_{f_\infty} \hookrightarrow E_\infty$ is continuous*

which is the domain of the following norm (and endowed with it):

$$\|x\|_{f_\infty} = \sup\left\{\frac{\|x_i\|}{c_i} | i \in \mathbb{N}\right\}.$$

Then D_∞ contains $\mathcal{B}$, the unit ball (of radius 1 centered at 0) of B_{f_∞}.

In [**56**], the question of the regularity of the implicit function is left open, because the domain D_∞ is not a priori open in O_∞. This lack of regularity induces a critical breakdown in generalizing the classical proof of the Frobenius theorem to this setting. This gap is filled in [**55**] mostly by the following result:

THEOREM 3.28. *Let*

$$f_i : O_i \to F_i, \quad i \in \mathbb{N} \cup \{\infty\}$$

be a family of maps, let u_∞ the implicit function defined on the domain D_∞, as in Theroem 3.27. Then, there exists a domain D such that $\mathcal{B} \subset D \subset D_\infty$ such that the function u_∞ is smooth for the subset diffeology of D.

The same way, we can state the corresponding Frobenius theorem from the same reference:

THEOREM 3.29. *Let*

$$f_i : O_i \to L(E_i, F_i), \quad i \in \mathbb{N}$$

be a collection of smooth maps satisfying the following condition:

$$i > j \Rightarrow f_j|_{O_i} = f_i$$

and such that,

$$\begin{gathered} \forall (x,y) \in O_i, \forall a, b \in E_i \\ (D_1 f_i(x,y)(a)(b) + (D_2 f_i(x,y))(f_i(x,y)(a))(b) = \\ (D_1 f_i(x,y)(b)(a) + (D_2 f_i(x,y))(f_i(x,y)(b))(a). \end{gathered}$$

Then, $\forall (x_0, y_0) \in O_\infty$*, there exists a diffeological subspace* D *of* O_∞ *that contains* (x_0, y_0) *and a smooth map* $J : D \to F_\infty$ *such that*

$$\forall (x,y) \in D, \quad D_1 J(x,y) = f_i(x, J(x,y))$$

and, if D_{x_0} *is the connected component of* (x_0, y_0) *in* $\{(x,y) \in D \,|\, x = x_0\}$*,*

$$J_i(x_0, .) = Id_{D_{x_0}}.$$

Moreover, there exists a sequence $(c_i)_{i \in \mathbb{N}} \in (\mathbb{R}_+^*)^{\mathbb{N}}$ *and a Banach space* B_{f_∞} *such that*

- $B_{f_\infty} \subset E_\infty \times F_\infty$ *(as a subset)*
- *the canonical inclusion map* $B_{f_\infty} \hookrightarrow E_\infty \times F_\infty$ *is continuous*

which is the domain of the following norm (and endowed with it):

$$\|x\|_{f_\infty} = \sup \left\{ \frac{\|x\|_{E_i \times F_i}}{c_i} | i \in \mathbb{N} \right\}.$$

Then D_∞ *contains* $\mathcal{B}$*, the unit ball (of radius* 1 *centered at 0) of* B_{f_∞}*.*

References

[1] S. Albeverio, A. Daletskii, and E. Lytvynov, *Laplace operators on differential forms over configuration spaces*, J. Geom. Phys. **37** (2001), no. 1-2, 15–46, DOI 10.1016/S0393-0440(00)00031-0. MR1806439

[2] S. Albeverio, Y. Kondratiev, E. Lytvynov, and G. Us, *Analysis and geometry on marked configuration spaces*, Infinite dimensional harmonic analysis (Kyoto, 1999), Gräbner, Altendorf, 2000, pp. 1–39. MR1851911

[3] I. M. Anderson, *Introduction to the variational bicomplex*, Mathematical aspects of classical field theory (Seattle, WA, 1991), Contemp. Math., vol. 132, Amer. Math. Soc., Providence, RI, 1992, pp. 51–73, DOI 10.1090/conm/132/1188434. MR1188434

[4] I. M. Anderson and N. Kamran, *Conservation laws and the variational bicomplex for second-order scalar hyperbolic equations in the plane*, Acta Appl. Math. **41** (1995), no. 1-3, 135–144, DOI 10.1007/BF00996109. Geometric and algebraic structures in differential equations. MR1362123

[5] M. Bauer, P. Harms, and P. W. Michor, *Sobolev metrics on shape space of surfaces*, J. Geom. Mech. **3** (2011), no. 4, 389–438, DOI 10.3934/jgm.2011.3.389. MR2888014

[6] M. Bauer, P. Harms, and P. W. Michor, *Sobolev metrics on shape space, II: Weighted Sobolev metrics and almost local metrics*, J. Geom. Mech. **4** (2012), no. 4, 365–383, DOI 10.3934/jgm.2012.4.365. MR3011892

[7] Bergmann, R., Herzog, R., Loayza, E., and Welker, K. Shape optimization: what to do first, optimize or discretize? Advantages and disadvantages for PDE-constrained problems. In *Proceedings in Applied Mathematics and Mechanics* (2019), Wiley.

[8] D. Canarutto, *Gauge field theory in natural geometric language—a revisitation of mathematical notions of quantum physics*, Oxford University Press, Oxford, 2020, DOI 10.1093/oso/9780198861492.001.0001. MR4261567

[9] K. T. Chen, *Iterated path integrals*, Bull. Amer. Math. Soc. **83** (1977), no. 5, 831–879, DOI 10.1090/S0002-9904-1977-14320-6. MR454968

[10] P. Cherenack, *Applications of Frölicher spaces to cosmology*, Ann. Univ. Sci. Budapest. Eötvös Sect. Math. **41** (1998), 63–91 (1999). MR1691919

[11] Cherenack, P. Frölicher versus differential spaces: A prelude to cosmology, *J. Categorical Structures* **1-2** (2000), reprinted in: Kozma, L. (ed.) et al., Steps in differential geometry. Proceedings of the colloquium on differential geometry, Debrecen, Hungary, July 25-30, 2000. Debrecen: Univ. Debrecen, Institute of Mathematics and Informatics, 85-94 (2001).

[12] J. D. Christensen and E. Wu, *Tangent spaces and tangent bundles for diffeological spaces* (English, with English and French summaries), Cah. Topol. Géom. Différ. Catég. **57** (2016), no. 1, 3–50. MR3467758

[13] J. D. Christensen, G. Sinnamon, and E. Wu, *The D-topology for diffeological spaces*, Pacific J. Math. **272** (2014), no. 1, 87–110, DOI 10.2140/pjm.2014.272.87. MR3270173

[14] J. D. Christensen and E. Wu, *The homotopy theory of diffeological spaces*, New York J. Math. **20** (2014), 1269–1303. MR3312059

[15] J. D. Christensen and E. Wu, *Exterior bundles in diffeology*, Israel J. Math. **253** (2023), no. 2, 673–713, DOI 10.1007/s11856-022-2372-9. MR4575422

[16] Cootes, T., Taylor, C., Cooper, D., and Graham, J. Active shape models-their training and application. *Computer Vision and Image Understanding* **61** no 1 (1995), 38–59.

[17] Droske, M., and Rumpf, M. Multiscale joint segmentation and registration of image morphology. *IEEE Transactions on Pattern Analysis and Machine Intelligence* **29** no 12 (2007), 2181–2194.

[18] B. Dugmore and P. P. Ntumba, *On tangent cones of Frölicher spaces*, Quaest. Math. **30** (2007), no. 1, 67–83, DOI 10.2989/16073600780205747. MR2309243

[19] D. S. Freed, *The geometry of loop groups*, J. Differential Geom. **28** (1988), no. 2, 223–276. MR961515

[20] A. Frölicher, *Smooth structures*, Category theory (Gummersbach, 1981), Lecture Notes in Math., vol. 962, Springer, Berlin-New York, 1982, pp. 69–81. MR682945

[21] A. Frölicher and A. Kriegl, *Linear spaces and differentiation theory*, Pure and Applied Mathematics (New York), John Wiley & Sons, Ltd., Chichester, 1988. A Wiley-Interscience Publication. MR961256

[22] M. Fuchs, B. Jüttler, O. Scherzer, and H. Yang, *Shape metrics based on elastic deformations*, J. Math. Imaging Vision **35** (2009), no. 1, 86–102, DOI 10.1007/s10851-009-0156-z. MR2515445

[23] Gangl, P., Langer, U., Laurain, A., Meftahi, H., and Sturm, K. Shape optimization of an electric motor subject to nonlinear magnetostatics. B1002–B1025.

[24] Glöckner, H.; Aspects of differential calculus related to infinite-dimensional vector bundles and Poisson vector spaces. *Axioms* **11** no 5 (2022), articleID 221.

[25] Goldammer, N.; Welker, K.; Towards optimization techniques on diffeological spaces by generalizing Riemannian concepts. Preprint.

[26] M. Golubitsky and V. Guillemin, *Stable mappings and their singularities*, Graduate Texts in Mathematics, Vol. 14, Springer-Verlag, New York-Heidelberg, 1973. MR0341518

[27] M. Gromov, *Partial differential relations*, Ergebnisse der Mathematik und ihrer Grenzgebiete (3) [Results in Mathematics and Related Areas (3)], vol. 9, Springer-Verlag, Berlin, 1986, DOI 10.1007/978-3-662-02267-2. MR864505

[28] Hafner, B. J., Zachariah, S. G., and Sanders, J. E. Characterisation of three-dimensional anatomic shapes using principal components: Application to the proximal tibia. *Medical and Biological Engineering and Computing* **38**, 1 (2000), 9–16.

[29] R. S. Hamilton, *The inverse function theorem of Nash and Moser*, Bull. Amer. Math. Soc. (N.S.) **7** (1982), no. 1, 65–222, DOI 10.1090/S0273-0979-1982-15004-2. MR656198

[30] M. Hintermüller and A. Laurain, *Optimal shape design subject to elliptic variational inequalities*, SIAM J. Control Optim. **49** (2011), no. 3, 1015–1047, DOI 10.1137/080745134. MR2806573

[31] M. Hintermüller and W. Ring, *A second order shape optimization approach for image segmentation*, SIAM J. Appl. Math. **64** (2003/04), no. 2, 442–467, DOI 10.1137/S0036139902403901. MR2049659

[32] M. W. Hirsch, *Differential topology*, Graduate Texts in Mathematics, vol. 33, Springer-Verlag, New York, 1994. Corrected reprint of the 1976 original. MR1336822

[33] P. Iglesias-Zemmour, *Diffeology*, Mathematical Surveys and Monographs, vol. 185, American Mathematical Society, Providence, RI, 2013, DOI 10.1090/surv/185. MR3025051

[34] P. Iglesias, Y. Karshon, and M. Zadka, *Orbifolds as diffeologies*, Trans. Amer. Math. Soc. **362** (2010), no. 6, 2811–2831, DOI 10.1090/S0002-9947-10-05006-3. MR2592936

[35] R. S. Ismagilov, *Representations of infinite-dimensional groups*, Translations of Mathematical Monographs, vol. 152, American Mathematical Society, Providence, RI, 1996. Translated from the Russian manuscript by D. Deart, DOI 10.1090/mmono/152. MR1393939

[36] J. Eells Jr., *A setting for global analysis*, Bull. Amer. Math. Soc. **72** (1966), 751–807, DOI 10.1090/S0002-9904-1966-11558-6. MR203742

[37] D. G. Kendall, *Shape manifolds, Procrustean metrics, and complex projective spaces*, Bull. London Math. Soc. **16** (1984), no. 2, 81–121, DOI 10.1112/blms/16.2.81. MR737237

[38] B. Khesin and R. Wendt, *The geometry of infinite-dimensional groups*, Ergebnisse der Mathematik und ihrer Grenzgebiete. 3. Folge. A Series of Modern Surveys in Mathematics [Results in Mathematics and Related Areas. 3rd Series. A Series of Modern Surveys in Mathematics], vol. 51, Springer-Verlag, Berlin, 2009. MR2456522

[39] H. Kihara, *Model category of diffeological spaces*, J. Homotopy Relat. Struct. **14** (2019), no. 1, 51–90, DOI 10.1007/s40062-018-0209-3. MR3913971

[40] Kilian, M., Mitra, N., and Pottmann, H. Geometric modeling in shape space. *ACM Transactions on Graphics (SIGGRAPH)* **26** no 3 (2007), #64, 1–8.

[41] I. Kolář, P. W. Michor, and J. Slovák, *Natural operations in differential geometry*, Springer-Verlag, Berlin, 1993, DOI 10.1007/978-3-662-02950-3. MR1202431

[42] A. Kriegl and P. W. Michor, *The convenient setting of global analysis*, Mathematical Surveys and Monographs, vol. 53, American Mathematical Society, Providence, RI, 1997, DOI 10.1090/surv/053. MR1471480

[43] A. Kriegl, P. W. Michor, and A. Rainer, *An exotic zoo of diffeomorphism groups on* $\mathbb{R}^n$, Ann. Global Anal. Geom. **47** (2015), no. 2, 179–222, DOI 10.1007/s10455-014-9442-0. MR3313140

[44] K. Kuribayashi, *Simplicial cochain algebras for diffeological spaces*, Indag. Math. (N.S.) **31** (2020), no. 6, 934–967, DOI 10.1016/j.indag.2020.08.002. MR4170260

[45] K. Kuribayashi, *A comparison between two de Rham complexes in diffeology*, Proc. Amer. Math. Soc. **149** (2021), no. 11, 4963–4972, DOI 10.1090/proc/15622. MR4310119

[46] Kurtek, S., Klassen, E., Ding, Z., and Srivastava, A. A novel Riemannian framework for shape analysis of 3D objects. In *2010 IEEE Computer Society Conference on Computer Vision and Pattern Recognition* (2010), IEEE.

[47] M. Laubinger, *A Lie algebra for Frölicher groups*, Indag. Math. (N.S.) **21** (2011), no. 3-4, 156–174, DOI 10.1016/j.indag.2011.04.001. MR2835925

[48] J. Leslie, *On a diffeological group realization of certain generalized symmetrizable Kac-Moody Lie algebras*, J. Lie Theory **13** (2003), no. 2, 427–442. MR2003152

[49] Ling, H., and Jacobs, D. Shape classification using the inner-distance. *IEEE Transactions on Pattern Analysis and Machine Intelligence* **29**, 2 (2007), 286–299.

[50] J.-P. Magnot, *Difféologie du fibré d'holonomie d'une connexion en dimension infinie* (French, with English and French summaries), C. R. Math. Acad. Sci. Soc. R. Can. **28** (2006), no. 4, 121–128. MR2340641

[51] J.-P. Magnot, *Ambrose-Singer theorem on diffeological bundles and complete integrability of the KP equation*, Int. J. Geom. Methods Mod. Phys. **10** (2013), no. 9, 1350043, 31, DOI 10.1142/S0219887813500436. MR3103144

[52] J.-P. Magnot, *Remarks on the geometry and the topology of the loop spaces* $H^s(S^1, N)$*, for* $s \leq 1/2$, Int. J. Maps Math. **2** (2019), no. 1, 14–37. MR4162164

[53] Magnot, J.-P. Differentiation on spaces of triangulations and optimized triangulations. *Journal of Physics: Conference Series* **738** (2016) article ID 012088.

[54] J.-P. Magnot, *The group of diffeomorphisms of a non-compact manifold is not regular*, Demonstr. Math. **51** (2018), no. 1, 8–16, DOI 10.1515/dema-2018-0001. MR3769995

[55] J.-P. Magnot, *On the differential geometry of numerical schemes and weak solutions of functional equations*, Nonlinearity **33** (2020), no. 12, 6835–6867, DOI 10.1088/1361-6544/abaa9f. MR4173564

[56] J.-P. Magnot, *On the domain of implicit functions in a projective limit setting without additional norm estimates*, Demonstr. Math. **53** (2020), no. 1, 112–120, DOI 10.1515/dema-2020-0008. MR4121350

[57] J.-P. Magnot and E. G. Reyes, *Well-posedness of the Kadomtsev-Petviashvili hierarchy, Mulase factorization, and Frölicher Lie groups*, Ann. Henri Poincaré **21** (2020), no. 6, 1893–1945, DOI 10.1007/s00023-020-00896-3. MR4100916

[58] J.-P. Magnot and E. G. Reyes, *The Cauchy problem of the Kadomtsev-Petviashvili hierarchy and infinite-dimensional groups*, Nonlinear systems and their remarkable mathematical structures. Vol. 2, CRC Press, Boca Raton, FL, 2020, pp. 382–414. MR4591211

[59] Magnot, J.-P., Reyes, E., and Rubtsov, V. Infinite order structures on differential equations. Preprint.

[60] Magnot, J.-P., Reyes, E., and Rubtsov, V. Frölicher structures, diffieties, and a formal Kadomtsev–Petviashvili hierarchy. to appear in the proceedings of A. Vinogradov's memorial conference

[61] J.-P. Magnot and J. Watts, *The diffeology of Milnor's classifying space*, Topology Appl. **232** (2017), 189–213, DOI 10.1016/j.topol.2017.10.011. MR3720892

[62] P. W. Michor and D. Mumford, *Vanishing geodesic distance on spaces of submanifolds and diffeomorphisms*, Doc. Math. **10** (2005), 217–245. MR2148075

[63] P. W. Michor and D. Mumford, *Riemannian geometries on spaces of plane curves*, J. Eur. Math. Soc. (JEMS) **8** (2006), no. 1, 1–48, DOI 10.4171/JEMS/37. MR2201275

[64] L. Younes, P. W. Michor, J. Shah, and D. Mumford, *A metric on shape space with explicit geodesics*, Atti Accad. Naz. Lincei Rend. Lincei Mat. Appl. **19** (2008), no. 1, 25–57, DOI 10.4171/RLM/506. MR2383560

[65] P. W. Michor and D. Mumford, *An overview of the Riemannian metrics on spaces of curves using the Hamiltonian approach*, Appl. Comput. Harmon. Anal. **23** (2007), no. 1, 74–113, DOI 10.1016/j.acha.2006.07.004. MR2333829

[66] Mio, W., Srivastava, A., and Joshi, S.; On shape of plane elastic curves. *International Journal of Computer Vision* **73** no 3 (2006), 307–324.

[67] P. Ntumba, *DW complexes and their underlying topological spaces*, Quaest. Math. **25** (2002), no. 1, 119–134, DOI 10.2989/16073600209486006. MR1897316

[68] P. P. Ntumba, *Sikorski and Frölicher CW-complexes compared*, Demonstratio Math. **38** (2005), no. 1, 207–221. MR2123735

[69] A. Nägel, V. Schulz, M. Siebenborn, and G. Wittum, *Scalable shape optimization methods for structured inverse modeling in 3D diffusive processes*, Comput. Vis. Sci. **17** (2015), no. 2, 79–88, DOI 10.1007/s00791-015-0248-9. MR3411257

[70] P. J. Olver, *Applications of Lie groups to differential equations*, Graduate Texts in Mathematics, vol. 107, Springer-Verlag, New York, 1986, DOI 10.1007/978-1-4684-0274-2. MR836734

[71] H. Omori, *Groups of diffeomorphisms and their subgroups*, Trans. Amer. Math. Soc. **179** (1973), 85–122, DOI 10.2307/1996492. MR377975

[72] H. Omori, *Infinite-dimensional Lie groups*, Translations of Mathematical Monographs, vol. 158, American Mathematical Society, Providence, RI, 1997. Translated from the 1979 Japanese original and revised by the author, DOI 10.1090/mmono/158. MR1421572

[73] A. Paganini, *Approximate shape gradients for interface problems*, New trends in shape optimization, Internat. Ser. Numer. Math., vol. 166, Birkhäuser/Springer, Cham, 2015, pp. 217–227, DOI 10.1007/978-3-319-17563-8_9. MR3467382

[74] Perperidis, D., Mohiaddin, R., and Rueckert, D. Construction of a 4D statistical atlas of the cardiac anatomy and its use in classification. In *Lecture Notes in Computer Science*, J. Duncan and G. Gerig, Eds. Springer Berlin Heidelberg, 2015, pp. 402–410.

[75] E. Pervova, *Diffeological vector pseudo-bundles*, Topology Appl. **202** (2016), 269–300, DOI 10.1016/j.topol.2016.01.014. MR3464165

[76] E. Pervova, *Diffeological gluing of vector pseudo-bundles and pseudo-metrics on them*, Topology Appl. **220** (2017), 65–99, DOI 10.1016/j.topol.2017.02.002. MR3619281

[77] E. Pervova, *Diffeological Clifford algebras and pseudo-bundles of Clifford modules*, Linear Multilinear Algebra **67** (2019), no. 9, 1785–1828, DOI 10.1080/03081087.2018.1472202. MR3980666

[78] A. Pressley and G. Segal, *Loop groups*, Oxford Mathematical Monographs, The Clarendon Press, Oxford University Press, New York, 1986. Oxford Science Publications. MR900587

[79] T. Robart, *Sur l'intégrabilité des sous-algèbres de Lie en dimension infinie* (French, with French summary), Canad. J. Math. **49** (1997), no. 4, 820–839, DOI 10.4153/CJM-1997-042-7. MR1471060

[80] M. Rumpf and B. Wirth, *A nonlinear elastic shape averaging approach*, SIAM J. Imaging Sci. **2** (2009), no. 3, 800–833, DOI 10.1137/080738337. MR2540171

[81] S. Schmidt, E. Wadbro, and M. Berggren, *Large-scale three-dimensional acoustic horn optimization*, SIAM J. Sci. Comput. **38** (2016), no. 6, B917–B940, DOI 10.1137/15M1021131. MR3566914

[82] V. Schulz and M. Siebenborn, *Computational comparison of surface metrics for PDE constrained shape optimization*, Comput. Methods Appl. Math. **16** (2016), no. 3, 485–496, DOI 10.1515/cmam-2016-0009. MR3518366

[83] V. H. Schulz, M. Siebenborn, and K. Welker, *Efficient PDE constrained shape optimization based on Steklov-Poincaré-type metrics*, SIAM J. Optim. **26** (2016), no. 4, 2800–2819, DOI 10.1137/15M1029369. MR3582826

[84] Siebenborn, M., and Welker, K. Computational aspects of multigrid methods for optimization in shape spaces.

[85] Söhn, M., Birkner, M., Yan, D., and Alber, M. Modelling individual geometric variation based on dominant eigenmodes of organ deformation: implementation and evaluation. *Physics in Medicine and Biology* **50** no 24 (2005), 5893–5908.

[86] Souriau, J. Un algorithme g'en'erateur de structures quantiques. *Astérisque* **25** (1985), 64-117 (2011).

[87] A. Stacey, *Comparative smootheology*, Theory Appl. Categ. **25** (2011), No. 4, 64–117. MR2805746

[88] A. Stacey, *The smooth structure of the space of piecewise-smooth loops*, Glasg. Math. J. **59** (2017), no. 1, 27–59, DOI 10.1017/S0017089516000033. MR3576326

[89] C. T. J. Dodson, G. Galanis, and E. Vassiliou, *Geometry in a Fréchet context: A projective limit approach*, London Mathematical Society Lecture Note Series, vol. 428, Cambridge University Press, Cambridge, 2016, DOI 10.1017/CBO9781316556092. MR3469847

[90] Vincent, M. Diffeological differential geometry. Master's thesis, University of Copenhagen, (2008).

[91] J. Watts, *Diffeologies, Differential Spaces, and Symplectic Geometry*, ProQuest LLC, Ann Arbor, MI, 2012. Thesis (Ph.D.)–University of Toronto (Canada). MR3153238

[92] K. Welker, *Suitable spaces for shape optimization*, Appl. Math. Optim. **84** (2021), no. suppl. 1, S869–S902, DOI 10.1007/s00245-021-09788-2. MR4316804

[93] B. Wirth, L. Bar, M. Rumpf, and G. Sapiro, *A continuum mechanical approach to geodesics in shape space*, Int. J. Comput. Vis. **93** (2011), no. 3, 293–318, DOI 10.1007/s11263-010-0416-9. MR2787012

[94] Zolésio, J.-P. Control of moving domains, shape stabilization and variational tube formulations. In *International Series of Numerical Mathematics*, K. Kunisch, J. Sprekels, G. Leugering, and F. Tröltzsch, Eds. Birkhäuser Basel (2007), pp. 329–382.

Helmut-Schmidt-University/University of the Federal Armed Forces Hamburg, Holstenhofweg 85, 22043 Hamburg, Germany
Email address: goldammer@hsu-hh.de

LAREMA - UMR CNRS 6093, Université d'Angers, 2 Boulevard Lavoisier 49045 Angers cedex 01 and Lycée Jeanne dArc 30 avenue de Grande Bretagne F-63000 Clermont-Ferrand http://orcid.org/0000-0002-3959-3443
Email address: jean-pierr.magnot@ac-clermont.fr

TU Bergakademie Freiberg, Akademiestr. 6, D-09599 Freiberg, Germany
Email address: Kathrin.Welker@math.tu-freiberg.de

Contemporary Mathematics
Volume **794**, 2024
https://doi.org/10.1090/conm/794/15925

Elastic diffeological spaces

Christian Blohmann

ABSTRACT. We introduce a class of diffeological spaces, called elastic, on which the left Kan extension of the tangent functor of smooth manifolds defines an abstract tangent functor in the sense of Rosický. On elastic spaces there is a natural Cartan calculus, consisting of vector fields and differential forms, together with the Lie bracket, de Rham differential, inner derivative, and Lie derivative, satisfying the usual graded commutation relations. Elastic spaces are closed under arbitrary coproducts, finite products, and retracts. Examples include manifolds with corners and cusps, diffeological groups and diffeological vector spaces with a mild extra condition, mapping spaces between smooth manifolds, and spaces of sections of smooth fiber bundles.

1. Introduction

1.1. The quest for a Cartan calculus on diffeological spaces. A category that contains smooth manifolds as a full subcategory but has better properties, such as having all limits, colimits, and exponential objects, is often called a convenient setting for differential geometry. The price to be paid for this convenience is that such categories are usually too large as to allow for strong geometric results that hold for all its objects. A typical example is the category of diffeological spaces. It is a quasi-topos with all its good categorical properties. But since it contains arbitrary quotients, arbitrary subsets, and arbitrary intersections of smooth manifolds, topological spaces, vector spaces of arbitrary cardinality, and much more, a theorem that holds for all diffeological spaces must hold in all these cases. So instead of trying to prove statements that would have to cover this impossible generality of situations, the task is often to identify conditions that are strong enough to prove a desired result, but weak enough to allow for a wide range of examples and applications.

A considerable part of the infinitesimal differential geometric computations on a smooth manifold M can be carried out in its Cartan calculus, which consists of the tangent bundle $TM \to M$, the Lie bracket of vector fields, the graded algebra of differential forms $\Omega(M)$, together with the de Rham differential d, the inner derivative ι_v and the Lie derivative $\mathcal{L}_v$ for every vector field v, which satisfy the relations

$$[d, d] = 0, \quad [\iota_v, \iota_w] = 0, \quad [\iota_v, d] = \mathcal{L}_v,$$
$$[\mathcal{L}_v, \iota_w] = \iota_{[v,w]}, \quad [\mathcal{L}_v, d] = 0, \quad [\mathcal{L}_v, \mathcal{L}_w] = \mathcal{L}_{[v,w]},$$

2020 *Mathematics Subject Classification.* Primary 58A40; Secondary 58A03, 18F15, 18F40.
Key words and phrases. Diffeological space, tangent structure, Cartan calculus.

where the bracket is the graded commutator of graded derivations of $\Omega(M)$. For example, local definitions and calculations of symplectic geometry can typically be worked out in the Cartan calculus, such as hamiltonian vector fields, Poisson brackets, hamiltonian actions, Dirac structures, generalized complex geometry, contact structures, the L_∞-algebra of a multisymplectic structure, homotopy momentum maps, infinitesimal models for equivariant cohomology, etc. Another example is local Lagrangian Field Theory, where the derivation of the Euler-Lagrange equations, local symmetries, Noether's theorems, the theory of Jacobi fields, etc. take place in the Cartan calculus of the infinite jet bundle, also known as the variational bicomplex [**DF99**]. In fact, the question we will address in this paper was motivated by current developments in Lagrangian Field Theory and geometric deformation theory, areas where the basic geometric objects, spaces of fields and paths in a moduli space of structures, are naturally equipped with diffeologies.

QUESTION 1.1. What are the conditions a diffeological space must satisfy so that it is equipped with a natural Cartan calculus?

Of course, there are always the tautological conditions which promote the desired outcome to axioms, in our case the existence of a Cartan calculus. The task is to identify a set of conditions that is minimal or at least so small that it can be verified in a wide range of cases.

The basic structures of the Cartan calculus on a smooth manifold M, the differential graded algebra of differential forms $\Omega(M)$ and the tangent bundle $TM \to M$, are local, that is, they can be defined first on the open subsets $U \subset \mathbb{R}^n$ of a chart $U \subset M$ and then glued together on an atlas. More precisely, the functor $U \mapsto \Omega(U)$ is a sheaf and the tangent functor $U \mapsto TU$ a cosheaf, so that

$$\begin{aligned} \Omega(M) &\cong \lim_{U \to M} \Omega(U) \\ TM &\cong \operatorname*{colim}_{U \to M} TU, \end{aligned}$$

where the limit in differential graded algebras and the colimit in manifolds are taken over the category of charts of a maximal atlas of M. For a diffeological space X we simply replace the category of charts by the category of plots and obtain the definitions

$$\begin{aligned} \Omega(X) &:= \lim_{U \to X} \Omega(U) \\ TX &:= \operatorname*{colim}_{U \to X} TU \end{aligned} \tag{1}$$

of the de Rham complex and the tangent diffeological space of X, where the colimit is taken in the category of diffeological spaces (cf. Section 1.5).

The maps $X \mapsto \Omega(X)$ and $X \mapsto TX$ are the pointwise left Kan extensions from smooth manifolds to diffeological spaces. It follows that $\Omega(X)$ and TX are functorial in X. More importantly, the left Kan extension applies also to natural transformations, such as the bundle projection $\pi_U : TU \to U$ and the zero section $0_U : U \to TU$, which suggests that this is the natural method to generalize the Cartan calculus of smooth manifolds to diffeological spaces.

How do we define the Lie bracket of vector fields on a diffeological space? The first guess is to start from the Lie algebras $\mathcal{X}(U) = \Gamma(U, TU)$ of vector fields on all plots $U \to X$. However, $U \mapsto \mathcal{X}(U)$ is not a functor, so that the left Kan extension cannot be applied. We could map the vector fields to the space of derivations of $C^\infty(X) = \Omega^0(X)$, which is equipped with the commutator bracket. However, this map is generally not injective, and even if it is, its image may not be closed under the bracket. Worse, the map $X \mapsto \mathrm{Der}(C^\infty(X))$ is still not a functor, so that this does not solve the problem of naturality. The conclusion is that the spaces of vector fields on plots are not a good starting point for the construction of a natural Cartan calculus on diffeological spaces.

Fortunately, the situation has been analyzed carefully by Rosický who has identified the natural structure of the tangent functor that is needed to define the Lie bracket of vector fields. He defines an abstract tangent structure on a category $\mathcal{C}$ to be an endofunctor $T : \mathcal{C} \to \mathcal{C}$ together with the natural transformations of the bundle projection $\pi_X : TX \to X$, zero section $0_X : X \to TX$, fiberwise addition $+_X : TX \times_X TX \to TX$, exchange of order of differentiation $\tau_X : T^2X \to T^2X$, and inclusion of the tangent fibers into the vertical tangent space $\lambda_X : TX \to T^2X$, which have to satisfy a rather long list of axioms (Definition 2.7). It is instructive to see how all these structures come together to define the Lie bracket (9) of vector fields, avoiding any reference to the commutator bracket of derivations of some structure ring.

For us, the main advantage of Rosický's approach is that all the structure is given by functors and natural transformations, to which we can apply the left Kan extension. However, this does not yield an abstract tangent structure on all diffeological spaces. The main issue is that the pointwise left Kan extension, which is given by a colimit, does not preserve limits, in particular the pullback on which the fiberwise addition of tangent vectors is defined. More precisely, the natural morphism

$$\operatorname*{colim}_{U \to X} TU \times_U TU \longrightarrow TX \times_X TX, \tag{2}$$

is not an isomorphism for all diffeological spaces X. In fact, this map is generally neither surjective nor injective, as the following two examples show.

EXAMPLE 1.2 (Axis cross of the plane). Consider the subset $\{(x,y) \in \mathbb{R}^2 \mid xy = 0\} \subset \mathbb{R}^2$ with the subspace diffeology. The two tangent vectors at the orgin in the direction of the x-axis and the y-axis cannot be represented on the same plot (Figure 1). It follows that (2) is not surjective.

EXAMPLE 1.3 (Folded line). Consider the diffeological quotient space of the action $\mathbb{Z}_2 \times \mathbb{R} \to \mathbb{R}$, $(k,x) \mapsto kx$, where $\mathbb{Z}_2 = \{1,-1\}$. The quotient map $\mathbb{R} \to \mathbb{R}/\mathbb{Z}_2$ is a plot. The tangent vectors $(0,1)$ and $(0,-1)$ on its domain represent the same tangent vector on $\mathbb{R}/\mathbb{Z}_2$. This implies that the pairs $\zeta = \big((0,1),(0,1)\big)$ and $\eta = \big((0,1),(0,-1)\big)$ in $T\mathbb{R} \times_{\mathbb{R}} T\mathbb{R}$ represent the same pair of tangent vectors in $T(\mathbb{R}/\mathbb{Z}_2) \times_{\mathbb{R}/\mathbb{Z}_2} T(\mathbb{R}/\mathbb{Z}_2)$. Since the tangent morphism of every morphism of plots preserves the sum of a pair of tangent vectors at a point and since the sum of ζ is zero but that of η is not, the two pairs cannot be equivalent in $\operatorname{colim}_{U \to \mathbb{R}/\mathbb{Z}_2} TU \times_U TU$. We conclude that (2) is not injective.

1.2. The axiom of elasticity. Only if (2) is an isomorphism, the left Kan extension of the addition $+_U$ on on plots is a morphism $+_X : TX \times_X TX \to TX$ that can be viewed as a fiberwise addition of tangent vectors on the diffeological space X. Therefore, requiring (2) to be an isomorphism is the first condition we have to impose for a diffeological space to have a natural Cartan calculus.

A k-form in $\Omega(X)$ is a family of k-forms on all plots $U \to X$ that are compatible with the pullbacks along morphisms of plots. A vector field, however, is not represented by a family of vector fields on the plots. For this reason, there is no natural operation of inner derivative on $\Omega(X)$. For the inner derivative, we have to define a k-form as a fiberwise multilinear and antisymmetric morphism

$$\alpha : \underbrace{TX \times_X \ldots \times_X TX}_{=:T_k X} \longrightarrow \mathbb{R}.$$

(We avoid defining a tensor product, which would entail the usual technical issues of completion when the fibers are infinite-dimensional.) The notation $T_k X$ for the k-fold fiber product is standard in the literature on abstract tangent structures. The inner derivative of α with respect to a vector field $v : X \to TX$ is then given by precomposition

$$\iota_v \alpha : T_{k-1}X \xrightarrow{\cong} X \times_X T_{k-1}X \xrightarrow{v \times_X \mathrm{id}} T_k X \xrightarrow{\alpha} \mathbb{R}.$$

If we define forms as maps $T_k X \to \mathbb{R}$, how can we define the differential? The differential of a function $f : X \to \mathbb{R}$ is given by the tangent map,

$$df : TX \xrightarrow{Tf} T\mathbb{R} \xrightarrow{\cong} \mathbb{R} \times \mathbb{R} \xrightarrow{\mathrm{pr}_2} \mathbb{R}.$$

However, the functions and exact 1-forms do not generate the ring of forms, so that this construction cannot be extended to higher forms.

We are now in the following dilemma. Either we define differential forms as families of forms on the plots, in which case we have a differential but no inner derivative. Or we define them as fiberwise multilinear and antisymmetric morphisms $T_k X \to \mathbb{R}$, in which case we have an inner derivative, but no differential. The way out is to require that the two notions of differential forms coincide.

We have already imposed the condition that (2) is an isomorphism, which induces an isomorphism

$$\mathrm{Hom}(TX \times_X TX, \mathbb{R}) \xrightarrow{\cong} \lim_{U \to X} \mathrm{Hom}(TU \times_U TU, \mathbb{R}). \tag{3}$$

It is easy to see that this isomorphism is equivariant with respect to the exchange of the two factors for the fiber product. Moreover, the maps are fiberwise multilinear on $T_k X$ if and only if they are on all $T_k U$. This shows that the isomorphism (3) induces an isomorphism from fiberwise multilinear and antisymmetric morphisms on $TX \times_X TX$ to $\Omega^2(X)$. Since we need such an isomorphism for forms of arbitrary degree k, we have to impose the following axiom:

AXIOM (E1). The natural morphisms

$$\theta_{k,X} : \operatorname*{colim}_{U \to X} T_k U \longrightarrow T_k X,$$

are isomorphisms for all $k > 1$.

This axiom has the following geometric interpretation. Every tangent vector $v_x \in T_x X$ is represented by a path. One can picture this by stretching out x in the direction of v_x to a smooth path $\gamma : (-\varepsilon, \varepsilon) \to X$ of short but non-zero length through $\gamma(0) = x$, such that the coordinate tangent vector $\frac{\partial}{\partial t}$ at the origin of the interval is mapped by $T_0\gamma$ to v_x. In this sense, every point of a diffeological space has some elasticity in a single infinitesimal direction.

However, we generally cannot simultaneously stretch out x in the directions of several tangent vectors $v_x^1, \ldots, v_x^k \in TX_x$. That is, we cannot always find a plot $p : U \to X$ with $p(0) = x$ such that $(T_0 p)\frac{\partial}{\partial t^i} = v_x^i$, where $(t^1, \ldots, t^k)$ are the

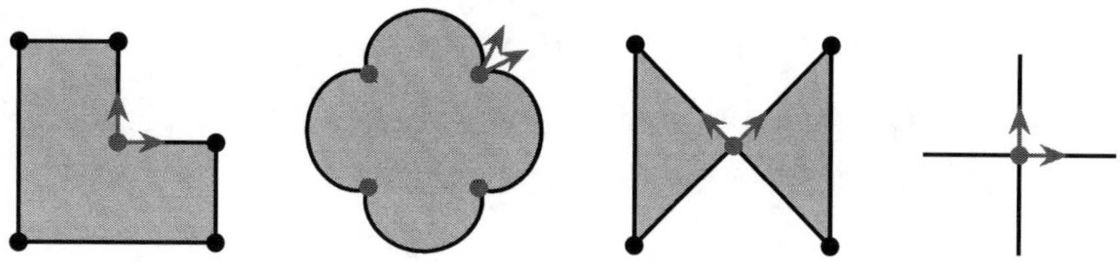

FIGURE 1. Diffeological subspaces of $\mathbb{R}^2$ with non-elastic points marked in red, at which two tangent directions cannot be represented on the same plot.

canonical coordinates of $U \subset \mathbb{R}^k$. And even if we can find such a plot, it may happen that the tangent map Tp is not injective at 0, so that we cannot identify the tangent vectors on X with the coordinate vectors on U. This identification is possible at every point $x \in X$ if and only if the morphism $\theta_{k,X}$ is a bijection. If in addition we want this condition to be compatible with the smooth structure, then we have to make the stronger assumption that $\theta_{k,X}$ is an isomorphism of diffeological spaces. In this sense, Axiom (E1) captures the geometric idea of the "elasticity" of a diffeological space in which any finite set of tangent directions can be streched out to a smooth "membrane" given by the image of a plot.

EXAMPLE 1.4 (Pasta diffeologies). We can equip a smooth manifold M with an alternative diffeology by defining the plots be all smooth maps $p : U \to M$ such that the rank of $Tp : TU \to TM$ is everywhere less than or equal to r. Since (i) the precomposition of p with a smooth function f does not increase the rank, (ii) the rank is a local property, and (iii) the rank of constant maps is zero, this defines a diffeology, which we call the rank-r-restricted diffeology. The rank-r-restricted diffeology is r-dimensional in the sense of Definition 1.12 in [**Mag13**].

For $r = 0$ we obtain the discrete diffeology. If $r = 1$, then every plot factors through $\mathbb{R}$, so that we obtain the **Spaghetti diffeology** [**IZ13**, Sec. 1.10, footnote 1]. The case $r = 2$ might then be called the **Fettuccine diffeology**. It was suggested by the participants of the AMS-EMS-SMF meeting 2022 in Grenoble that the case $r = 3$ should be called the **Gnocchi diffeology**. For the rank-r-restricted diffeology the morphism $\theta_{k,M}$ of Axiom (E1) is an isomorphism for all $k \leq r$ but not for $r < k < \dim M$.

1.3. The additional axioms. So far we have the Axiom (E1) that ensures that we have a fiberwise addtion on TX and an inner derivative on differential forms. For the definition of the Lie bracket we need more structure. In particular, we need a natural morphism $\tau_X : T^2X \to T^2X$ that exchanges the order of differentiation when we apply the tangent functor twice. On a euclidean space $U \subset \mathbb{R}^n$, every tangent vector is represented by a path $\mathbb{R} \to U$ on some plot, so that a tangent vector on the manifold of tangent vectors is represented by a smooth path of smooth paths, which is the same as a smooth map $\mathbb{R}^2 \to U$. Exchanging the order of differentiation is achieved by exchanging the parameters,

$$\begin{aligned} \tau_{1\leftrightarrow 2} : \mathbb{R}^2 &\longrightarrow \mathbb{R}^2 \\ (t_1, t_2) &\longmapsto (t_2, t_1), \end{aligned}$$

which descends by the commutative diagram

$$
\begin{array}{ccc}
C^\infty(\mathbb{R}^2, U) & \xrightarrow{\tau^*_{1\leftrightarrow 2}} & C^\infty(\mathbb{R}^2, U) \\
\downarrow & & \downarrow \\
T^2U & \xrightarrow[\tau_U]{} & T^2U
\end{array}
\tag{4}
$$

to an endomorphism of T^2U.

When we extend τ_U to diffeological spaces, the problem arises that the left Kan extension does not preserve the product of endofunctors, that is, the natural morphism

$$\theta^2_X : \operatorname*{colim}_{U\to X} T^2U \longrightarrow T^2X$$

is generally not an isomorphism. We could impose the condition that θ^2_X is an isomorphism, but this would be unnecessarily strong. It suffices to require the left Kan extension of τ_U to descend to a morphism $\tau_X : T^2X \to T^2X$. Since θ^2_X is a subduction for all X (Proposition 3.26), such a τ_X is unique. This condition can be expressed more intuitively in terms of the smooth families in the same way as for euclidean spaces. We can show that we can represent elements in T^2X by plots $\mathbb{R}^2 \to X$. More precisely, we have a subduction

$$\underline{\mathcal{D}\mathrm{flg}}(\mathbb{R}^2, X) \longrightarrow T^2X,$$

where $\underline{\mathcal{D}\mathrm{flg}}$ denotes the inner hom of diffeological spaces, that is, the set of morphisms equipped with the functional diffeology. The second axiom can now be expressed in a way that is completely analogous to diagram (4).

Axiom (E2). There is a natural morphism $\tau_X : T^2X \to T^2X$, such that the diagram

$$
\begin{array}{ccc}
\underline{\mathcal{D}\mathrm{flg}}(\mathbb{R}^2, X) & \xrightarrow{\tau^*_{1\leftrightarrow 2}} & \underline{\mathcal{D}\mathrm{flg}}(\mathbb{R}^2, U) \\
\downarrow & & \downarrow \\
T^2X & \xrightarrow[\tau_X]{} & T^2X
\end{array}
$$

commutes.

Next, consider the natural morphism $\lambda_X : TX \to T^2X$ that maps $v \in TX$ to the vertical tangent vector on TX represented by the path $t \mapsto tv$. On a smooth manifold, this morphism induces an isomorphism between every tangent space and the tangent space of the tangent space. For diffeologial vector spaces this can fail, as the following example shows.

Example 1.5. Consider $\mathbb{R}^n$ equipped with k-times differentiable maps as plots. This is a diffeological vector space that we denote by $\mathbb{R}^n_{C^k}$. Its tangent diffeological space is given for $k > 0$ by

$$T\mathbb{R}^n_{C^k} \cong \mathbb{R}^n_{C^k} \times \mathbb{R}^n_{C^{k-1}},$$

which shows that the vector space and its tangent fiber are not isomorphic. Assume that $k > 1$, so that we can apply the tangent functor twice. The vertical lift,

$$
\begin{aligned}
\lambda_{\mathbb{R}^n_{C^k}} : \mathbb{R}^n_{C^k} \times \mathbb{R}^n_{C^{k-1}} &\longrightarrow \mathbb{R}^n_{C^k} \times \mathbb{R}^n_{C^{k-1}} \times \mathbb{R}^n_{C^{k-1}} \times \mathbb{R}^n_{C^{k-2}} \\
(x, v) &\longmapsto (x, 0, 0, v),
\end{aligned}
$$

is not a subduction.

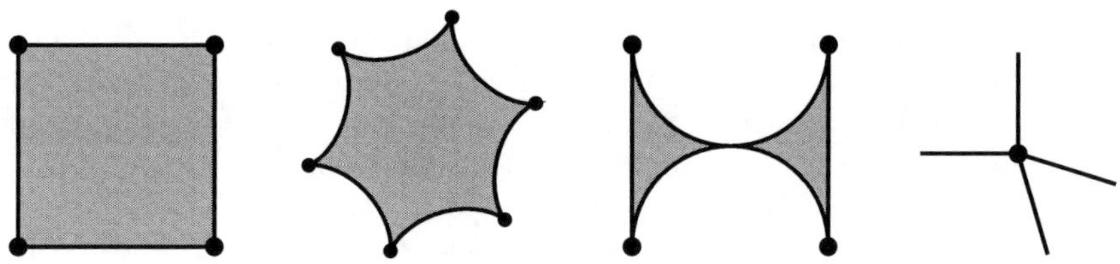

FIGURE 2. Elastic diffeological subspaces of $\mathbb{R}^2$. The tangent spaces are 0 at the marked points, $\mathbb{R}$ at points on the black lines, and $\mathbb{R}^2$ at gray points in the interior.

The definition of the Lie bracket in terms of the tangent structure yields a map from X to the vertical subbundle of T^2X restricted to the zero section of TX. We have to be able to identify this bundle with TX for the bracket to be again a vector field. This condition is not specific to diffeological spaces. A vector field on a C^k-manifold is a C^k-map. The commutator of two such vector fields is a C^{k-1}-map which is, therefore, not a vector field on the C^k-manifold. To exclude such phenomena we have to impose the following axiom:

AXIOM (E3). The vertical lift $\lambda_X : TX \to T^2X$ is an induction.

There are two more axioms. For smooth manifolds the tangent functor commutes with pullbacks over submersions. This follows from the local standard form of submersions, which is proved using the implicit function theorem. Such a genuinely analytic result cannot hold for all diffeological spaces, which is why we need to impose the following axiom:

AXIOM (E4). The tangent functor commutes with fiber products of the tangent bundle, $TT_kX \cong T_kTX$.

Finally, we want the diffeological spaces that satisfy our axioms to form a category. This requires the collection of diffeological spaces that satisfy the axioms to be closed under the functors T_k, which leads to the following axiom:

AXIOM (E5). For every finite set of positive integers $k_1, \dots, k_n$ the diffeological space $X' := T_{k_1} \cdots T_{k_n} X$ satisfies axioms (E1) through (E4).

A diffeological space that satisfies Axioms (E1)-(E5) will be called **elastic**. If we drop Axiom (E5), then we still have a natural Cartan calculus on X. We call a diffeological space that satisfies Axioms (E1)-(E4) **weakly elastic**. The category of weakly elastic spaces is not closed under the functors T_k.

1.4. Summary. The main result about elastic diffeological spaces is the following: The left Kan extension of the tangent structure on euclidean spaces defines a tangent structure with scalar $\mathbb{R}$-multiplication on the category of elastic spaces (Theorem 4.2). The proof of this statement, which is very long and technical, is carried out in detail in a much longer paper [**Blo**]. Here we will present an overview of the conceptual framework, the main properties, and important examples of elastic spaces.

In Section 2 we review Rosický's concept of abstract tangent structures. We also spell out the conditions for a compatible scalar multiplication, which is only mentioned briefly in the original paper [**Ros84**]. For clarity, we spell out the tangent structure of euclidean spaces explicitly.

In Section 3 we state some new results about the left Kan extension from euclidean spaces to diffeological spaces, that will be needed later. We use the definition of diffeological spaces as concrete sheaves on the site of euclidean spaces, which the best approach for the categorical constructions we will study. We then state, without proof, our results on the compatibility of the left Kan extension with products, coproducts, subductions, composition of endofunctors, and the D-topology of diffeological spaces, which are all needed for the proof of the main Theorem 4.2.

Section 4 contains the formal definition of elastic spaces and the statement, without proof, of the main Theorem 4.2. We discuss some alternative choices of axioms for which the theorem remains true. First, we observe that we can drop Axiom (E5) from the definition of elastic spaces and still obtain a Cartan calculus on the space X. This category of weaker diffeological spaces will no longer be closed under the tangent functor and its fiber products, so that we no longer have a category with an abstract tangent structure in the sense of Rosický. We can also slightly relax the Axiom (E1) of elasticity by requiring that the condition only holds for $k = 2$. This weaker condition will be sufficient to prove Theorem 4.2. As explained in the introduction, our choice of the stronger Axiom (E1) is motivated by our wish to obtain not just a tangent structure, but a full-fledged Cartan calculus. In an earlier version of the notion of elastic spaces presented in talks, Axioms (E1), (E2), and (E4) were replaced by a single condition, which later turned out to be unnecessarily strong. Finally, we state, without proof, that the category of elastic spaces is closed under restrictions to open subsets, coproducts, finite products, and retracts.

In Section 5 we will give a number of examples for elastic diffeological spaces that show that, while the conditions of elasticity are quite strong, they still allow for an interesting range of applications. The first main result, Theorem 5.2, shows that a diffeological Lie group G is elastic if and only if the natural map $\mathfrak{g} \to T_0\mathfrak{g}$ from the vector space $\mathfrak{g} = T_eG$ to its tangent space at 0 is an induction. This is a surprisingly weak condition, which is not particular to diffeological spaces. For example, it is not satisfied by differentiable group structure on a C^k-manifold, $k < \infty$. The proof of Theorem 5.2, which is long and technical, will be given in [**Blo**]. The second main result, Theorem 5.13, states that the diffeological space of sections of a smooth fiber bundle $F \to M$ is elastic. The proof of this theorem is again long and involved, so that it will be given in [**Blo**]. More examples of elastic spaces in this sections are: manifolds with corners and cusps, diffeological vector spaces satisfying a mild extra condition, manifolds modelled on elastic vector spaces, mapping spaces $\underline{\mathcal{D}\mathrm{flg}}(X, A)$ to diffeological vector spaces satisfying $A \cong T_0A$, the space of smooth $\mathbb{R}$-valued functions on a diffeological space, the mapping space of smooth manifolds.

1.5. Prior work.

1.5.1. *The tangent functor.* We have defined in (1) the de Rham complex $\Omega(X)$ and the tangent functor TX by the pointwise left Kan extension. $\Omega(X)$ is the standard de Rham complex as defined in [**IZ13**, Sec. 6.28], which can be seen by spelling out the limit in terms of products and equalizers [**ML98**, Sec.V, Thm. 2]. For the tangent functor a number of inequivalent definitions have been used in the literature, some of which were reviewed, compared, and improved in the invaluable paper of Christensen and Wu [**CW16**]. The authors show in [**CW16**, Example 4.3] that for the internal tangent bundle of [**Hec95**] the fibers are generally not diffeological vector spaces and the fiber-wise addition generally not smooth, contrary to what is claimed in [**Hec95**]. Christensen and Wu observe that the tangent bundle of [**Hec95**] can be viewed as left Kan extension, albeit with values in the wrong target category, and resolve the issue of smoothness by defining the internal tangent bundle of X as pointwise left Kan extension in the appropriate category of vector space objects in the overcategory $\mathcal{D}\mathrm{flg}/X$ [**CW16**, p. 40].

Our definition (1) of TX is very close to the one of Christensen and Wu. The difference is that we take the left Kan extension in $\mathcal{D}\mathrm{flg}$, so that the tangent fibers are cones, rather than vector spaces (cf. [**BFW13**, Sec. A.1]). There are two reasons for this choice.

The first, technical reason is that our definition of Lie bracket uses Rosický's notion of abstract tangent structure [**Ros84**] that is described in terms of natural transformations of endofunctors of diffeological spaces. To obtain these natural transformations we have to take our Kan extension in $\mathcal{D}\mathrm{flg}$. The condition that the tangent cone T_xX is a vector space (with final sums defined on the plots) is equivalent to the Axiom (E1) of elasticity.

The second, geometric reason is that the tangent fiber at x should be an infinitesimal model of the space at x. More precisely, T_xX should describe the directions in which smooth paths can leave or enter x with finite velocity. From this point of view, the vector space structure of the tangent fibers of a manifold reflects the fact that the space itself looks locally like a vector space. For a diffeological space that is not a manifold, forcing the tangent fiber to be a vector space obscures this local geometric information. For example, the tangent vector space at 0 of three smooth lines through the origin of $\mathbb{R}^2$ is $\mathbb{R}^3$, whereas the tangent cone is an accurate local model of the diffeological space.

There is another class of definitions of tangent vectors and tangent bundles that involve the smooth real-valued functions on the diffeological space. An *external* tangent vector is a derivation of the germ of smooth functions at a point. Many definitions require such a tangent vector to be represented by a smooth path and use its action on functions to define the tangency condition of two paths [**Vin08**, **GW21**]. This includes the original definition of Souriau [**Sou70**] and definitions that are used in the context of diffeological groups [**Les03**, **Mag18**]. Note that so-defined tangent fibers are often called "internal tangent cones", we follow the terminology of [**CW16**, Sec. 3.4.3] calling these approaches "mixed", that is, a combination of the strictly internal and the strictly external definition.

We avoid involving smooth functions on the diffeological spaces in our strictly internal definition of tangent vectors for two reasons. The first reason is that this would be at odds with the conceptually simple categorical approach we use. The second reason is that there are many interesting spaces, such as noncommutative tori [**DI85**], that have a rich diffeological structure, but a poor ring of functions on them. In fact, non-commutative tori turn out to be elastic, which shows that there is no reason to exclude them here by using a definition of tangent space that is best suited for Fröhlicher spaces.

1.5.2. *Lie brackets for groups.* It is standard textbook material to obtain the bracket of the Lie algebra of a Lie group by differentiating the adjoint group action twice: once to obtain the adjoint action of the group on the tangent space at the identity and a second time to obtain the adjoint action of the Lie algebra. While [**Hec95**] has used this approach for for diffeological groups to define a bilinear operation, Leslie was the first to give sufficient conditions for it to be a Lie bracket satisfying the Jacobi identity [**Les03**, Thm. 1.14]. However, the assumptions for Leslie's result are quite strong.

Firstly, the tangent cones are defined only for diffeologies of "lattice type" [**Les03**, Def. 1.1], which means that any two plots factor locally through a common third plot. Since the diffeology of every diffeological group is of lattice type, as is stated without proof after Definition 1.3, this does not affect the main results, but limits their scope considerably. For example, the assumption that the spaces are of lattice type must be added implicitly in Lemma 1.4 of [**Les03**].

Secondly, the tangent vectors are defined in terms of their action as directional derivatives on real-valued smooth functions on the diffeological space, which requires a good supply of such functions. For this reason, the strong condition [**Les03**, Def. 1.13 i)] must be made that the tangent space at the identity is separated by linear functions (cf. [**CW19**]).

Thirdly, [**Les03**, Def. 1.13 ii)] requires all linear plots of the tangent fiber to factor through a complete Hausdorff locally convex topological vector space. This condition puts the diffeological setup squarely back in a functional analytic context which is then used to proof the main Theorem 1.14.

It was remarked in [**Mag18**, p. 12] that the conditions of Definition 1.13 and Theorem 1.14 of [**Les03**] are sufficient, but hardly necessary for a Lie bracket of a diffeological Lie group to exist. In Definition 1.10 of [**Mag18**], diffeological Lie groups are defined as those diffeological groups that possess a smooth Lie bracket. This is the natural tautological definition that promotes the desired property of the group to an axiom, rather than spelling out conditions that are sufficient but unnecessarily strong.

[**Mag18**] uses the definition of tangent cones of [**DN07**], which is essentially the same as in [**Les03**] without the assumption that the diffeology is of "lattice type. In the terminology of [**CW16**], this is a "mixed" approach, where tangent vectors are represented, internally, by paths, which are defined to be equivalent if they have the same action, externally, on smooth real-valued functions on the diffeological space.

1.5.3. *Lie brackets and diffeological Lie groupoids.* In [**BFW13**] the construction of the Lie bracket by the adjoint group action was successfully generalized to compute the Lie algebroid of an infinite dimensional diffeological groupoid that is relevant in general relativity. This worked due to the special nature of the groupoid. For example, the tangent bundle of the Lie algebroid is trivial. In more general situations, the construction of the Lie bracket of sections of the Lie algebroid via the adjoint action of the bisection group [**CdSW99**, Chap. 15] will fail if there is no good supply of (local) bisections of the groupoid. In this case, the Lie bracket should be defined by the Lie bracket of invariant vector fields on the groupoid, for which an ambient Lie algebra of vector fields on the diffeological groupoid is needed. This was one of the original motivations of the present work.

1.6. Outlook. One of the motivations to develop the concept of elastic spaces came from classical field theory, where the basic structure, the "space" of fields is the set of sections of a smooth fiber bundle $F \to M$ leaving it often unclear or implicit what "space" means mathematically. It is often observed that $\mathcal{F} = \Gamma(M, F)$ is a Fréchet manifold, which is subsequently viewed as blanket license to treat $\mathcal{F}$ as if it were an ordinary finite-dimensional smooth manifold. For example, it is often an implicit assumption that there is a natural differential bigraded algebra of smooth forms on $\mathcal{F} \times M$ that restricts to the variational bicomplex on $J^\infty F$ [**DF99**]. In the same vein, in the study of symmetries, such as Noether's theorems or BV-theory, the constructions are often explained rigorously only for finite Lie groups acting on finite-dimensional manifolds and then generalized with a leap of faith to group-valued functions or diffeomorphisms acting on the spaces of fields. A closer analysis shows that only the diffeological structure is being used. For example, a variation of a field is a smooth path of sections and a tangent vector to the space of solutions of the field equations (i.e. a generalized Jacobi field) is represented by a smooth path of fields. All this suggests that there is a diffeological construction of a Cartan calculus on $\mathcal{F}$. This approach is validated by Theorem 5.13.

Another motivation comes from geometric deformation theory. Conceptually, a deformation is a path in the moduli space of structures, such as the morphisms of an algebraic structure, riemannian metrics, or complex structures, all of which are equipped with a natural functional diffeology. This suggests that the geometric moduli spaces can be conceptualized by (higher) stacks that are presented by (higher) diffeological groupoids. The infinitesimal deformation theory should then be given by the fibers of the tangent bundle of the moduli space, which should be presented by the corresponding (higher) Lie algebroid. In order to define this procedure rigorously, a Lie theory for diffeological groupoids has to be developed. This has lead us to the conclusion that we need a tangent structure on the diffeological spaces of the groupoid. It is encouraging that for diffeological Lie groups the condition of elasticity is surprisingly weak. Lie theory for diffeological groupoids and their application to geometric deformation theory is work in progress.

2. Abstract tangent structures

It is fairly straightforward to generalize the de Rham complex $\Omega(M)$, which is a contravariant functor from smooth manifolds to differential graded algebras, to other categories. While vector fields are in some sense dual to differential forms, their generalization is a much more difficult problem. An immediate obstacle is that the map $M \to \mathcal{X}(M)$ that sends a manifold to its Lie algebra of vector fields is not a functor. It is the tangent bundle $TM \to M$ that is functorial in M and, therefore, lends itself easily to generalizations. Given a generalized tangent bundle $\pi_X : TX \to X$ in some category, the vector fields are naturally defined as the sections of the morphism π_X. The obvious question is now the following:

QUESTION 2.1. Given a morphism $TX \to X$ in some catgory, what is the natural structure needed to equip its space of sections with the structure of a Lie algebra?

This question turns out to be quite involved. A vector field on a smooth manifold can be identified with a derivation of its ring of smooth functions $C^\infty(M)$, which is closed under the commutator bracket of the ambient ring of endomorphisms of $C^\infty(M)$. However, $M \mapsto \mathrm{Der}(C^\infty(M))$ is still no functor. Moreover, in a generalized setting, the identification of sections of $TX \to X$ with derivations on some structure ring on X seems to be an overly strong requirement that is extraneous to differential geometric considerations. So how can we define the Lie bracket of vector fields on M directly in terms of the tangent functor using a categorical approach that lends itself to generalizations?

This issue has been solved by Rosický in [**Ros84**]. In a first step, we observe that on manifolds the vector space structure on vector fields is induced by the vector space structure on the tangent fibers. To allow for categories that do not contain the real numbers as object, we relax the structure of $\mathbb{R}$-vector space to that of an abelian group. The structure we then need on a generalized tangent bundle $TX \to X$ is the natural transformations of addition $+_X : TX \times_X TX \to TX$ and a zero section $0_X : X \to TX$ that equip $TX \to X$ with the structure of an abelian group over X.

For the definition of the Lie bracket, we start from the coordinate formula

$$\Big[v^i \frac{\partial}{\partial x^i} w^j \frac{\partial}{\partial x^j}\Big] = \Big(v^i \frac{\partial w^j}{\partial x^i} - w^i \frac{\partial v^j}{\partial x^i}\Big) \frac{\partial}{\partial x^j}$$

for vector fields on $\mathbb{R}^n$. The right hand side is given by the derivation of w with respect to v minus the derivation of v with respect to w. In order to generalize this formula we must make sense of the differentiation of one vector field with respect to another and the subtraction of the two terms.

The derivation of $w : X \to TX$ with respect to $v : X \to TX$ is given by the composition

$$X \xrightarrow{v} TX \xrightarrow{Tw} T^2X.$$

However, we cannot yet subtract $Tw(v_x)$ and $Tv(w_x)$ since the basepoint of $Tw(v_x)$ in $\pi_{TX} : T(TX) \to TX$ is w_x, whereas that of $Tv(w_x)$ is v_x. We first have to exchange the order of differentiation of the twofold tangent bundle. That is, we need a natural transformation $\tau_X : T^2X \to T^2X$ that satisfies $\tau_X \circ \pi_{TX} = T\pi_X$. Then the basepoint of $\tau_X(Tv(w_x))$ is also w_x, so that we can take the difference

$$Tw(v_x) - \tau_X\big(Tv(w_x)\big) = \big(+_{TX} \circ (Tw \circ v, -\tau_X \circ Tv \circ w)\big)_x. \tag{5}$$

The result lies in the vertical tangent bundle of $T^2X \to TX$, that is, the kernel of $T\pi_{TX}$. In the last step we have to be able to identify at every point $w_x \in TX$ the vertical tangent space with T_xX itself.

In a smooth manifold, there is a morphism $\lambda_{2,M} : TM \times_M TM \to T^2M$ that maps the pair (w_x, u_x) to the vertical tangent vector in T^2M that is represented by the path $t \mapsto w_x + tu_x$. This map induces an isomorphism $TM \times_M TM \cong \ker T\pi_{TM}$. In the generalized setting this structure is promoted to an axiom: We require a natural morphism $\lambda_{2,X} : TX \times_X TX \to T^2X$ that induces an isomorphism $TX \times_X TX \cong \ker T\pi_{TX}$. Using this isomorphism, the expression (5) can be viewed as an element in $TX \times_X TX$ that can be projected onto the second factor which produces an element in TX, which is the value of the bracket $[v, w]$ at x.

For this bracket to satisfy the Jacobi identity, a number of compatibility relations and properties of the various structures, bundle projection, zero section, fiberwise addition, exchange of the order of differentiation, and the vertical lift have to be required [**Ros84**]. Rosicky's axiomatization of all this structure is the basis of this paper.

2.1. Preliminary remarks on terminology and notation.

TERMINOLOGY 2.2. Let Wibble be an algebraic theory. Let X be an object in a category $\mathcal{C}$ such that the overcategory $\mathcal{C}\downarrow X$ has all finite products (i.e. pullbacks over X). A Wibble object in $\mathcal{C}\downarrow X$ will be called a **bundle of Wibbles over** X.

In this paper Wibble will be one of: monoid, group, abelian group, module, $\mathbb{R}$-vector space (for categories containing $\mathbb{R}$ as an object). If $W \to X$ is a bundle of Wibbles and if the pullback $W_x = * \times_X W$ over a point $x : * \to X$ exists in $\mathcal{C}$, then W_x is a Wibble object in $\mathcal{C}$. In other words, every fiber of a bundle of Wibbles is a Wibble object in $\mathcal{C}$, which justifies the terminology. Note, that the notion of bundle of Wibbles does not make any assumptions on local trivializations, whatsoever. So a bundle of vector spaces over a manifold M is more general than a vector bundle over M.

REMARK 2.3. The main purpose of Terminology 2.2 is to unify (for the purpose of this paper) the varied terminology found in the literature and to use a term that is self-explanatory for a category theorist. In [**Ros84**, p. 1] a bundle of (abelian) groups over an endofunctor $F : \mathcal{C} \to \mathcal{C}$ is called an "natural (abelian) group bundle over F". A bundle of vector spaces over a diffeological space X is called a "regular vector bundle" in [**Vin08**], a "diffeological vector space over X" in [**CW16**], and a "diffeological vector pseudo-bundle" in [**Per16**].

NOTATION 2.4. We will follow [**Ros84**] for the notation of the compositions of functors and natural transformations. The composition of functors $G : \mathcal{A} \to \mathcal{B}$ and $F : \mathcal{B} \to \mathcal{C}$ will be denoted by juxtaposition $FG : \mathcal{A} \to \mathcal{C}$. Therefore, the horizontal composition of natural transformations $\alpha : F \to F'$ and $\beta : G \to G'$ (the Godement product) will also be denoted by juxtaposition $\alpha\beta : FG \to F'G'$. Its components are given by the following commutative diagram:

$$\begin{array}{ccc} FG(A) & \xrightarrow{F(\beta_A)} & FG'(A) \\ {\scriptstyle \alpha_{G(A)}}\downarrow & \searrow^{(\alpha\beta)_A} & \downarrow{\scriptstyle \alpha_{G'(A)}} \\ F'G(A) & \xrightarrow[F'(\beta_A)]{} & F'G'(A) \end{array}$$

The identity natural transformation $F \to F$ will be denoted by F, so that

$$(F\beta)_A = F(\beta_A)$$
$$(\alpha G)_A = \alpha_{G(A)}.$$

The vertical composition of α with a natural transformation $\alpha' : F' \to F''$ will be denoted by $\alpha' \circ \alpha : F \to F''$. Its components are given by $(\alpha' \circ \alpha)_A = \alpha'_A \circ \alpha_A$. The monoidal category of endofunctors will be denoted by $\mathrm{End}(\mathcal{C})$.

2.2. Rosický's axioms. In [**Ros84**], Rosický introduced the notion of **abstract tangent functor**, which captures much of the categorical structure of the tangent functor of manifolds. The following notion is implicit in Rosický's definition:

DEFINITION 2.5. Let $F : \mathcal{C} \to \mathcal{C}$ be a functor and $\tau : F^2 \to F^2$ a natural transformation. Let $\tau_{12} := \tau F$ and $\tau_{23} := F\tau$ be the two trivial extensions of τ to natural transformations $F^3 \to F^3$. We call τ a **braiding on** F if it satisfies the braid relations $\tau_{12} \circ \tau_{23} \circ \tau_{12} = \tau_{23} \circ \tau_{12} \circ \tau_{23}$. A braiding τ is called a **symmetric structure on** F if it satisfies $\tau \circ \tau = F^2$.

REMARK 2.6. A symmetric structure on F defines an action of the symmetric group S_n on F^n.

A bundle of groups over X consists of a morphism $\pi : A \to X$, the bundle projection, together with the morphisms $0 : X \to A$ and $+ : A \times_X A \to A$ of the group structure. Let $\pi' : A' \to X'$, $0' : X' \to A'$, $+' : A' \times_{X'} A' \to A'$ be another bundle of groups. A **morphism of bundles** is a commutative diagram

$$\begin{array}{ccc} A & \xrightarrow{\varphi} & A' \\ {\scriptstyle\pi}\downarrow & & \downarrow{\scriptstyle\pi'} \\ X & \xrightarrow[\psi]{} & X' \end{array}$$

There is a unique morphism $\varphi \times_\psi \varphi : A \times_X A \to A' \times_{X'} A'$ that makes the following diagram commutative:

$$\begin{array}{ccc} A \times_X A & \xrightarrow{\varphi \times_\psi \varphi} & A \times_X A \\ \downarrow & & \downarrow \\ A \times A & \xrightarrow[\varphi \times \varphi]{} & A' \times A' \end{array}$$

The pair (φ, ψ) is a **morphism of bundles of groups** if the diagram

$$\begin{array}{ccc} A \times_X A & \xrightarrow{\varphi \times_\psi \varphi} & A' \times_X A' \\ {\scriptstyle +}\downarrow & & \downarrow{\scriptstyle +'} \\ A & \xrightarrow[\varphi]{} & A' \end{array}$$

commutes. An endofunctor F **preserves the fiber product** if the natural morphism of bundles over FX,

$$\nu_{k,X} : F(A \times_X^{\pi,\pi} \ldots \times_X^{\pi,\pi} A) \longrightarrow FA \times_{FX}^{F\pi,F\pi} \ldots \times_{FX}^{F\pi,F\pi} FA, \tag{6}$$

where both sides have the same number k of factors, is an isomorphism for all k.

DEFINITION 2.7 (Sec. 2 in[**Ros84**], Def. 2.3 in [**CC14**]). A **tangent structure** of a category $\mathcal{C}$ consists of a functor $T : \mathcal{C} \to \mathcal{C}$ together with natural transformations $\pi : T \to 1$, $0 : 1 \to T$, $+ : T_2 \to T$, $\lambda : T \to T^2$, and $\tau : T^2 \to T^2$, such that the following axioms hold:

- **Fiber products:** The pullbacks
$$T_k := \underbrace{T \times_1 T \times_1 \ldots \times_1 T}_{k \text{ factors}}$$
over $T \xrightarrow{\pi} 1$ exist for all $k \geq 1$, are pointwise, and preserved by T.
- **Bundle of abelian groups:** $T \xrightarrow{\pi} 1$ with neutral element 0 and addition $+$ is a bundle of abelian groups over 1 (Terminology 2.2).
- **Symmetric structure:** $\tau : T^2 \to T^2$ is a symmetric structure on T (Definition 2.5). Moreover, τ is a morphism of bundles of groups. That is, the diagrams
$$\begin{array}{ccc} T^2 & \xrightarrow{\tau} & T^2 \\ & {\scriptstyle T\pi}\searrow \quad \swarrow{\scriptstyle \pi T} & \\ & T & \end{array}$$

and

$$\begin{array}{ccc} TT_2 \xrightarrow{\nu_2} T^2 \times_T^{T\pi,T\pi} T^2 \xrightarrow{\tau\times_T\tau} & & T_2T \\ {\scriptstyle T+}\downarrow & & \downarrow{\scriptstyle +T} \\ T^2 & \xrightarrow[\tau]{} & T^2 \end{array}$$

commute, where ν_2 is morphism (6) for $A = TX \xrightarrow{\pi} X$, $F = T$, and $k = 2$.

- **Vertical lift:** The diagrams

$$\begin{array}{ccc} T & \xrightarrow{\lambda} & T^2 \\ {\scriptstyle\pi}\downarrow & & \downarrow{\scriptstyle\pi T} \\ 1 & \xrightarrow[0]{} & T \end{array} \qquad \begin{array}{ccc} T & \xrightarrow{\lambda} & T^2 \\ {\scriptstyle\lambda}\downarrow & & \downarrow{\scriptstyle\lambda T} \\ T^2 & \xrightarrow[T\lambda]{} & T^3 \end{array}$$

commute. Moreover, the first diagram is a morphism of bundles of groups, that is $(+T) \circ (\lambda \times_0 \lambda) = \lambda \circ +$.

- **Compatibility of vertical lift and symmetric structure:** The diagrams

$$\begin{array}{ccc} & T & \\ {\scriptstyle\lambda}\swarrow & & \searrow{\scriptstyle\lambda} \\ T^2 & \xrightarrow[\tau]{} & T^2 \end{array} \qquad \begin{array}{ccc} T^2 \xrightarrow{T\lambda} T^3 \xrightarrow{\tau T} & & T^3 \\ {\scriptstyle\tau}\downarrow & & \downarrow{\scriptstyle T\tau} \\ T^2 & \xrightarrow[\lambda T]{} & T^3 \end{array}$$

commute.

- **The vertical lift is a kernel:** The diagram

(7)
$$\begin{array}{ccc} T & \xrightarrow{\lambda} & T^2 \\ {\scriptstyle\pi}\downarrow & & \downarrow{\scriptstyle(\pi T, T\pi)} \\ 1 & \xrightarrow[0\times_1 0]{} & T_2 \end{array}$$

is a pointwise pullback.

TERMINOLOGY 2.8. In [**CC14**] and subsequent work, Rosický's original condition that $T \to 1$ be a bundle of abelian groups was relaxed to a bundle of abelian monoids. In this terminology, Rosicky's stronger notion is called a tangent structure with negatives. All tangent structures in this paper will be with negatives.

REMARK 2.9. Diagram (7) is a pointwise pullback if and only if

$$T \xrightarrow{\lambda} T^2 \overset{\xrightarrow{\pi T}}{\underset{\xrightarrow[0\circ\pi\circ\pi T]{}}{\xrightarrow{T\pi}}} T$$

is a pointwise triple equalizer. This condition is the original axiom in [**Ros84**].

REMARK 2.10. The vertical lift can be extended by the additive bundle structure to the map

(8)
$$\lambda_2 : T_2 \xrightarrow{T0\times_0\lambda} T_2T \xrightarrow{+T} T^2 \xrightarrow{\tau} T^2.$$

It was shown in [**CC14**, Lem. 3.10], assuming all other axioms of a tangent structure (with negatives), that axiom (7) is satisfied if and only if

$$\begin{array}{ccc} T_2 & \xrightarrow{\lambda_2} & T^2 \\ {\scriptstyle \pi\circ\mathrm{pr}_1}\downarrow & & \downarrow{\scriptstyle T\pi} \\ 1 & \xrightarrow[0]{} & T \end{array}$$

is a pointwise pullback.

2.3. Scalar multiplication. Let R be a ring object in the category $\mathcal{C}$. This gives rise to an endofunctor $R \times 1 : \mathcal{C} \to \mathcal{C}$, $X \mapsto R \times X$, which is equipped with the projection $\mathrm{pr}_1 : R \times 1 \to 1$. The ring structure of R equips $R \times 1 \to 1$ with the structure of a ring internal to endofunctors over 1. Let $\pi : T \to 1$ be an abelian group object in the category of endofunctors over 1. An $(R \times 1 \to 1)$-module structure on $T \to 1$ is given explicitly by a natural transformation

$$\kappa_X : R \times TX \longrightarrow TX,$$

such that the following diagrams commute for all $X \in \mathcal{C}$:

(i) Morphism of bundles:

$$\begin{array}{ccc} R \times TX & \xrightarrow{\kappa_X} & TX \\ & {\scriptstyle \pi_X \circ \mathrm{pr}_2}\searrow \quad \swarrow{\scriptstyle \pi_X} & \\ & X & \end{array}$$

(ii) Associativity:

$$\begin{array}{ccc} R \times R \times TX & \xrightarrow{\mathrm{id}_R \times \kappa_X} & R \times TX \\ {\scriptstyle \cdot\times\mathrm{id}_{R\times TX}}\downarrow & & \downarrow{\scriptstyle \kappa_X} \\ R \times TX & \xrightarrow[\kappa_X]{} & TX \end{array}$$

(iii) Unitality:

$$\begin{array}{ccc} \{1\} \times TX & \hookrightarrow & R \times TX \\ & {\scriptstyle \cong}\searrow & \downarrow{\scriptstyle \kappa_X} \\ & & TX \end{array}$$

(iv) Linearity in R:

$$\begin{array}{ccc} R \times R \times TX & \xrightarrow{+\times\mathrm{id}_{TX}} & R \times TX \\ {\scriptstyle \mathrm{id}_{R^2}\times\Delta_{TX}}\downarrow & & \\ (R \times TX) \times_X (R \times TX) & & \downarrow{\scriptstyle \kappa_X} \\ {\scriptstyle \kappa_X \times_X \kappa_X}\downarrow & & \\ TX \times_X TX & \xrightarrow[+_X]{} & TX \end{array}$$

Here $\Delta_{TX} : TX \to TX \times_X TX$ is the diagonal morphism and the factors of the codomain are reordered.

(v) Linearity in TX:

$$\begin{array}{ccc}
R \times TX \times_X TX & \xrightarrow{\mathrm{id} \times +_X} & R \times TX \\
\Big\downarrow{\scriptstyle \Delta_R \times \mathrm{id}_{T_2 X}} & & \\
(R \times TX) \times_X (R \times TX) & & \Big\downarrow{\scriptstyle \kappa_X} \\
\Big\downarrow{\scriptstyle \kappa_X \times_X \kappa_X} & & \\
TX \times_X TX & \xrightarrow[+_X]{} & TX
\end{array}$$

Here $\Delta_R : R \to R \times R$ is the diagonal morphisms and the factors of the codomain are reordered.

We will call this structure more succinctly an R-**module structure** on $T \to 1$ and $T \to 1$ a **bundle of** R-**modules** (Terminology 2.2). If T is part of a tangent structure on $\mathcal{C}$, then we also have to require the compatibility with the symmetric structure and the vertical lift.

DEFINITION 2.11. Let R be a ring internal to a category $\mathcal{C}$ with a tangent structure. An R-module structure $\kappa_X : R \times TX \to TX$ will be called a **scalar multiplication** of the tangent structure if the following diagrams commute for all $X \in \mathcal{C}$:

(vi) Compatibility with the symmetric structure:

$$\begin{array}{ccc}
R \times T^2X & \xrightarrow{\mathrm{id}_R \times \tau_X} & R \times T^2X \\
& {\scriptstyle \kappa_{TX}} \searrow \quad \swarrow {\scriptstyle T\kappa_X} & \\
& T^2X &
\end{array}$$

(vii) Compatibilty with the vertical lift:

$$\begin{array}{ccc}
R \times TX & \xrightarrow{\mathrm{id}_R \times \lambda_X} & R \times T^2X \\
\Big\downarrow{\scriptstyle \kappa_X} & & \Big\downarrow{\scriptstyle \kappa_{TX}} \\
TX & \xrightarrow[\lambda_X]{} & T^2X
\end{array}$$

The tangent structures we consider here will all be equipped with an $\mathbb{R}$-scalar multiplication.

REMARK 2.12. As is the case for any module structure, the commutative diagram (v) implies that the scalar multiplication by $0 \in R$ sends TX to the zero section, that is, the diagram

$$\begin{array}{ccc}
\{0\} \times TX & \hookrightarrow & R \times TX \\
\Big\downarrow{\scriptstyle \pi_X \circ \mathrm{pr}_2} & & \Big\downarrow{\scriptstyle \kappa_X} \\
X & \xrightarrow[0_X]{} & TX
\end{array}$$

is commutative. If κ is such that this diagram and diagrams (i)-(iii) are commutative, then κ will be called an R-**cone** structure and $T \to 1$ a **bundle of** R-**cones**.

2.4. The Lie bracket of vector fields.

DEFINITION 2.13. Let $\mathcal{C}$ be a category with a tangent structure. A **vector field** on $X \in \mathcal{C}$ is a section of $\pi_X : TX \to X$.

The bracket of two vector fields $v, w : X \to TX$ is defined as follows. The composition of v and $Tw : TX \to T^2X$ satisfies

$$\begin{aligned}\pi_{TX} \circ Tw \circ v &= w \circ \pi_X \circ v = w \circ \mathrm{id}_X \\ &= w.\end{aligned}$$

When we exchange v and w, we have $\pi_X \circ Tv \circ w = v$. In order to be able to subtract the two terms in the fiber product $T^2X \times_{TX} T^2X$, we have two apply the symmetric structure on T^2, so that we obtain

$$\begin{aligned}\pi_{TX} \circ \tau_X \circ Tv \circ w &= T\pi_X \circ Tv \circ w = T\mathrm{id}_X \circ w = \mathrm{id}_{TX} \circ w \\ &= w.\end{aligned}$$

This shows that $Tw \circ v$ and $\tau_X \circ Tv \circ w$ project to the same fiber of $\pi_{TX} : T^2X \to TX$, so that we can take the difference

$$\delta(v, w) := Tw \circ v - \tau_X \circ Tv \circ w,$$

where the minus denotes the difference in the bundle of abelian groups $\pi_{TX} : T^2X \to TX$. We have

$$\begin{aligned}T\pi_X \circ \delta(v, w) &= T\pi_X \circ Tw \circ v - T\pi_X \circ \tau_X \circ Tv \circ w \\ &= T(\pi_X \circ w) \circ v - \pi_{TX} \circ Tv \circ w \\ &= T\mathrm{id}_X \circ v - v \circ \pi_X \circ w \\ &= 0,\end{aligned}$$

so that the map $\delta(v, w) : X \to T^2X$ takes values in the kernel of $T\pi_X : T^2X \to TX$, which is isomorphic to $TX \times_X TX$. By projecting on the second factor we thus obtain the vector field $[v, w] : X \to TX$. This construction can be summarized by the following commutative diagram:

(9)
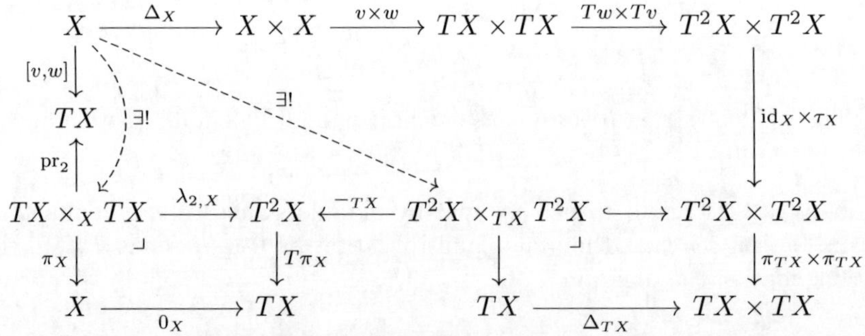

This shows that all of the tangent structure is needed for the definition of the bracket of vector fields. It was announced in [**Ros84**] and proved in [**CC15**] with the input of Rosický that $[v, w]$ satisfies the Jacobi relation.

The set of vector fields $\Gamma(X, TX)$ has the structure of an abelian group with addition

$$v + w := +_X \circ (v \times w) \circ \Delta_X,$$

where $\Delta_X : X \to X \times X$ is the diagonal morphism. When the tangent structure has a scalar multiplication by R, then $\Gamma(X, TX)$ is a module over the ring $\mathcal{C}(X, R)$ of R-valued functions, given by

$$\kappa(f, v) = \kappa_X \circ (f \times v) \circ \Delta_X.$$

2.5. The tangent structure of euclidean spaces. The eponymous example for tangent structures is the tangent functor of open subsets of real vector spaces, which is the local model for the tangent functor of smooth manifolds. Let $\mathcal{E}$ucl denote the category which has open subsets of $\mathbb{R}^n$, $n \geq 0$ as objects and smooth maps as morphisms. $\mathcal{E}$ucl will be called the category of **euclidean spaces**. Its tangent functor will be denoted by

$$T : \mathcal{E}\mathrm{ucl} \longrightarrow \mathcal{E}\mathrm{ucl}.$$

On an open subset $U \subset \mathbb{R}^n$, the functors that appear in the definition 2.7 of a tangent category are given explicitly by

$$\begin{aligned} TU &= U \times \mathbb{R}^n \\ T^2U &= U \times \mathbb{R}^n \times \mathbb{R}^n \times \mathbb{R}^n \\ T^kU &= U \times (\mathbb{R}^n)^{2^k-1} \\ T_2U &= U \times \mathbb{R}^n \times \mathbb{R}^n \\ T_kU &= U \times (\mathbb{R}^n)^k. \end{aligned}$$

On a smooth map $f : U \to V \subset \mathbb{R}^m$ the functors are given by

$$\begin{aligned} Tf : (u, u_0^i) &\longmapsto \Big(f(u), \frac{\partial f^a}{\partial x^i} u_0^i\Big) \\ T^2f : (u, u_0^i, u_1^i, u_{01}^i) &\longmapsto \Big(f(u), \frac{\partial f^a}{\partial x^i} u_0^i, \frac{\partial f^a}{\partial x^i} u_1^i, \frac{\partial f^a}{\partial x^i} u_{01}^i + \frac{\partial^2 f^a}{\partial x^i \partial x^j} u_0^i u_1^j\Big) \\ T_2f : (u, u_0^i, u_1^i) &\longmapsto \Big(f(u), \frac{\partial f^a}{\partial x^i} u_0^i, \frac{\partial f^a}{\partial x^i} u_1^i\Big). \end{aligned}$$

The formulas for T^k and T_k are analogous. The natural transformations of the tangent category structure are given by

$$\begin{aligned} \pi_U : (u, u_0) &\longmapsto u \\ 0_U : u &\longmapsto (u, 0) \\ +_U : (u, u_0, v_0) &\longmapsto (u, u_0 + v_0) \\ \lambda_U : (u, u_0) &\longmapsto (u, 0, 0, u_0) \\ \tau_U : (u, u_0, u_1, u_{01}) &\longmapsto (u, u_1, u_0, u_{01}). \end{aligned}$$

The commutativity of T_2 and T is given by the isomorphism

$$\begin{aligned} T(T_2U) &\longrightarrow T_2(TU) \\ \big((u, u_0, v_0), (u_1, u_{01}, v_{01})\big) &\longmapsto \big((u, u_1), (u_0, u_{01}), (v_0, v_{01})\big) \end{aligned}$$

The bundle projection extends to T^2 as

$$\begin{aligned} (\pi T)_U = \pi_{TU} : (u, u_0, u_1, u_{01}) &\longmapsto (u, u_0) \\ (T\pi)_U = T\pi_U : (u, u_0, u_1, u_{01}) &\longmapsto (u, u_1). \end{aligned}$$

The other natural transformations that appear in the definition, $+T$, $T+$, λT, $T\lambda$, τT, and $T\tau$, are obtained in a similar way. The extension (8) of the vertical lift is given by

$$\lambda_2 : (u, u_0, v_0) \longmapsto (u, u_0, 0, v_0).$$

The following propositions can be checked by explicit elementary calculation:

PROPOSITION 2.14. *$\mathcal{E}$ucl with the tangent functor T and the natural transformations π, 0, $+$, τ, and λ is a tangent structure on $\mathcal{E}$ucl.*

PROPOSITION 2.15. *The fiberwise multiplication by real numbers,*

$$\begin{aligned} \kappa_U : \mathbb{R} \times TU &\longrightarrow TU \\ (r, (u, u_0)) &\longmapsto (u, r u_0), \end{aligned}$$

is a scalar multiplication of the tangent structure (Definition 2.11).

3. Left Kan extension to diffeological spaces

The structures on smooth manifolds that we wish to generalize to diffeological spaces, such as the tangent bundle and the algebra of differential forms, are local and universal in the sense that they are defined on all open subsets of $\mathbb{R}^n$ and then glued together along an atlas. In categorial terms, the local structure is given by a functor $F : \mathcal{E}\mathrm{ucl} \to \mathcal{C}$ and the glueing operation by the colimit

$$FM := \operatorname*{colim}_{U \to M} FU$$

over a maximal atlas. In categorical terms, FM is the pointwise left Kan extension of F along the inclusion of euclidean spaces into smooth manifolds. For the generalization of this construction to diffeological spaces we replace the charts of the smooth manifold M with the plots of the diffeological space X,

$$FX := \operatorname*{colim}_{U \to X} FU.$$

This is the pointwise left Kan extension of F along the inclusion $y : \mathcal{E}\mathrm{ucl} \to \mathcal{D}\mathrm{flg}$ of euclidean spaces into diffeological spaces, that is, the colimit of the functor

$$y \downarrow X \longrightarrow \mathcal{E}\mathrm{ucl} \xrightarrow{F} \mathcal{C},$$

where $y \downarrow X$ is the category of plots (Terminology 3.13) and where the first arrow maps a plot to its domain [**ML98**, Sec. X.5].

In order to use the left Kan extension as universal tool to generalize differential geometric structures to diffeological spaces, we have to study its categorical properties in some detail. For this it is best to work with the definition of diffeological spaces as concrete sheaves on euclidean spaces, which was elucidated in [**BH11**]. The site $\mathcal{E}$ucl of euclidean spaces is concrete, which means that there is a faithful functor $\mathcal{E}\mathrm{ucl} \to \mathcal{S}\mathrm{et}$, $U \to |U|$ that maps covers to surjective maps. A sheaf $D : \mathcal{E}\mathrm{ucl}^{\mathrm{op}} \to \mathcal{S}\mathrm{et}$ is concrete if it is a subsheaf of $U \mapsto \mathcal{S}\mathrm{et}(|U|, D(*))$. Explicitly, this means that $D(U)$ is a collection of maps of sets $|U| \to X := D(*)$, called plots, that satisfy the properties of a sheaf. In this way, we recover the original definition of diffeology.

As is the case for any category of concrete sheaves, the category of diffeological spaces $\mathcal{D}$flg is a quasi-topos, a category with a classifier for strong subobjects, which has a number of good properties (Proposition 3.6). Other good properties of $\mathcal{D}$flg are inherited from the site $\mathcal{E}$ucl. For example, the category of plots of a diffeological space, the index category used to compute the pointwise left Kan extension, is sifted, which suggest a compatibility with finite products.

Every representable presheaf is a concrete sheaf, so that the Yoneda embedding factors through a full and faithful functor

$$y : \mathcal{E}\mathrm{ucl} \longrightarrow \mathcal{D}\mathrm{flg}.$$

While y is simply the Yoneda embedding restricted on its codomain, the restriction changes the computation of colimits. A colimit in $\mathcal{D}\mathrm{flg}$ is given by first taking the pointwise colimit in the functor category $\mathcal{S}\mathrm{et}^{\mathcal{E}\mathrm{ucl}^{\mathrm{op}}}$ and then applying the left adjoint of the inclusion $\mathcal{D}\mathrm{flg} \to \mathcal{S}\mathrm{et}^{\mathcal{E}\mathrm{ucl}^{\mathrm{op}}}$. This second step has consequences for the properties of the left Kan extension, which is given by a colimit.

Most structures that we will consider are given by endofunctors $F : \mathcal{E}\mathrm{ucl} \to \mathcal{E}\mathrm{ucl}$. Since $\mathcal{E}\mathrm{ucl}$ is not cocomplete and since we want to obtain an endofunctor of diffeological spaces, we have to place the codomain of F in diffeological spaces by composing with y before taking the Kan extension. The Kan extended endofunctor will be denoted by

$$\mathbb{L}F := \mathrm{Lan}_y\, yF : \mathcal{D}\mathrm{flg} \longrightarrow \mathcal{D}\mathrm{flg}.$$

The main question we will address in this section is the following.

QUESTION 3.1. What are the categorical properties of $\mathbb{L} : \mathrm{End}(\mathcal{E}\mathrm{ucl}) \to \mathrm{End}(\mathcal{E}\mathrm{ucl})$ and which properties of F are preserved by $\mathbb{L}$?

It follows from the naturality of the Kan extension that $\mathbb{L}$ is a functor

$$\mathbb{L} : \mathrm{End}(\mathcal{E}\mathrm{ucl}) \longrightarrow \mathrm{End}(\mathcal{D}\mathrm{flg})$$

of categories of endofunctors, which means that the vertical composition of natural transformations and, therefore, commutative diagrams of natural transformations are preserved. This is the most important property for our purposes, since it implies that the commutative diagrams that appear in the Definition 2.7 of tangent structures carry over to the diffeological setting.

Unfortunately, this is where the good news about the left Kan extension to diffeological spaces end. Unlike the Kan extension to presheaves, the Kan extension of a functor to concrete sheaves does generally not preserve colimits, not even finite coproducts. Nor does $\mathbb{L}$ preserve the composition of endofunctors. Fortunately, the functors F we want to extend have some good categorical properties, that entail good properties of their Kan extensions $\mathbb{L}F$.

Using that the category of plots $U \to X$ is sifted, we show that if F preserves finite products, then so does its Kan extension $\mathbb{L}F$ (Proposition 3.19). It also implies that $\mathbb{L}$ preserves finite products of endofunctors (Proposition 3.20). For the compatibility with coproducts we have to assume that F is a cosheaf, that is, $F : \mathcal{E}\mathrm{ucl}^{\mathrm{op}} \to \mathcal{E}\mathrm{ucl}^{\mathrm{op}}$ is a sheaf. Then $\mathbb{L}F$ preserves coproducts (Corollary 3.23) and subductions (Proposition 3.24).

While $\mathbb{L}$ does not preserve the composition of endofunctors, there is a natural morphism

$$\mathbb{L}(FG) \longrightarrow (\mathbb{L}F)(\mathbb{L}G), \tag{10}$$

which is generally not an isomorphism. We can show that if F is a cosheaf, then (10) applied to a diffeological space is a subduction (Proposition 3.26). The morphism

$$\mathbb{L}(FGH) \longrightarrow \mathbb{L}(F)\mathbb{L}(G)\mathbb{L}(H)$$

we obtain by applying (10) twice does not depend on whether we first apply it to $(FG)H$ or to $F(GH)$ (Proposition 3.27). In this sense μ is associative. It is straightforward to see that (10) is natural in F and G (Proposition 3.28).

Finally, we study the compatibility of the Kan extension with the D-topology of diffeological spaces. We observe that many of the functors on euclidean spaces we are interested in come with a natural transformation $F \to 1$ such that the pullback along any open embedding $V \to U$ satisfies $FV \cong V \times_U FU$. We call such an $F \to 1$ a local bundle, by default of a better term. We can show that if $F \to 1$ is local, then so is its Kan extension $\mathbb{L}F \to 1$ (Proposition 3.35). Then we prove that a morphism $FX \to GX$ of local bundles is an induction, subduction, epimorphism, or isomorphism if all restrictions to the open subsets of a cover of X are (Proposition 3.34).

3.1. Diffeological spaces as concrete sheaves. Recall that $\mathcal{E}\mathrm{ucl}$ denotes the category which has all open subsets of euclidean spaces $\mathbb{R}^n$, $n \geq 0$ as objects and all smooth maps as morphisms. Open covers define a Grothendieck pretopology.

DEFINITION 3.2. The small category $\mathcal{E}\mathrm{ucl}$ together with the Grothendieck topology generated by the pretopology of open covers will be called the **site of euclidean spaces**.

The terminal object in $\mathcal{E}\mathrm{ucl}$ is $* := \mathbb{R}^0$. The functor of points,

$$\begin{aligned} |\_| : \mathcal{E}\mathrm{ucl} &\longrightarrow \mathcal{S}\mathrm{et} \\ U &\longmapsto \mathcal{E}\mathrm{ucl}(*, U), \end{aligned}$$

is faithful, so that it equips $\mathcal{E}\mathrm{ucl}$ with the structure of a concrete category. Moreover, every cover $\{U_i \to U\}$ is surjective on the underlying sets. A site with these properties is called **concrete**.

Let $X : \mathcal{E}\mathrm{ucl}^{\mathrm{op}} \to \mathcal{S}\mathrm{et}$ be a presheaf. Then there is a morphism of presheaves defined by

$$\begin{aligned} X(U) &\longrightarrow \mathcal{S}\mathrm{et}\big(|U|, |X|\big) \\ p &\longmapsto \big((* \xrightarrow{u} U) \mapsto X_u(p)\big), \end{aligned} \tag{11}$$

where $X_u \equiv X(u) : X(U) \to X(*)$ is the restriction of $X(U)$ to the point u.

DEFINITION 3.3. A presheaf $X : \mathcal{E}\mathrm{ucl}^{\mathrm{op}} \to \mathcal{S}\mathrm{et}$ on a concrete site is **concrete** if (11) is a monomorphism, i.e. if the maps defined in (11) are injective for all $U \in \mathcal{E}\mathrm{ucl}$. A sheaf is concrete if it is concrete as a presheaf. A morphism between concrete sheaves is a morphism of the underlying presheaves.

DEFINITION 3.4. A concrete sheaf on the site of euclidean spaces is called a **diffeological space**. A morphism of diffeological spaces is a morphism of sheaves. The category of diffeological spaces will be denoted by $\mathcal{D}\mathrm{flg}$.

THEOREM 3.5 (Thm. 5.25 in [**BH11**]). *The category of diffeological spaces is a quasitopos with small limits and small colimits.*

Theorem 3.5 implies that the category of diffeological spaces has a number of convenient properties. For clarity and later reference, we will spell out some of them.

PROPOSITION 3.6. *The category $\mathcal{D}$flg of diffeological spaces has the following properties:*

- *$\mathcal{D}$flg is locally cartesian closed, i.e. for every object X in $\mathcal{D}$flg the overcategory $\mathcal{D}\text{flg}\downarrow X$ is cartesian closed.*
- *Strong monomorphisms and strong epimorphisms are effective.*
- *(Strong) monomorphisms and (strong) epimorphisms are stable under pullback.*
- *$\mathcal{D}$flg is quasiadhesive, that is, the pushout of a strong monomorphism is a strong monomorphism and the pushout square is a pullback square.*
- *The initial object is strict, i.e. every morphism $X \to \emptyset$ is an isomorphism.*
- *Coproducts are disjoint, i.e. $X \to X \sqcup Y \leftarrow Y$ are monomorphisms and $X \times_{X \sqcup Y} Y \cong \emptyset$.*
- *The functor of points $\mathcal{D}\text{flg} \to \mathcal{S}\text{et}$, $X \to \mathcal{D}\text{flg}(*, X)$ is faithful. It has a left and a right adjoint, so that it preserves limits and colimits.*

TERMINOLOGY 3.7. The strong epimorphisms in $\mathcal{D}$flg are called **subductions**, the strong monomorphisms **inductions**.

EXAMPLE 3.8. Here are some of the most basic examples for diffeologies:

(a) The **fine diffeology** or **discrete diffeology** on a set S is the diffeology for which the plots are the locally constant maps.[1]

(b) The **coarse diffeology**, or **indiscrete diffeologoy**, or **trivial diffeology** on a set S is given by $U \mapsto \mathcal{S}\text{et}\big(|U|, S\big)$, i.e. all maps are plots.

(c) Every topological space X is equipped with the **continuous diffeology** given by $U \mapsto \mathcal{T}\text{op}(U, X)$, i.e. the plots are the continuous maps.

(d) Every smooth finite-dimensional manifold M is equipped with the **natural diffeology** given by $U \mapsto \mathcal{M}\text{fld}(U, M)$, i.e. the plots are the infinitely often differentiable maps.

(e) We will denote the exponential objects in $\mathcal{D}$flg by

$$\underline{\mathcal{D}\text{flg}}(X, Y) \equiv Y^X$$

and call them the **diffeological mapping spaces**. The diffeology, which is given by the universal property

$$\mathcal{D}\text{flg}\big(U, \underline{\mathcal{D}\text{flg}}(X, Y)\big) \cong \mathcal{D}\text{flg}(U \times X, Y),$$

is called the **functional diffeology**.

REMARK 3.9. By Proposition 3.6, the forgetful functor $\mathcal{D}\text{flg} \to \mathcal{S}\text{et}$, $X \mapsto |X|$ has a left and a right adjoint. The right adjoint equips a set S with the coarse diffeology. The left adjoint equips it with the fine diffeology. In other words, the fine diffeology on a set is the free diffeology, the coarse diffeology is the cofree diffeology.

REMARK 3.10. The map that sends a smooth manifold M to its natural diffeology $U \mapsto \mathcal{M}\text{fld}(U, M)$ defines a full, faithful, and injective functor $\mathcal{M}\text{fld} \to \mathcal{D}\text{flg}$.

REMARK 3.11. Every diffeological space X is naturally equipped with the finest topology such that all plots are continuous, which is called the D-topology. This topology is determined by the smooth curves only, so that many different diffeologies

[1] In [**BH11**, Example (2), p. 5794] it is stated incorrectly that the discrete diffeology is given by the constant maps.

induce the same topology [**CSW14**, Thm. 3.7]. Mapping a diffeology on X to the induced topology is left adjoint to mapping a topology to the continuous diffeology [**CSW14**, Prop. 3.3]. The topology induced by the discrete (trivial) diffeology is the discrete topology. In general, however, neither the unit nor the counit of the adjunction is an isomorphism.

The maps $p \in X(U) \subset \mathcal{S}\mathrm{et}(|U|, |X|)$ of a diffeological space X are called **plots**. If we spell out the defining conditions of a concrete sheaf, we obtain the traditional definition of diffeological spaces in terms of plots.

For every open cover $\{U_i \to U\}$ in $\mathcal{E}\mathrm{ucl}$, U is the coequalizer of $\coprod_{i,j} U_i \cap U_j \rightrightarrows \coprod U_i$. In other words, the site of euclidean spaces is subcanonical, so that every representable presheaf is a sheaf. Since every representable presheaf is concrete, it follows that all representable presheaves on $\mathcal{E}\mathrm{ucl}$ are concrete sheaves. We conclude that the Yoneda embedding $Y : X \mapsto \mathcal{D}\mathrm{flg}(\_, X)$ factors as

$$\begin{array}{ccc} & & \mathcal{S}\mathrm{et}^{\mathcal{E}\mathrm{ucl}^{\mathrm{op}}} \\ & \overset{Y}{\nearrow} & \uparrow I \\ \mathcal{E}\mathrm{ucl} & \xrightarrow[y]{} & \mathcal{D}\mathrm{flg} \end{array}$$

through a functor y. Since Y and I are full and faithful, so is y. Since I is full and faithful, the Yoneda lemma implies that the evaluation of the concrete sheaf $X \in \mathcal{D}\mathrm{flg}$ on $U \in \mathcal{E}\mathrm{ucl}$ is given by

$$\begin{aligned} X(U) &\cong \mathcal{S}\mathrm{et}^{\mathcal{E}\mathrm{ucl}^{\mathrm{op}}}(YU, IX) \cong \mathcal{S}\mathrm{et}^{\mathcal{E}\mathrm{ucl}^{\mathrm{op}}}(IyU, IX) \\ &\cong \mathcal{D}\mathrm{flg}(yU, X). \end{aligned}$$

It follows that limits in $\mathcal{D}\mathrm{flg}$ are computed pointwise and that I preserves limits. By the adjoint functor theorem, I has a left adjoint,

$$K : \mathcal{S}\mathrm{et}^{\mathcal{E}\mathrm{ucl}^{\mathrm{op}}} \rightleftarrows \mathcal{D}\mathrm{flg} : I,$$

which was computed and studied in [**BH11**, Sec. 5.3]. Explicitly, K is given by a procedure called concretization followed by the Grothendieck plus construction.

The left adjoint K is a retract, $KI \cong \mathrm{id}_{\mathcal{D}\mathrm{flg}}$, which implies that the colimit of a diagram in $X : \mathcal{I} \to \mathcal{D}\mathrm{flg}$ can be computed as

$$\begin{aligned} \operatorname*{colim}_{i \in \mathcal{I}} X_i &\cong \operatorname*{colim}_{i \in \mathcal{I}} KIX_i \\ &\cong K \operatorname*{colim}_{i \in \mathcal{I}} IX_i, \end{aligned}$$

that is, by first computing the colimit in presheaves and then applying K. As a further consequence, it can be shown that y is dense:

PROPOSITION 3.12 (Prop. 51 in [**BH11**]). *Every $X \in \mathcal{D}\mathrm{flg}$ is the colimit of $y \downarrow X \to \mathcal{E}\mathrm{ucl} \to \mathcal{D}\mathrm{flg}$, which we will write as*

$$X \cong \operatorname*{colim}_{yU \to X} yU.$$

TERMINOLOGY 3.13. The comma category $y \downarrow X$ is called the **category of plots** of X.

REMARK 3.14. It is customary and convenient to identify notationally the domain of a plot $U \in \mathcal{E}\mathrm{ucl}$ with the diffeological space $yU \in \mathcal{D}\mathrm{flg}$. In this paper, however, we deal with a number of subtleties of Kan extensions along y where this identification would invite wrong proofs by notation (a trap the author has fallen into more than once). Therefore, we will always spell out the embedding y.

3.2. Left Kan extension to diffeological spaces.

NOTATION 3.15. Let $F : \mathcal{E}\mathrm{ucl} \to \mathcal{E}\mathrm{ucl}$ be an endofunctor. The left Kan extension of yF along the embedding $y : \mathcal{E}\mathrm{ucl} \hookrightarrow \mathcal{D}\mathrm{flg}$ will be denoted by

$$\mathbb{L}F := \mathrm{Lan}_y\, yF, \tag{12}$$

which is an endofunctor of $\mathcal{D}\mathrm{flg}$.

The left Kan extension will be our device to extend the tangent structure of euclidean to diffeological spaces. $\mathbb{L}$ is functorial, which means that $\mathbb{L}$ preserves the vertical compostion of natural transformations $\hat{\alpha} : \hat{F} \to \hat{F}'$ and $\hat{\alpha}' : \hat{F}' \to \hat{F}''$ between endofunctors $\hat{F}, \hat{F}', \hat{F}'' \in \mathrm{End}(\mathcal{E}\mathrm{ucl})$, that is,

$$\mathbb{L}(\alpha' \circ \alpha) = (\mathbb{L}\alpha') \circ (\mathbb{L}\alpha).$$

Since $\mathcal{E}\mathrm{ucl}$ is small and $\mathcal{D}\mathrm{flg}$ is cocomplete, $\mathbb{L}F$ exists and is pointwise, that is, it can be computed by the colimit [**ML98**, Thm. X.5.3]

$$\begin{aligned}(\mathbb{L}F)(X) &\cong \mathrm{colim}(y \downarrow X \longrightarrow \mathcal{E}\mathrm{ucl} \xrightarrow{yF} \mathcal{D}\mathrm{flg}) \\ &= \operatorname*{colim}_{yU \to X} yFU,\end{aligned}$$

for all $X \in \mathcal{D}\mathrm{flg}$. Let $\alpha : F \to G$ be a natural transformation of functors $\mathcal{E}\mathrm{ucl} \to \mathcal{E}\mathrm{ucl}$. Then $y\alpha : yF \to yG$ is a natural transformation of functors $\mathcal{E}\mathrm{ucl} \to \mathcal{D}\mathrm{flg}$. The left Kan extension $\mathrm{Lan}_y\, yF$ is functorial in yF, so that we have a natural transformation

$$\mathbb{L}\alpha := \mathrm{Lan}_y\, y\alpha : \mathbb{L}F \longrightarrow \mathbb{L}G. \tag{13}$$

Together (12) and (13) define a functor

$$\mathbb{L} : \mathrm{End}(\mathcal{E}\mathrm{ucl}) \longrightarrow \mathrm{End}(\mathcal{D}\mathrm{flg}), \tag{14}$$

where $\mathrm{End}(\mathcal{C})$ denotes the category of endofunctors and natural transformations of the category $\mathcal{C}$.

PROPOSITION 3.16. *The diagram*

$$\begin{array}{ccc} \mathcal{E}\mathrm{ucl} & \xrightarrow{\quad F \quad} & \mathcal{E}\mathrm{ucl} \\ {\scriptstyle y}\big\downarrow & & \big\downarrow{\scriptstyle y} \\ \mathcal{D}\mathrm{flg} & \xrightarrow[\quad \mathbb{L}F \quad]{} & \mathcal{D}\mathrm{flg} \end{array}$$

commutes for all endofunctors F, that is,

$$(\mathbb{L}F)yU = yFU$$

for all $U \in \mathcal{E}\mathrm{ucl}$.

PROOF. Since y is full and faithful, the statement follows from [**ML98**, Cor. 3, Sec. X.3] or from [**Kel05**, Prop. 4.23]. □

COROLLARY 3.17. *The functor* (14) *is full and faithful.*

Corollary 3.17 means that the natural transformations $\mathbb{L}F \to \mathbb{L}G$ of left Kan extended functors are in natural bijection with the natural transformations $F \to G$ of the underlying functors of euclidean spaces. It is ultimately a consequence of the fact that the inclusion $\mathcal{E}\mathrm{ucl} \hookrightarrow \mathcal{D}\mathrm{flg}$ is full and faithful (cf. [**Wat13**]).

3.3. Compatibility with products, coproducts, and subductions. Since $y : \mathcal{E}\mathrm{ucl} \to \mathcal{D}\mathrm{flg}$ is full and faithful, a smooth map $f : U \to V$ of euclidean spaces is a strong epimorphism if and only if $yf : yU \to yV$ is a strong epimorphism, which is the same thing as a subduction. For this reason we will call a strong epimorphism f a subduction. This is the case if every point $v_0 \in V$ has an open neighborhood $V_0 \subset V$ such that there is a smooth map $g_0 : V_0 \to U$ satisfying $f \circ g_0 = \mathrm{id}_{V_0}$. In short, f is a subduction if it has local sections. In particular, every surjective submersion is a subduction.

PROPOSITION 3.18. *Let $\alpha : F \to G$ be a natural transformation of endofunctors of $\mathcal{E}\mathrm{ucl}$. If $\alpha_U : FU \to GU$ is a subduction for all $U \in \mathcal{E}\mathrm{ucl}$, then $(\mathbb{L}\alpha)_X : (\mathbb{L}F)X \to (\mathbb{L}G)X$ is a subduction for all $X \in \mathcal{D}\mathrm{flg}$.*

PROOF. Since α_U is a subduction, so is $y\alpha_U$. Subductions in $\mathcal{D}\mathrm{flg}$ are the same as regular epimorphisms, so that $y\alpha_U$ is a regular epimorphism for all $U \in \mathcal{E}\mathrm{ucl}$. The left Kan extension $\alpha_X = (\mathbb{L}\alpha)_X$ is given by the colimit over the category of plots $y{\downarrow}X$. Since colimits preserve regular epimorphisms, α_X is a regular epimorphism, that is, a subduction. □

PROPOSITION 3.19. *If a functor $F : \mathcal{E}\mathrm{ucl} \to \mathcal{E}\mathrm{ucl}$ preserves finite products, then so does $\mathbb{L}F : \mathcal{D}\mathrm{flg} \to \mathcal{D}\mathrm{flg}$.*

Let $F : \mathcal{I} \to \mathrm{End}(\mathcal{E}\mathrm{ucl})$, $i \mapsto F_i$ be a functor. Due to the universal properties of colimits and limits, we have for every $X \in \mathcal{D}\mathrm{flg}$ the natural morphism

$$\operatorname*{colim}_{yU \to X} \lim_{i \in \mathcal{I}} yF_iU \longrightarrow \lim_{i \in \mathcal{I}} \operatorname*{colim}_{yU \to X} yF_iU.$$

Assuming that the limit $\lim_{i \in \mathcal{I}} F_i$ exists in $\mathrm{End}(\mathcal{E}\mathrm{ucl})$, it can be written as the natural transformation

$$\mathbb{L} \lim_i F_i \longrightarrow \lim_i \mathbb{L}F_i,$$

where we have used that y preserves limits. This is not an isomorphism unless the colimit and the limit commute.

Let $G : \mathcal{J} \to \mathrm{End}(\mathcal{E}\mathrm{ucl})$ be another diagram, such that $\lim_i G_i$ exists. Any natural transformation $\alpha_i : F_i \to G_i$ induces a commutative diagram

$$\begin{array}{ccc} \mathbb{L}\lim_i F_i & \xrightarrow{\mathbb{L}(\lim_i \alpha_i)} & \mathbb{L}\lim_i G_i \\ \downarrow & & \downarrow \\ \lim_i \mathbb{L}F_i & \xrightarrow[\lim_i \mathbb{L}\alpha_i]{} & \lim_i \mathbb{L}G_i \end{array}$$

PROPOSITION 3.20. *Let $F_1, \ldots, F_k \in \mathrm{End}(\mathcal{E}\mathrm{ucl})$ be a finite family of endofunctors. Then we have an isomorphism*

$$\mathbb{L}(F_1 \times \ldots \times F_k) \cong \mathbb{L}F_1 \times \ldots \times \mathbb{L}F_k.$$

It is well-known, that the left Kan extension of an arbitrary functor along the Yoneda embedding $Y : \mathcal{E}\mathrm{ucl} \to \mathcal{S}\mathrm{et}^{\mathcal{E}\mathrm{ucl}^{\mathrm{op}}}$ preserves all colimits. This is not true for Kan extensions along $y : \mathcal{E}\mathrm{ucl} \to \mathcal{D}\mathrm{flg}$. Already for the preservation of coproducts we have to make additional assumptions. Recall that a functor $F : \mathcal{E}\mathrm{ucl} \to \mathcal{C}$ is a **cosheaf** if $F^{\mathrm{op}} : \mathcal{E}\mathrm{ucl}^{\mathrm{op}} \to \mathcal{C}^{\mathrm{op}}$ is a sheaf.

EXAMPLE 3.21. The following functors on $\mathcal{E}\mathrm{ucl}$ are cosheaves:

(a) the tangent functor $T : \mathcal{E}\mathrm{ucl} \to \mathcal{E}\mathrm{ucl}$;

(b) fiber products of the tangent functor $T_k : \mathcal{E}\mathrm{ucl} \to \mathcal{E}\mathrm{ucl}$;
(c) if $F : \mathcal{E}\mathrm{ucl} \to \mathcal{C}$ and $G : \mathcal{E}\mathrm{ucl} \to \mathcal{E}\mathrm{ucl}$ are cosheaves, then so is their composition FG;
(d) the de Rham functor $\Omega : \mathcal{E}\mathrm{ucl} \to \mathrm{dg}\mathcal{A}\mathrm{lg}^{\mathrm{op}}$, which maps U to the differential graded algebra of differential forms and smooth maps to the pullback of forms;
(d) if $F : \mathcal{M}\mathrm{fld}^{\mathrm{op}} \to \mathcal{C}$ is a sheaf on the big site of manifolds and open covers, then the restriction $F : \mathcal{E}\mathrm{ucl} \hookrightarrow \mathcal{M}\mathrm{fld} \to \mathcal{C}^{\mathrm{op}}$ is a cosheaf;

PROPOSITION 3.22. *If $F : \mathcal{E}\mathrm{ucl} \to \mathcal{C}$ is a cosheaf, then its left Kan extension along $y : \mathcal{E}\mathrm{ucl} \to \mathcal{D}\mathrm{flg}$ preserves coproducts.*

COROLLARY 3.23. *If $F : \mathcal{E}\mathrm{ucl} \to \mathcal{E}\mathrm{ucl}$ is a cosheaf, then $\mathbb{L}F$ preserves coproducts.*

PROPOSITION 3.24. *If $F : \mathcal{E}\mathrm{ucl} \to \mathcal{E}\mathrm{ucl}$ is a cosheaf, then $\mathbb{L}F$ preserves subductions.*

PROPOSITION 3.25. *Let $X : \mathcal{I} \to \mathcal{D}\mathrm{flg}$ be a functor. If $F : \mathcal{E}\mathrm{ucl} \to \mathcal{E}\mathrm{ucl}$ is a cosheaf, then the natural morphism*

$$\operatorname{colim}(\mathbb{L}F)X \longrightarrow (\mathbb{L}F)\operatorname{colim} X$$

is a subduction.

3.4. Compatibility with the composition of endofunctors. Let $F, G : \mathcal{E}\mathrm{ucl} \to \mathcal{E}\mathrm{ucl}$ be endofunctors. The left Kan extension of their product is given by the colimit

$$\begin{aligned} \mathbb{L}(FG)X &\cong \operatorname*{colim}_{yU \to X} yFGU \\ &\cong \operatorname*{colim}_{yU \to X} (\mathbb{L}F)yGU, \end{aligned} \tag{15}$$

where we have used Proposition 3.16. The product of the Kan extensions can be written as

$$(\mathbb{L}F)(\mathbb{L}G)X \cong (\mathbb{L}F)\operatorname*{colim}_{yU \to X} yGU. \tag{16}$$

Due to the universal property of the colimit, we have a natural morphism

$$\operatorname*{colim}_{yU \to X} (\mathbb{L}F)yGU \longrightarrow (\mathbb{L}F)\operatorname*{colim}_{yU \to X} yGU. \tag{17}$$

Composing this morphism with the isomorphism (15) on its domain and with isomorphism (16) on the codomain, we obtain a natural morphism

$$\mu_{F,G} : \mathbb{L}(FG)X \longrightarrow (\mathbb{L}F)(\mathbb{L}G)X. \tag{18}$$

This is an isomorphism if and only if (17) is, which is generally not the case.

PROPOSITION 3.26. *If F is a cosheaf, then* (18) *is a subduction for all $X \in \mathcal{D}\mathrm{flg}$.*

PROOF. Since F is a cosheaf, it follows from Proposition 3.25 that (17) is a subduction, which implies that (18) is a subduction. $\square$

The morphism (18) is associative in the followoing sense.

PROPOSITION 3.27. *Let $F, G, H : \mathcal{E}\mathrm{ucl} \to \mathcal{E}\mathrm{ucl}$ be endofunctors. Then the diagram of natural transformations*

$$\begin{array}{ccc} \mathbb{L}(FGH) & \xrightarrow{\mu_{F,GH}} & (\mathbb{L}F)(\mathbb{L}GH) \\ {\scriptstyle \mu_{FG,H}}\downarrow & & \downarrow{\scriptstyle (\mathbb{L}F)\mu_{G,H}} \\ \mathbb{L}(FG)(\mathbb{L}H) & \xrightarrow[\mu_{F,G}(\mathbb{L}H)]{} & (\mathbb{L}F)(\mathbb{L}G)(\mathbb{L}H) \end{array}$$

is commutative.

PROPOSITION 3.28. *Let $\alpha : F \to F'$ and $\beta : G \to G'$ be natural transformations of endofunctors of $\mathcal{E}\mathrm{ucl}$. Then the diagram*

$$\begin{array}{ccc} \mathbb{L}(FG) & \xrightarrow{\mathbb{L}(\alpha\beta)} & \mathbb{L}(F'G') \\ {\scriptstyle \mu_{F,G}}\downarrow & & \downarrow{\scriptstyle \mu_{F',G'}} \\ (\mathbb{L}F)(\mathbb{L}G) & \xrightarrow[(\mathbb{L}\alpha)(\mathbb{L}\beta)]{} & (\mathbb{L}F')(\mathbb{L}G') \end{array}$$

is commutative.

3.5. Compatibility with the D-topology.

DEFINITION 3.29 (Sec. 2.8 in [**IZ13**]). The **D-topology** on a diffeological space is the finest topology (on the underlying set) such that every plot is continuous.

Explicitly, a subset $Y \subset X$ is open in the D-topolgy if and only if for every plot $p : yU \to X$, the preimage $p^{-1}(Y) \subset U$ is open. Every morphism of diffeological spaces is continuous with respect to the D-topologies. In the following "open" and "continuous" are always meant with respect to the D-topology. An open subset $S \subset X$ of a diffeological space is naturally equipped with the subspace diffeology, so that the inclusion $i : S \to X$ is an open induction.

Let $\{S_i \subset X\}_{i \in I}$ be an open cover of a diffeological space X. Then the diagram

$$\coprod_{i,j} S_{ij} \rightrightarrows \coprod_i S_i \longrightarrow X \,, \tag{19}$$

where $S_{ij} = S_i \times_X S_j$, is a coequalizer. A functor $F : \mathcal{D}\mathrm{flg} \to \mathcal{D}\mathrm{flg}$ is a **cosheaf** if F preserves the coequalizer (19), that is, if

$$\coprod_{i,j} FS_{ij} \rightrightarrows \coprod_i FS_i \longrightarrow FX$$

is a coequalizer for every open cover. A special kind of cosheaf is given by a natural bundle that satisfies the following locality condition.

DEFINITION 3.30. A bundle $\pi : F \to 1$ of endofunctors of diffeological spaces will be called **local** if for every open induction $i : S \to X$ the commutative diagram

$$\begin{array}{ccc} FS & \xrightarrow{Fi} & FX \\ {\scriptstyle \pi_S}\downarrow & & \downarrow{\scriptstyle \pi_X} \\ S & \xrightarrow[i]{} & X \end{array}$$

is a pullback.

PROPOSITION 3.31. *If a bundle $\pi : F \to 1$ of endofunctors of diffeological spaces is local, then F is a cosheaf.*

PROPOSITION 3.32. *If a bundle $\pi : F \to 1$ of endofunctors of diffeological spaces is local, then F preserves open inductions.*

PROPOSITION 3.33. *Let $\pi : F \to 1$ and $\rho : G \to 1$ be bundles of endofunctors of diffeological spaces. If both bundles are local, then the fiber product $F \times_1 G \to 1$ and the composition $FG \to 1$ is local.*

PROPOSITION 3.34. *Let $\pi : F \to 1$ and $\rho : G \to 1$ be bundles of endofunctors of diffeological spaces and $\alpha : F \to G$ a morphism of bundles, $\rho \circ \alpha = \pi$. If both bundles are local, then the following are equivalent:*

(i) *$\alpha_X : FX \to GX$ is an induction (subduction, epimorphism, isomorphism).*
(ii) *There is an open cover $\{S_i \to X\}$, such that $\alpha_{S_i} : FS_i \to GS_i$ is an induction (subduction, epimorphism, isomorphism) for all i.*

Definition 3.30 applies also to bundles of endomorphisms of euclidean spaces. Explicitly, a bundle $\pi : F \to 1$ is local if $\pi_U^{-1}(V) \cong FV$ for every open subset $V \subset U$. The next proposition shows that this notion locality is preserved by the left Kan extension to diffeological spaces.

PROPOSITION 3.35. *If a bundle $\pi : F \to 1$ of endofunctors of euclidean spaces is local, then so is its left Kan extension $\mathbb{L}\pi : \mathbb{L}F \to 1$.*

4. Elastic diffeological spaces

The main result for the category of elastic spaces is that the left Kan extension of the tangent structure on euclidean spaces defines a tangent structure with scalar $\mathbb{R}$-multiplication (Theorem 4.2). The proof of this statement is quite long and will be given in [**Blo**]. It uses a variety of techniques, some categorical, some geometric, and all of the results about the left Kan extension given in the preceeding Section 3.2.

4.1. The tangent structure of elastic spaces. The tangent structure of euclidean spaces consists of the tangent functor $\hat{T} : \mathcal{E}\text{ucl} \to \mathcal{E}\text{ucl}$ together with the natural transformations of the bundle projection $\hat{\pi} : \hat{T} \to 1$, the zero section $\hat{0} : 1 \to \hat{T}$, the addition $\hat{+} : \hat{T}_2 \to \hat{T}$, the symmetric structure $\hat{\tau} : \hat{T}^2 \to \hat{T}^2$, and the vertical lift $\hat{\lambda} : \hat{T} \to \hat{T}^2$, which we from now one decorate with hats in order to distinguish them from their Kan extensions to diffeological spaces. The structure was spelled out explicitly in Section 2.5.

We would like to extend this structure to diffeological spaces by applying the Kan extension functor $\mathbb{L} = \operatorname{Lan}_y y\,\_$. The Kan extension of the tangent functor will be denoted by

$$T := \mathbb{L}\hat{T} : \mathcal{D}\text{flg} \to \mathcal{D}\text{flg}.$$

DEFINITION 4.1. A diffeological space X is called **elastic** if the following axioms hold:

(E1) The natural morphisms

$$\theta_{k,X} : (\mathbb{L}\hat{T}_k)X \longrightarrow T_k X$$

are isomorphisms for all $k > 1$.

(E2) There is a natural morphism $\tau_X : T^2X \to T^2X$, such that the diagram

$$\begin{array}{ccc} (\mathbb{L}\hat{T}^2)X & \xrightarrow{(\mathbb{L}\hat{\tau})_X} & (\mathbb{L}\hat{T}^2)X \\ \theta^2_X \big\downarrow & & \big\downarrow \theta^2_X \\ T^2X & \xrightarrow[\tau_X]{} & T^2X \end{array}$$

commutes.

(E3) The natural morphism

$$\lambda_X : TX \xrightarrow{(\mathbb{L}\hat{\lambda})_X} (\mathbb{L}\hat{T}^2)X \xrightarrow{\theta^2_X} T^2X$$

is an induction.

(E4) The natural morphisms

$$\nu_{k,X} : TT_kX \longrightarrow T^2X \times_{TX}^{T\pi_X, T\pi_X} \dots \times_{TX}^{T\pi_X, T\pi_X} T^2X$$

are injective for all $k > 1$.

(E5) For every finite set of positive integers $k_1, \dots, k_n$ the diffeological space $X' := T_{k_1} \cdots T_{k_n} X$ satisfies axioms (E1) through (E4).

The full subcategory of elastic diffeological spaces will be denoted by $\mathcal{E}\mathrm{lst} \subset \mathcal{D}\mathrm{flg}$.

THEOREM 4.2. *The category of elastic diffeological spaces has a tangent structure given by the Kan extended tangent functor $T = \mathbb{L}\hat{T}$, bundle projection $\pi_X := (\mathbb{L}\hat{\pi})_X$, and zero section $0_X := (\mathbb{L}\hat{0})_X$, the addition*

$$+_X := \theta_{2,X}^{-1} \circ (\mathbb{L}\hat{+})_X \circ \theta_{2,X},$$

the symmetric structure τ_X of Axiom (E2) of Definition 4.1, *and the vertical lift λ_X of Axiom (E3). Moreover, the Kan extension of the scalar $\mathbb{R}$-multiplication is a scalar $\mathbb{R}$-multiplication in the sense of Definition* 2.11.

4.2. Alternative axioms. The axioms for abstract tangent structures are to hold for the entire category. This is why the Axiom (E5) has to be included in the Definition 4.1 of elastic spaces, so that the application of T_k does not lead out of the subcategory of elastic spaces. However, Rosický's axioms still make sense pointwise for a single object X that lies in an ambient category where T_k is defined. This is our situation, which suggests the following more general concept.

DEFINITION 4.3. A diffeological space X is called **weakly elastic** if the Axioms (E1)-(E4) of Definition 4.1 hold.

On a weakly elastic space we still have most of the structure of differential calculus, like a Lie algebra of vector fields. We only have to observe that TX and its fiber products may no longer share the same good properties.

REMARK 4.4. Theorem 4.2 still holds if Axiom (E1) of Definition 4.1 is replaced with the following weaker version:

(E1') The natural morphism

$$\theta_{k,X} : (\mathbb{L}\hat{T}_k)X \longrightarrow T_kX$$

is an isomorphism for $k = 2$ and an epimorphism for all $k > 2$.

As explained in the introduction, we need the stronger Axiom (E1) so that we obtain a Cartan calculus.

In earlier versions of the definition of elastic space, we have used the following axiom:

(E0) The natural morphisms

$$\mathbb{L}(\hat{T}_k\hat{T})X \longrightarrow T_kTX$$

are isomorphisms for all $k \geq 1$.

PROPOSITION 4.5. *The Axiom (E0) implies the Axioms (E1), (E2), and (E4) of Definition* 4.1.

While this proposition shows that Axiom (E0) is logically stronger than (E1), (E2), and (E4), we currently do not know an example of an elastic space that does not satisfy the stronger Axiom (E0).

4.3. Stability properties of elastic spaces. The conditions for a diffeological space to be elastic are quite strong. Limits, colimits, and subspaces of elastic spaces are generally no longer elastic. There are still some categorical constructions which preserve the elasticity of a space.

The first is the restriction to open subsets in the D-topology. The functors $\mathbb{L}\hat{T}_k$ and $\mathbb{L}\hat{T}^2$ that appear in the definition of elastic spaces are the Kan extension of natural fiber bundles in $\mathcal{E}$ucl. If $\hat{F}U \to U$ is such a fiber bundle, its Kan extension $FX \to X$ is generally not a fiber bundle. But it can be shown that it is still compatible with the restriction to an open subset $i : S \hookrightarrow X$ on the base, in the sense that the commutative diagram

$$\begin{array}{ccc} FS & \xrightarrow{Fi} & FX \\ \downarrow & & \downarrow \\ S & \xrightarrow[i]{} & X \end{array}$$

is a pullback. Using this property, we can show the following result.

PROPOSITION 4.6. *Let X be a diffeological space. The following are equivalent:*

(i) *X is elastic.*
(ii) *X has an open cover $\{S_i \to X\}$ by elastic spaces S_i.*

In other words, Proposition 4.6 states that elasticity is a local property (with respect to the D-topology).

COROLLARY 4.7. *Let $\{X_i\}_{i\in I}$ be a small family of diffeological spaces. The following are equivalent:*

(i) *The coproduct $\sqcup_i X_i$ is elastic.*
(ii) *Every X_i is elastic.*

For the next property we use that the category of plots of a diffeological space is sifted [**ARV10**]. This means that colimits of functors on this category commute with finite products. This observation leads to the following result.

PROPOSITION 4.8. *Finite products of elastic spaces are elastic.*

In category theory, a **retract** is a pair of morphisms $i : X \to Y$ and $p : Y \to X$ such that $pi = \mathrm{id}_X$. In diffeology, a split monomorphism is a subduction, so that X

is a diffeological subspace of Y. A morphism $f : X \to X'$ is a retract of a morphism $g : Y \to Y'$ if there is a commutative diagram

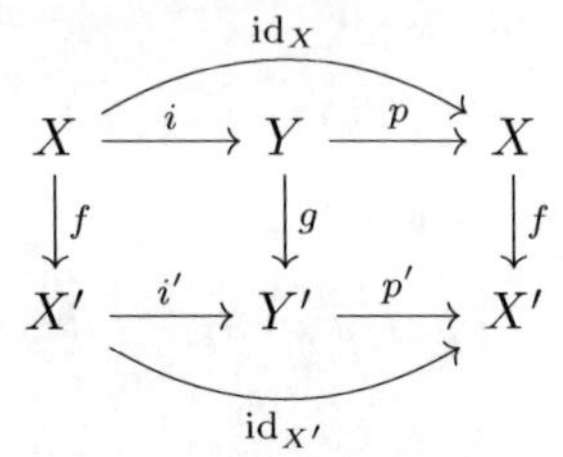

Strong monomorphisms and strong epimorphisms are stable under retracts in any category. In diffeological spaces this means that if g is a subduction or an induction, then so is f. The next statement is a consequence of this.

PROPOSITION 4.9. *Retracts of elastic spaces are elastic.*

5. Examples

The conditions for a diffeological space to be elastic are quite strong, so the question arises whether there are enough interesting examples and applications for the concept to be useful. We have just stated that finite products and small coproducts of elastic spaces are elastic. We have also seen that elastic spaces are stable under retracts, which allows for the construction of elastic spaces with certain kinds of singularities.

The original example that motivated the concept of diffeology by Souriau is that of diffeological groups. Our main result for groups states that a diffeological group is elastic if and only if the vertical lift $\lambda_G : TG \to T^2G$ is an induction (Theorem 5.2). This mild condition is needed to avoid that the bracket of vector fields takes values in a tangent bundle with a weaker diffeology, as is the case for vector fields on groups on a C^k-manifold. This shows that almost all diffeological groups that come to mind are elastic. Even a seemingly pathological group like the quotient $\mathbb{R}/\mathbb{Q}$ is elastic.

Every diffeological vector space has an underlying abelian group, so that we can apply the characterization of diffeological groups. It follows that a diffeological vector space A is elastic if the natural map $T_0A \to T_0T_0A$ that maps v_a to the tangent vector represented by the path $t \mapsto tv_a$ is an induction. A diffeological manifold modelled on A is elastic if and only if A is elastic. A stronger condition which ensures elasticity is that $A \to T_0A$ is an isomorphism, which yields a trivialization $TA \cong A \times A$ of the tangent bundle. We call such vector spaces tangent stable. All fine diffeological vector spaces have this property.

If A is tangent stable then the diffeological mapping space $\underline{\mathcal{D}\mathrm{flg}}(X, A)$ is elastic for all X. An important example is the algebra $C^\infty(X) \cong \underline{\mathcal{D}\mathrm{flg}}(X, \mathbb{R})$ of smooth functions on X. The diffeological space of sections $\Gamma(M, F)$ of a smooth fiber bundle $\rho : F \to M$ is elastic (Theorem 5.13). As expected, its tangent space is given by the sections of the vertical tangent bundle $\ker T\rho \to M$. As a corollary, the diffeological space $\underline{\mathcal{D}\mathrm{flg}}(M, N)$ of smooth maps of manifolds is elastic, the tangent space being given by

$$T\,\underline{\mathcal{D}\mathrm{flg}}(M, N) \cong \underline{\mathcal{D}\mathrm{flg}}(M, TN).$$

More examples can be constructed by forming retracts.

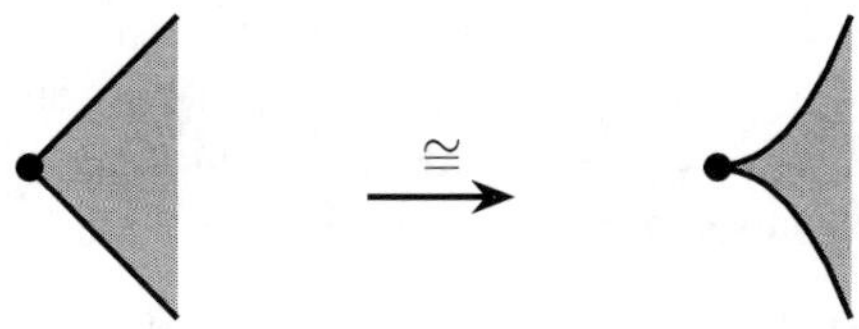

FIGURE 3. Squeezing a corner by multiplying the y-coordinate with the function $f(x) = x^{\frac{3}{2}}$. The boundary of the resulting diffeological subspace of $\mathbb{R}^2$ is the curve $x^2 = y^5$ with a cusp at $(0,0)$.

5.1. Manifolds with corners. Consider the set $[0,\infty) \subset \mathbb{R}$ equipped with the subspace diffeology. Every smooth map $p : U \to \mathbb{R}$ with image $p(U) \subset [0,\infty)$ has a vanishing first derivative at every point u_0 with $p(u_0) = 0$. It follows that

$$T_0[0,\infty) = 0.$$

The tangent space at an interior point $x \in (0,\infty)$ is given by $T_x[0,\infty) = \mathbb{R}$. We conclude that as a set $TX = \{(0,0)\} \cup (0,\infty) \times \mathbb{R}$. A map $U \to TX$ is a plot if and only if it factors locally through $Tq : TV \to TX$ for some plot $q : V \to X$. Using this explicit description of TX, we can show that X is elastic.

It follows from Proposition 4.8 that any finite product

$$\mathbb{R}^n_k := [0,\infty)^k \times \mathbb{R}^{n-k}$$

is elastic. Since the diffeological tangent functor commutes with products, the tangent spaces are given by

$$T_{(x_1,\ldots,x_n)}\mathbb{R}^n_k = T_{x_1}[0,\infty) \times \ldots T_{x_k}[0,\infty) \times \mathbb{R}^{n-k}.$$

Finally, it follows from Proposition 4.6 that every diffeological space modeled locally on $\mathbb{R}^n_k$ is elastic. Such spaces are called **manifolds with corners** [**Joy12**]. In the the context of diffeology, they were studied in [**GIZ19**]. We conclude:

PROPOSITION 5.1. *Manifolds with corners are elastic.*

5.2. Manifolds with cusps. Consider the following subset of $\mathbb{R}^2$,

$$X := \{(x,y) \mid x \geq 0 \ \wedge \ |y| \leq x\},$$

with the subspace diffeology. We can squeeze or stretch the corner at $(0,0)$ by multiplying the y coordinate by a smooth function $f \in C^\infty\big([0,\infty)\big)$ (see Figure 3),

$$\begin{aligned} \varphi : X &\longrightarrow X_f \\ (x,y) &\longmapsto \big(x, f(x)\,y\big), \end{aligned}$$

where

$$X_f := \{(x,y) \mid x \geq 0 \ \wedge \ |y| \leq f(x)\}.$$

Assume that $f(x) > 0$ for $x > 0$. Then φ has an inverse map given by $\varphi^{-1}(0,0) = (0,0)$ and

$$\varphi^{-1}(x,y) = \Big(x, \frac{y}{f(x)}\Big)$$

otherwise. Let us equip X_f with the pullback diffeology of φ^{-1}. Since φ^{-1} is surjective, φ is an isomorphism of diffeological spaces. X is $[0,\infty) \times [0,\infty)$ rotated by minus 45 degrees, so it is elastic by Proposition 5.1. Since φ is an isomorphism X_f is elastic, too.

We cannot describe here in general, what a manifold with cusps or similar defects is, since any squeezing operation that is a smooth retract will produce a new model for elastic subspaces of $\mathbb{R}^n$. In Figure 2 on page 55 we have given a few examples of elastic subspaces of $\mathbb{R}^2$ that can be obtained in this way.

5.3. Diffeological groups. A diffeological group is a group object (G, m, e) in $\mathcal{D}$flg. As for ordinary groups, the inverse $i : G \to G$, $g \mapsto g^{-1}$ is unique, so it is a property rather than a structure. By applying an endofunctor $F : \mathcal{D}\mathrm{flg} \to \mathcal{D}\mathrm{flg}$ that preserves products, we obtain a diffeological group (FG, Fm, Fe). Examples for F are T and T_k (see Proposition 3.19). (TG, Tm, Te) is called the tangent group.

From the multiplication of the tangent group we obtain the left G-translation of TG given by

$$L : G \times TG \xrightarrow{0_G \times \mathrm{id}_{TG}} TG \times TG \xrightarrow{Tm} TG.$$

If we restrict the second argument to the tangent fiber at e, we obtain the map

$$\begin{aligned} \varphi_G : G \times T_e G &\longrightarrow TG \\ (g, v_e) &\longmapsto L_g v_e, \end{aligned} \tag{20}$$

where $T_e G = \{e\} \times_G TG$ is the tangent fiber at the group identity. It has an inverse given by

$$\varphi_G^{-1} : TG \xrightarrow{(i \circ \pi_G, \mathrm{id}_{TG})} G \times TG \xrightarrow{L} T_e G,$$

which maps $v_g \mapsto L_{g^{-1}} v_g$. We conclude that (20) is an isomorphism. In analogy to Lie groups, we will denote

$$\mathfrak{g} := T_e G,$$

so that we obtain the trivialization of the tangent bundle,

$$TG \cong G \times \mathfrak{g}.$$

TG is itself a diffeological group with neutral element 0_e. Using the isomorphism

$$\begin{aligned} T_{(x,y)}(X, Y) &\cong \{(x, y)\} \times_{X \times Y} T(X \times Y) \\ &\cong (\{x\} \times_X TX) \times (\{y\} \times_Y TY) \\ &\cong T_x X \times T_y Y, \end{aligned}$$

we obtain the trivialization

$$\begin{aligned} T^2 G &\cong TG \times T_{(0_e)} TG \\ &\cong G \times \mathfrak{g} \times T_{(e,0)}(G \times \mathfrak{g}) \\ &\cong G \times \mathfrak{g} \times T_e G \times T_0 \mathfrak{g} \\ &\cong G \times \mathfrak{g} \times \mathfrak{g} \times T_0 \mathfrak{g}. \end{aligned}$$

In this trivialization the vertical lift $\lambda_G : TG \to T^2 G$ is given by

$$\begin{aligned} G \times \mathfrak{g} &\longrightarrow G \times \mathfrak{g} \times \mathfrak{g} \times T_0 \mathfrak{g} \\ (g, a) &\longmapsto (g, 0, 0, \lambda_{\mathfrak{g}}^{\perp}(a)) \end{aligned}$$

where

$$\lambda_{\mathfrak{g}}^{\perp} : \mathfrak{g} \longrightarrow T_0 \mathfrak{g},$$

maps $a \in \mathfrak{g}$ to the tangent vector represented by the path $t \mapsto ta$.

THEOREM 5.2. *A diffeological group G is elastic if and only if $\lambda_{\mathfrak{g}}^{\perp} : \mathfrak{g} \to T_0 \mathfrak{g}$ is an induction.*

It follows from Theorem 5.2 and Theorem 4.2 that $\mathfrak{g}$ and $T_0\mathfrak{g}$ are diffeological vector spaces. For an elastic group $\lambda_{\mathfrak{g}}^{\perp}$ is an isomorphism and $\mathfrak{g}$ is equipped with the Lie bracket of invariant vector fields.

EXAMPLE 5.3. Every Lie group G is elastic when G is a smooth manifold. However, when equip G with the C^k-diffeology, then $\mathfrak{g} = \mathbb{R}^n_{C^k}$ and $T_0\mathfrak{g} = \mathbb{R}^n_{C^{k-1}}$, so that $\lambda_{\mathfrak{g}}^{\perp}$ is no longer an induction (cf. Example 1.5).

EXAMPLE 5.4. The diffeomorphism group of a smooth manifold is elastic.

If a diffeological group G is elastic, then there is a Lie bracket on the space of *all* vector fields on G. This induces a Lie bracket on the much smaller space of right invariant vector fields, which is the Lie algebra $\mathfrak{g}$. Most work on diffeological Lie groups is only concerned with the bracket on $\mathfrak{g}$, such as [**Les03**]. (For a critical appreciation of Leslie's pioneering work and a comment on the technical differences, see Section 1.5.2.) The condition $\mathfrak{g} \cong T_0\mathfrak{g}$ of Theorem 5.2 is the same as the condition in [**Lau11**, Thm. 3.12] for groups in the category of Fröhlicher spaces.

5.4. Diffeological vector spaces. A diffeological vector space is an $\mathbb{R}$-vector space object in $\mathcal{D}$flg. Explicitly, this means that the addition and scalar multiplication are morphisms of diffeological spaces. Diffeological vector spaces are a rich and subtle structure [**CW19**].

PROPOSITION 5.5. *A diffeological vector space A is elastic if and only if the natural linear map $T_0A \to T_0T_0A$ is an induction.*

PROOF. This is Theorem 5.2 for the additive diffeological group A. □

In many cases, elastic diffeological vector spaces satisfy the stronger condition $A \cong T_0A$, which is equivalent to

$$TA \cong A \times A.$$

We will call diffeological vector spaces with this property **tangent stable**. Tangent stable vector spaces are elastic.

PROPOSITION 5.6. *All fine diffeological vector spaces are tangent stable (hence elastic).*

EXAMPLE 5.7. $\mathbb{R}^n$ with the C^k-diffeology is a diffeological vector space that is not elastic.

More interesting examples for elastic diffeological vector spaces will be given in Section 5.6 by mapping spaces.

REMARK 5.8. It would be interesting to compare elastic and tangent stable diffeological vector spaces to other classes of diffeological vector spaces, as for example studied in [**CW19**].

5.5. Diffeological manifolds. We recall that a **diffeological manifold** modelled on the diffeological vector space A is a diffeological space X such that every point of X has an open neighborhood that is isomorphic as diffeological space to an open subset of A.

PROPOSITION 5.9. *A diffeological manifold is elastic if and only if it is modelled on an elastic diffeological vector space.*

PROOF. This follows from Proposition 4.6. □

5.6. Mapping spaces. If A is diffeological vector space, then the mapping space $\underline{\mathcal{D}\mathrm{flg}}(X,A)$ is a diffeological vector space with pointwise addition and scalar multiplication.

PROPOSITION 5.10. *Let X be a diffeological space and A a diffeological vector space. If A is tangent stable, then so is $\underline{\mathcal{D}\mathrm{flg}}(X,A)$.*

This follows from the fact that the inner hom functor preserves limits in the second argument, so that $\underline{\mathcal{D}\mathrm{flg}}(X,A\times A)\cong\underline{\mathcal{D}\mathrm{flg}}(X,A)\times\underline{\mathcal{D}\mathrm{flg}}(X,A)$. Since a tangent stable vector space is a fortiori elastic, it follows that $\underline{\mathcal{D}\mathrm{flg}}(X,A)$ is elastic whenever A is tangent stable. Whether it suffices to assume that A is elastic is an open question. The following corollary generalizes Proposition 4.27 of [**CW16**].

COROLLARY 5.11. *If A is a tangent stable diffeological vector space then we have a natural isomorphisms*

$$T\,\underline{\mathcal{D}\mathrm{flg}}(X,A)\cong\underline{\mathcal{D}\mathrm{flg}}(X,TA)$$

for all $X\in\mathcal{D}\mathrm{flg}$.

COROLLARY 5.12. *The diffeological vector space*

$$C^\infty(X):=\underline{\mathcal{D}\mathrm{flg}}(X,\mathbb{R})$$

of smooth $\mathbb{R}$-valued functions on a diffeological space X is elastic.

THEOREM 5.13. *Let $\rho: F\to M$ be a fiber bundle of smooth manifolds. Then the diffeological space of sections $\Gamma(M,F)$ is elastic with tangent space*

$$T\Gamma(M,F)\cong\Gamma(M,VF),$$

the space of sections of the vertical tangent bundle $VF:=\ker T\rho$.

The proof of this theorem is long and involved.

COROLLARY 5.14. *The diffeological mapping space $\underline{\mathcal{D}\mathrm{flg}}(M,N)$ of smooth manifolds M and N is elastic with tangent space*

$$T\,\underline{\mathcal{D}\mathrm{flg}}(M,N)\cong\underline{\mathcal{D}\mathrm{flg}}(M,TN).$$

An analogous formula for the tangent space of the mapping space has been proved for the case that M is compact [**CW16**, Pro. 4.28].

COROLLARY 5.15. *The diffeological vector space of sections $\Gamma(M,A)$ of a smooth vector bundle $A\to M$ is tangent stable.*

References

[ARV10] J. Adámek, J. Rosický, and E. M. Vitale, *What are sifted colimits?*, Theory Appl. Categ. **23** (2010), No. 13, 251–260. MR2720191

[BFW13] Christian Blohmann, Marco Cezar Barbosa Fernandes, and Alan Weinstein, *Groupoid symmetry and constraints in general relativity*, Commun. Contemp. Math. **15** (2013), no. 1, 1250061, 25, DOI 10.1142/S0219199712500617. MR3022999

[BH11] John C. Baez and Alexander E. Hoffnung, *Convenient categories of smooth spaces*, Trans. Amer. Math. Soc. **363** (2011), no. 11, 5789–5825, DOI 10.1090/S0002-9947-2011-05107-X. MR2817410

[Blo] Christian Blohmann, *Tangent structure and Cartan calculus on elastic diffeological spaces*, In preparation.

[CC14] J. R. B. Cockett and G. S. H. Cruttwell, *Differential structure, tangent structure, and SDG*, Appl. Categ. Structures **22** (2014), no. 2, 331–417, DOI 10.1007/s10485-013-9312-0. MR3192082

[CC15] J. R. B. Cockett and G. S. H. Cruttwell, *The Jacobi identity for tangent categories* (English, with English and French summaries), Cah. Topol. Géom. Différ. Catég. **56** (2015), no. 4, 301–316. MR3445914

[CdSW99] Ana Cannas da Silva and Alan Weinstein, *Geometric models for noncommutative algebras*, Berkeley Mathematics Lecture Notes, vol. 10, American Mathematical Society, Providence, RI; Berkeley Center for Pure and Applied Mathematics, Berkeley, CA, 1999. MR1747916

[CSW14] J. Daniel Christensen, Gordon Sinnamon, and Enxin Wu, *The D-topology for diffeological spaces*, Pacific J. Math. **272** (2014), no. 1, 87–110, DOI 10.2140/pjm.2014.272.87. MR3270173

[CW16] J. Daniel Christensen and Enxin Wu, *Tangent spaces and tangent bundles for diffeological spaces* (English, with English and French summaries), Cah. Topol. Géom. Différ. Catég. **57** (2016), no. 1, 3–50. MR3467758

[CW19] J. Daniel Christensen and Enxin Wu, *Diffeological vector spaces*, Pacific J. Math. **303** (2019), no. 1, 73–92, DOI 10.2140/pjm.2019.303.73. MR4044857

[DF99] Pierre Deligne and Daniel S. Freed, *Classical field theory*, Quantum fields and strings: a course for mathematicians, Vol. 1, 2 (Princeton, NJ, 1996/1997), Amer. Math. Soc., Providence, RI, 1999, pp. 137–225. MR1701599

[DI85] Paul Donato and Patrick Iglésias, *Exemples de groupes difféologiques: flots irrationnels sur le tore* (French, with English summary), C. R. Acad. Sci. Paris Sér. I Math. **301** (1985), no. 4, 127–130. MR799609

[DN07] B. Dugmore and P. P. Ntumba, *On tangent cones of Frölicher spaces*, Quaest. Math. **30** (2007), no. 1, 67–83, DOI 10.2989/160736007780205747. MR2309243

[GIZ19] Serap Gürer and Patrick Iglesias-Zemmour, *Differential forms on manifolds with boundary and corners*, Indag. Math. (N.S.) **30** (2019), no. 5, 920–929, DOI 10.1016/j.indag.2019.07.004. MR3996773

[GW21] Nico Goldammer and Kathrin Welker, *Towards optimization techniques on diffeological spaces*, PAMM **20** (2021), no. 1, e202000040.

[Hec95] G. Hector, *Géométrie et topologie des espaces difféologiques* (French, with English summary), Analysis and geometry in foliated manifolds (Santiago de Compostela, 1994), World Sci. Publ., River Edge, NJ, 1995, pp. 55–80. MR1414196

[IZ13] Patrick Iglesias-Zemmour, *Diffeology*, Mathematical Surveys and Monographs, vol. 185, American Mathematical Society, Providence, RI, 2013, DOI 10.1090/surv/185. MR3025051

[Joy12] Dominic Joyce, *On manifolds with corners*, Advances in geometric analysis, Adv. Lect. Math. (ALM), vol. 21, Int. Press, Somerville, MA, 2012, pp. 225–258. MR3077259

[Kel05] G. M. Kelly, *Basic concepts of enriched category theory*, Repr. Theory Appl. Categ. **10** (2005), vi+137. Reprint of the 1982 original [Cambridge Univ. Press, Cambridge; MR0651714]. MR2177301

[Lau11] Martin Laubinger, *A Lie algebra for Frölicher groups*, Indag. Math. (N.S.) **21** (2011), no. 3-4, 156–174, DOI 10.1016/j.indag.2011.04.001. MR2835925

[Les03] Joshua Leslie, *On a diffeological group realization of certain generalized symmetrizable Kac-Moody Lie algebras*, J. Lie Theory **13** (2003), no. 2, 427–442. MR2003152

[Mag13] Jean-Pierre Magnot, *Ambrose-Singer theorem on diffeological bundles and complete integrability of the KP equation*, Int. J. Geom. Methods Mod. Phys. **10** (2013), no. 9, 1350043, 31, DOI 10.1142/S0219887813500436. MR3103144

[Mag18] Jean-Pierre Magnot, *The group of diffeomorphisms of a non-compact manifold is not regular*, Demonstr. Math. **51** (2018), no. 1, 8–16, DOI 10.1515/dema-2018-0001. MR3769995

[ML98] Saunders Mac Lane, *Categories for the working mathematician*, 2nd ed., Graduate Texts in Mathematics, vol. 5, Springer-Verlag, New York, 1998. MR1712872

[Per16] Ekaterina Pervova, *Diffeological vector pseudo-bundles*, Topology Appl. **202** (2016), 269–300, DOI 10.1016/j.topol.2016.01.014. MR3464165

[Ros84] J. Rosický, *Abstract tangent functors*, Diagrammes **12** (1984), JR1–JR11. MR800500

[Sou70] J.-M. Souriau, *Structure des systèmes dynamiques* (French), Dunod, Paris, 1970. Maîtrises de mathématiques. MR260238

[Vin08] Martin Vincent, *Diffeological differential geometry*, Master's thesis, Department of Mathematical Sciences, University of Copenhagen, 2008, Available at https://www.math.ku.dk/english/research/tfa/top/paststudents/ms-theses/martinvincent.msthesis.pdf, downloaded on 4/15/2019.

[Wat13] Jordan Watts, *Diffeologies, Differential Spaces, and Symplectic Geometry*, ProQuest LLC, Ann Arbor, MI, 2012. Thesis (Ph.D.)–University of Toronto (Canada). MR3153238

Max-Planck-Institut für Mathematik, Vivatsgasse 7, 53111 Bonn, Germany
Email address: blohmann@mpim-bonn.mpg.de

Contemporary Mathematics
Volume **794**, 2024
https://doi.org/10.1090/conm/794/15937

A remark on stability and the D-topology of mapping spaces

Alireza Ahmadi

ABSTRACT. We discuss how stability is related to the D-topology of mapping spaces, equipped with the functional diffeology. Indeed, we show that the stable classes of mapping spaces are D-open. After a reformulation of the classical stability theorem of manifolds with respect to the D-topology, we prove a version of the stability theorem in the class of diffeological étale manifolds.

1. Introduction

The stability of mapping spaces of manifolds[1] is an aspect of the classical differential geometry which concerns those properties that remain invariant under small deformations or perturbations in a smooth manner (see, e.g., [**4, 5**]). Let us first recall the classical definition of a stable property.

DEFINITION 1.1 ([**5**]). Let M and N be manifolds. A property P of smooth maps is *stable* on $C^\infty(M, N)$ under small deformations, whenever for any smooth map $h : M \times [0, 1] \to N$, where $[0, 1]$ is a manifold with boundary, if $h_0 : M \to N$, $x \mapsto h(x, 0)$ possesses the property P, then there exists a number $\epsilon > 0$ such that the map $h_t : M \to N$, $x \mapsto h(x, t)$ also possesses the property P, for all $t \in [0, \epsilon)$.

This article is aimed to study the stability of mapping spaces in the context of diffeology, which extends ordinary differential geometry by diffeological spaces. The reader can refer to the book [**10**] for a comprehensive introduction to diffeology. One advantage of working in this framework, among others, is that mapping spaces have a natural structure, called the functional diffeology, for which the natural map

$$C^\infty(X, C^\infty(Y, Z)) \longrightarrow C^\infty(X \times Y, Z)$$

taking $f \mapsto \tilde{f}$ with $\tilde{f}(x, y) = f(x)(y)$, is a diffeomorphism of diffeological spac es (see [**10**, §1.60]). This feature turns the category of diffeological spaces into a Cartesian closed one, which is very useful in smooth homotopy theory. This particularly enables us to define a coherent notion of stability in diffeology.

DEFINITION 1.2. Let P be a property for smooth maps between diffeological spaces X and Y, i.e., $C^\infty(X, Y)$. We say that the property P is *stable* on $C^\infty(X, Y)$ under small deformations whenever for any smooth homotopy $h : \mathbb{R} \to C^\infty(X, Y)$,

2020 *Mathematics Subject Classification.* Primary 58A05, 58C25; Secondary 57P99.
Key words and phrases. Mapping spaces, D-topology, stability theorem, diffeological étale manifolds.

[1]By a manifold we mean a second-countable Hausdorff finite-dimensional smooth manifold.

if $h(0)$ possesses the property P, then there exists a number $\epsilon > 0$ such that $h(t)$ also possesses the property P, for all $t \in (-\epsilon, \epsilon)$.

On the other hand, every diffeological space has an intrinsic topology, called the D-topology, in which a subset is D-open[2] if its preimage by any plot is open. Hence a mapping space obtain a topological structure from the functional diffeology.

PROPOSITION 1.3. *Let X and Y be diffeological spaces. A property P is stable on $\mathrm{C}^\infty(X, Y)$ if and only if*

$$\mathsf{class}(\mathsf{P}) = \{f \in \mathrm{C}^\infty(X, Y) \mid f \textit{ possesses the property } \mathsf{P}\}$$

is a D-open subset of $\mathrm{C}^\infty(X, Y)$, where $\mathrm{C}^\infty(X, Y)$ is equipped with the functional diffeology.

PROOF. As the D-topology of a diffeological space is determined via smooth paths by [**2**, Theorem 3.7], the proof is straightforward. □

COROLLARY 1.4. *It is immediate that for any stable property P,*

$$\mathsf{dim}(\mathsf{class}(\mathsf{P})) \leq \mathsf{dim}(\mathrm{C}^\infty(X, Y)),$$

where dim *denotes the diffeological dimension* [**10**, §1.78].

To compare the classical definition of stability with the diffeological one in the class of manifolds, we need the following lemma. But before that, consider the space $[0, 1]$ equipped with the diffeology of manifolds with boundary (see [**10**, §4.16] and [**6**]). The D-topology of $[0, 1]$ is the subspace topology inherited from $\mathbb{R}$.

LEMMA 1.5. *Let X be a diffeological space. A subset A of X is D-open if and only if for every $h \in \mathrm{C}^\infty([0, 1], X)$ with $h(0) \in A$, there exists a number $\epsilon > 0$ such that $h(t) \in A$, for all $t \in [0, \epsilon)$.*

PROOF. The proof is inspired by that of [**2**, Theorem 3.7]: Since smooth maps are D-continuous, if a subset A of X is D-open and $h \in \mathrm{C}^\infty([0, 1], X)$ with $h(0) \in A$, there exists a number $\epsilon > 0$ such that $h(t) \in A$, for all $t \in [0, \epsilon)$.

Suppose that $A \subseteq X$ and for every $h \in \mathrm{C}^\infty([0, 1], X)$ with $h(0) \in A$, there exists a number $\epsilon > 0$ such that $h(t) \in A$, for all $t \in [0, \epsilon)$. To see that A is D-open in X, take any plot $P : U \to X$. Let $r \in P^{-1}(A)$ and $\{r_n\}$ be a sequence which converges fast to r in U, up to suitable subsequences. By the special curve lemma [**12**, p. 18], there is a smooth map $C : \mathbb{R} \to U$ with $C(\frac{1}{n}) = r_n$ and $C(0) = r$. Set $c = C|_{[0,1]}$. Then $P \circ c : [0, 1] \to X$ is an element of $\mathrm{C}^\infty([0, 1], X)$ with $P \circ c(0) \in A$. By hypothesis, there exists a number $\epsilon > 0$ such that $P \circ c(t) \in A$, for all $t \in [0, \epsilon)$. Thus, for sufficiently large n we get, $P \circ c(\frac{1}{n}) \in A$ and consequently, $r_n = c(\frac{1}{n}) \in P^{-1}(A)$. Therefore, $P^{-1}(A) \subseteq U$ is open. □

COROLLARY 1.6. *Let M and N be manifolds. A property P is stable on $\mathrm{C}^\infty(M, N)$ (according to Definition 1.1) if and only if*

$$\mathsf{class}(\mathsf{P}) = \{f \in \mathrm{C}^\infty(M, N) \mid f \textit{ possesses the property } \mathsf{P}\}$$

is a D-open subset of $\mathrm{C}^\infty(M, N)$, where $\mathrm{C}^\infty(M, N)$ is equipped with the functional diffeology.

[2]Throughout this article, the prefix D of a topological property indicates the same property in terms of the D-topology.

PROPOSITION 1.7. *In the class of manifolds, Definitions* 1.1 *and* 1.2 *are equivalent, meaning that a property* P *is stable according to Definition* 1.1 *if and only if it is stable according to Definition* 1.2.

PROOF. The proof is a consequence of Proposition 1.3 and Corollary 1.6. □

1.1. Stability theorem. By Corollary 1.6, the classical stability theorem of manifolds (see, e.g., [**5**, p. 35]) can be rephrased in terms of the D-topology in the following way.

THEOREM 1.8. *(Stability theorem of manifolds). The following classes of smooth maps from a compact manifold K to a manifold M are D-open subsets of* $C^\infty(K, M)$, *where* $C^\infty(K, M)$ *is equipped with the functional diffeology.*

(i) *diffeomorphisms,*
(ii) *étale maps (i.e., local diffeomorphisms),*
(iii) *submersions,*
(iv) *immersions,*
(v) *embeddings,*
(vi) *maps transversal to any specified closed submanifold* $N \subseteq M$.

REMARK 1.9. Besides the D-topology, there are other well-known topologies on mapping spaces of manifolds such as the compact-open topology, the weak topology, the strong topology, and the Whitney topology. In [**12**, Theorem 43.1] or [**15**, Section 5], the stability theorem is shown with respect to the Whitney topology. Furthermore, in [**9**, Chapter 2], results similar to the stability theorem is proved with respect to the strong topology. As mentioned in [**2**, Corollary 4.15], one can reach Theorem 1.8 by comparisons made in [**2**], among the D-topology and the strong topology, and also other topologies. Of course, it seems that way would be longer and more complicated than our approach.

As an application, this restatement of the stability theorem can be helpful in some computations.

EXAMPLE 1.10. Let K be a compact manifold. In light of [**3**, Propositions 3.6], also [**7**, p. 72] and [**8**, Propositions 6.3], or equivalently [**3**, Corollary 4.29], one can compute the internal tangent spaces of the following subspaces of $C^\infty(K, K)$ at id_K, which are all isomorphic to the vector space of all smooth vector fields on K:

(1) diffeomorphisms,
(2) étale maps (equivalently, submersions or immersions),
(3) embeddings,
(4) maps transversal to any specified closed submanifold $N \subseteq K$.

It is natural to suggest such a problem for mapping spaces of diffeological spaces.

QUESTION 1.11 (Stability problem). Let X and Y be diffeological spaces. Which of the following classes and under what conditions are D-open in $C^\infty(X, Y)$?

- diffeomorphisms,
- (local) subductions,
- (local) inductions,
- (diffeological) submersions,

- (diffeological) immersions,
- (diffeological) étale maps,
- (diffeological) embeddings, etc.

(See Section 2 for terminology.)

Although the stability problem remains open in general, thanks to some linear-algebraic tools and additional topological conditions, we are able to generalize the stability theorem to diffeological étale manifolds, which is the main result of this article. Diffeological étale manifolds, introduced by the author in [**1**], constitute a class of diffeological spaces that includes the usual manifolds and also irrational tori. We briefly review this kind of diffeological space in Section 2.

THEOREM 1.12 (Stability theorem of diffeological étale manifolds). *Suppose that $\mathcal{K}$ and $\mathcal{M}$ are diffeological étale manifolds such that $\mathcal{K}$ is D-compact (i.e., compact with respect to the D-topology). Also,* $\mathrm{C}^\infty(\mathcal{K},\mathcal{M})$ *is equipped with the functional diffeology.*

(**a**) *The following classes of* $\mathrm{C}^\infty(\mathcal{K},\mathcal{M})$ *are D-open:*
  (a1) *diffeological submersions,*
  (a2) *diffeological immersions,*
  (a3) *diffeological étale maps,*
  (a4) *local subductions (provided that $\mathcal{M}$ is D-Hausdorff).*

(**b**) *If $\mathcal{M}$ is a usual manifold, then the following classes of* $\mathrm{C}^\infty(\mathcal{K},\mathcal{M})$ *are D-open:*
  (b1) *injective diffeological immersions,*
  (b2) *diffeological embeddings,*
  (b3) *diffeomorphisms.*

(**c**) *Let* $\mathsf{Inj}^\infty(\mathcal{K},\mathcal{M}) \subseteq \mathrm{C}^\infty(\mathcal{K},\mathcal{M})$ *be the diffeological subspace of injective smooth maps from $\mathcal{K}$ to $\mathcal{M}$. If $\mathcal{M}$ is D-Hausdorff, then the following classes of* $\mathsf{Inj}^\infty(\mathcal{K},\mathcal{M})$ *are D-open subsets of* $\mathsf{Inj}^\infty(\mathcal{K},\mathcal{M})$.
  (c1) *diffeological embeddings,*
  (c2) *diffeomorphisms.*

REMARK 1.13. Notice that we have to consider requirements that guarantee the injectivity remains stable in the cases of diffeological embeddings and diffeomorphisms. We can take two ways to handle this difficulty: 1) to consider the restrictive condition that $\mathcal{M}$ to be a usual manifold in item (**b**) so that diffeological immersions are locally injective (see Proposition 2.4) and follow the classical method of proof, or 2) to impose directly injectivity condition in (**c**). We take into account diffeomorphisms just as a subset in (**b**) and (**c**), without any refined diffeological structure.

2. Diffeological étale manifolds

In this section, we briefly recall the needed definitions and results about diffeological étale manifolds (see [**1**] for more details).

DEFINITION 2.1. We call a smooth map $f : X \to Y$ between diffeological spaces a *submersion* if for each x_0 in X, there exists a smooth local section $\sigma : O \to X$ of f passing through x_0 defined on a D-open subset $O \subseteq Y$ such that $f \circ \sigma(y) = y$ for all $y \in O$. A smooth map $f : X \to Y$ is said to be a *diffeological submersion* if the pullback of f by every plot in Y is a submersion.

PROPOSITION 2.2. *If $f : X \to M$ is a diffeological submersion into a manifold M, then it is a submersion. In particular, a map between manifolds is a diffeological submersion if and only if it is a submersion of manifolds.*

DEFINITION 2.3. A smooth map $f : X \to Y$ between diffeological spaces is an *immersion* if for each x_0 in X, there exist a D-open neighborhood $O \subseteq X$ of the point x_0, a D-open neighborhood $O' \subseteq Y$ of the set $f(O)$, and a smooth map $\varrho : O' \to X$ such that $\varrho \circ f(x) = x$ for all $x \in O$. A smooth map $f : X \to Y$ is a *diffeological immersion* if for any plot $P : U \to Y$ in Y, for each (r_0, x_0) in P^*X, there exist a D-open neighborhood O of (r_0, x_0) in P^*X, an open neighborhood $V \subseteq U$ of $P^*f(O)$ and a smooth map $\varrho : V \to U \times X$ such that $\varrho \circ P^*f(r, x) = (r, x)$ for all $(r, x) \in O$.

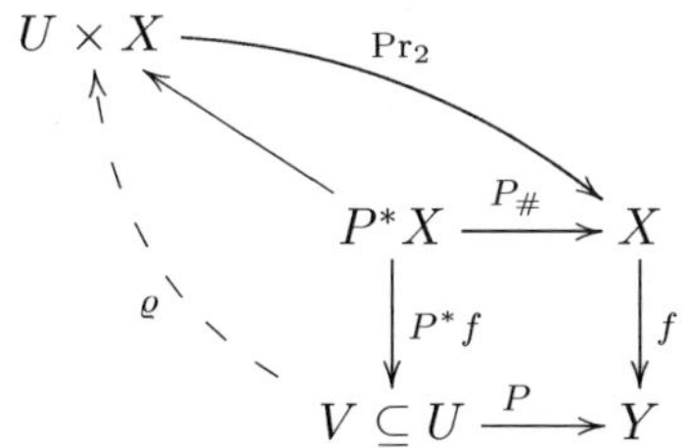

Any immersion of diffeological spaces is a local induction.

PROPOSITION 2.4. *If $f : X \to M$ is a diffeological immersion into a manifold M, then it is an immersion. In particular, a map between manifolds is a diffeological immersion if and only if it is an immersion of manifolds.*

DEFINITION 2.5. A map $f : X \to Y$ between diffeological spaces is called a *diffeological embedding* if it is both a diffeological immersion and a D-embedding (i.e., a topological embedding with respect to the D-topology).

Any diffeological embedding is an induction.

DEFINITION 2.6. A map $f : X \to Y$ between diffeological spaces is *étale* if for every x in X, there are D-open neighborhoods $O \subseteq X$ and $V \subseteq Y$ of x and $f(x)$, respectively, such that $f|_O : O \to O'$ is a diffeomorphism. A smooth map $f : X \to Y$ is a *diffeological étale map* if the pullback P^*f by every plot P in X is étale.

PROPOSITION 2.7. *If $f : X \to M$ is a diffeological étale map into a manifold M, then it is étale. In particular, a map between manifolds is a diffeological étale map if and only if it is a local diffeomorphism of manifolds.*

REMARK 2.8. Any diffeological étale map and more generally, any diffeological submersion is a D-open map.

DEFINITION 2.9. A diffeological space $\mathcal{M}$ is said to be a *diffeological étale n-manifold* if there exists a parametrized cover $\mathfrak{A}$, called an *atlas* for $\mathcal{M}$ consisting of diffeological étale maps from n-domains into $\mathcal{M}$. We call the elements of $\mathfrak{A}$*diffeological étale charts.*

In this situation, $\mathfrak{A}$ is actually a covering generating family for $\mathcal{M}$ and the diffeological dimension of $\mathcal{M}$ is equal to n, $\mathsf{dim}(\mathcal{M}) = n$. Moreover, at each point x of $\mathcal{M}$, the internal tangent space $T_x\mathcal{M}$ is isomorphic to $\mathbb{R}^n$. Therefore, for a diffeological étale n-manifold, the diffeological dimension and the dimension of the tangent spaces are the same.

If $\varphi : U \to \mathcal{M}$ are $\psi : V \to \mathcal{M}$ are two diffeological étale charts in $\mathcal{M}$ with $\varphi(r) = \psi(s)$, then we can find a unique smooth map $h : U' \to V$ defined on an open neighborhood $U' \subseteq U$ of r such that $h(r) = s$ and the following diagram commutes:

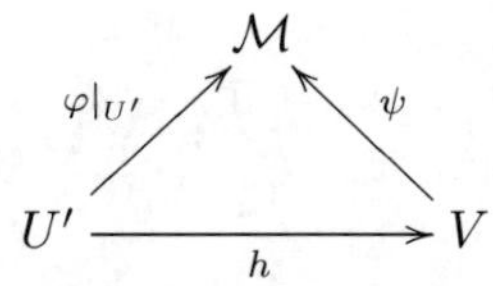

Although diffeological étale charts φ are ψ may not locally injective, h is indeed an étale map and uniquely determined by φ are ψ. Locally, h plays the role of a transition map.

EXAMPLE 2.10. Diffeological manifolds are diffeological étale manifolds but not conversely. The irrational torus $\mathbb{T}_\alpha = \mathbb{R}/(\mathbb{Z} + \alpha\mathbb{Z})$, $\alpha \notin \mathbb{Q}$, is a D-compact diffeological étale 1-manifold which is not a diffeological manifold. More generally, $\mathbb{T}^n_\Gamma = \mathbb{R}^n/\Gamma$, where Γ is a discrete subgroup of $\mathbb{R}^n$, is a diffeological étale n-manifold.

EXAMPLE 2.11. Any open subset of a diffeological étale n-manifold itself is a diffeological étale n-manifold in a natural way.

EXAMPLE 2.12. (Product étale manifolds). Suppose $\mathcal{M}_1, \ldots, \mathcal{M}_k$ are diffeological étale manifolds of dimensions , respectively. The product space $\mathcal{M}_1 \times \cdots \times \mathcal{M}_k$ is a diffeological étale manifold of dimension $n_1 + \cdots + n_k$ with diffeological étale charts of the form $\varphi_1 \times \cdots \times \varphi_k$, where φ_i is a diffeological étale chart of $\mathcal{M}_i$, $i = 1, \ldots, k$.

Recall that the internal tangent map of a smooth map is the linear map between the internal tangent spaces induced by the universal property (see [**3**, **7**, **8**]).

THEOREM 2.13. *Let $\mathcal{M}$ and $\mathcal{N}$ be diffeological étale manifolds and $f : \mathcal{M} \to \mathcal{N}$ be a smooth map.*

(i) *f is a diffeological submersion if and only if the internal tangent map $df_x : T_x\mathcal{M} \to T_{f(x)}\mathcal{N}$ is an epimorphism at each point $x \in \mathcal{M}$.*
(ii) *f is a diffeological immersion if and only if the internal tangent map $df_x : T_x\mathcal{M} \to T_{f(x)}\mathcal{N}$ is a monomorphism at each point $x \in \mathcal{M}$.*
(iii) *f is a diffeological étale map if and only if the internal tangent map $df_x : T_x\mathcal{M} \to T_{f(x)}\mathcal{N}$ is an isomorphism at each point $x \in \mathcal{M}$.*

DEFINITION 2.14. A smooth map $f : \mathcal{M} \to \mathcal{N}$ between diffeological étale manifolds has *internal rank* k at a point $x \in \mathcal{M}$ if the rank of the internal tangent map $df_x : T_x\mathcal{M} \to T_{f(x)}\mathcal{N}$ is equal to k. Moreover, f has *full rank* at x if its internal rank is equal to $\min\{\mathsf{dim}(\mathcal{M}), \mathsf{dim}(\mathcal{N})\}$.

PROPOSITION 2.15. *Let $f : \mathcal{M} \to \mathcal{N}$ be a smooth map and $x_0 \in \mathcal{M}$. If f has full rank k at x_0, then there is a D-open neighborhood $O \subseteq \mathcal{M}$ of x_0 such that f has full rank k on O.*

PROOF. Take any diffeological étale chart $\varphi : U \to \mathcal{M}$ with $\varphi(r_0) = x_0$ for some $r_0 \in U$. The composition $f \circ \varphi$ is a plot in $\mathcal{N}$. So there are diffeological étale chart $\psi : V \to \mathcal{N}$ and a smooth map $f : U' \to V$ defined on an open neighborhood of $r_0 \in U' \subseteq U$ such that $f \circ \varphi|_{U'} = \psi \circ F$. Then $df_{\varphi(r_0)} \circ d\varphi_{r_0} = d\psi_{F(r_0)} \circ dF_{r_0}$. Since $d\varphi_{r_0}$ and $d\psi_{F(r_0)}$ are isomorphisms, f has full rank k at r_0. By [**14**, Proposition 4.1], r_0 has a neighborhood $U'' \subseteq U'$ such that f has full rank k on U''. Therefore, f has full rank k on $O := \varphi(U'')$, which is a D-open neighborhood of x_0. □

LEMMA 2.16. *Let $f : \mathcal{L} \times \mathcal{N} \to \mathcal{M}$ be a smooth map between diffeological étale manifolds. For each $x \in \mathcal{L}$, define $f^x : \mathcal{N} \to \mathcal{M}$ by $f^x(y) = f(x, y)$. Then*

$$\mathsf{rank}\ d(\mathrm{Pr}_1, f)_{(x,y)} = \mathsf{dim}(\mathcal{L}) + \mathsf{rank}\ d\big(f^x\big)_y$$

for every $(x, y) \in \mathcal{L} \times \mathcal{N}$, where $\mathrm{Pr}_1 : \mathcal{L} \times \mathcal{N} \to \mathcal{L}$ is the projection on the first factor.

3. Proof of the stability theorem for diffeological étale manifolds

We first need some preliminary results.

LEMMA 3.1. *Let $\mathcal{M}$ and $\mathcal{N}$ be diffeological étale manifolds, and $\mathcal{L}$ be a usual manifold. Suppose that $h : \mathcal{L} \to \mathrm{C}^\infty(\mathcal{N}, \mathcal{M})$ be a smooth map. If $(x_0, y_0) \in \mathcal{L} \times \mathcal{N}$ and the map $h(x_0) : \mathcal{N} \to \mathcal{M}$ has full rank k at y_0, then there exist a D-open neighborhood $O_{x_0} \subseteq \mathcal{L}$ of x_0 and a D-open neighborhood $U_{y_0} \subseteq \mathcal{N}$ of y_0 such that for every $x \in O_{x_0}$, $h(x)|_{U_{y_0}} : U_{y_0} \to \mathcal{M}$ has full rank k on U_{y_0}.*

PROOF. Define $\hat{h} : \mathcal{L} \times \mathcal{N} \to \mathcal{M}$ by $\hat{h}(x, y) = h(x)(y)$. By Lemma 2.16, we have

$$\mathsf{rank}\ d(\mathrm{Pr}_1, \hat{h})_{(x_0,y_0)} = \mathsf{dim}(\mathcal{L}) + \mathsf{rank}\ d\big(h(x_0)\big)_{y_0} = \mathsf{dim}(\mathcal{L}) + k,$$

where $\mathrm{Pr}_1 : \mathcal{L} \times \mathcal{N} \to \mathcal{L}$ is the projection on the first factor. This means that $(\mathrm{Pr}_1, \hat{h}) : \mathcal{L} \times \mathcal{N} \to \mathcal{L} \times \mathcal{M}$ has full rank $\mathsf{dim}(\mathcal{L}) + k$ at (x_0, y_0). By Proposition 2.15, there exist a D-open neighborhood $O_{x_0} \subseteq \mathcal{L}$ of x_0 and a D-open neighborhood $U_{y_0} \subseteq \mathcal{N}$ of y_0 such that $(\mathrm{Pr}_1, \hat{h})$ has full rank $\mathsf{dim}(\mathcal{L}) + k$ on $O_{x_0} \times U_{y_0}$. Again by Lemma 2.16, we get

$$\mathsf{dim}(\mathcal{L}) + k = \mathsf{rank}\ d(\mathrm{Pr}_1, \hat{h})_{(x,y)} = \mathsf{dim}(\mathcal{L}) + \mathsf{rank}\ d\big(h(x)\big)_y$$

or

$$\mathsf{rank}\ d\big(h(x)\big)_y = k,$$

for all $(x, y) \in O_{x_0} \times U_{y_0}$. Hence for every $x \in O_{x_0}$, $h(x)|_{U_{y_0}} : U_{y_0} \to \mathcal{M}$ has full rank k on U_{y_0}. □

PROPOSITION 3.2. *Suppose that $\mathcal{L}$ is a usual manifold, and $\mathcal{M}, \mathcal{K}$ are diffeological étale manifolds such that $\mathcal{K}$ is D-compact. Let $h : \mathcal{L} \to \mathrm{C}^\infty(\mathcal{K}, \mathcal{M})$ be a smooth map. If $x_0 \in \mathcal{L}$ and the map $h(x_0) : \mathcal{K} \to \mathcal{M}$ has full rank k on $\mathcal{K}$, then there exists a D-open neighborhood $O \subseteq \mathcal{L}$ of x_0 such that for every $x \in O$, $h(x) : \mathcal{K} \to \mathcal{M}$ has full rank k on $\mathcal{K}$.*

PROOF. By Lemma 3.1, for each $y \in \mathcal{K}$, there exist a D-open neighborhood $O_{x_0,y} \subseteq \mathcal{L}$ of x_0 and a D-open neighborhood $U_y \subseteq \mathcal{K}$ of y such that for every $x \in O_{x_0,y}$, $h(x)|_{U_y} : U_y \to \mathcal{M}$ has full rank k on U_y. Since the collection $\{U_y\}_{y \in \mathcal{K}}$ covers $\mathcal{K}$ and $\mathcal{K}$ is D-compact, we conclude that there are finitely many points $y_1, \ldots, y_n$ in $\mathcal{K}$ for which $\bigcup_{i=1}^n U_{y_i} = \mathcal{K}$. Set $O = \bigcap_{i=1}^n O_{x_0,y_i}$, which is a D-open neighborhood of x_0 in $\mathcal{L}$. Therefore, $h(x) : \mathcal{K} \to \mathcal{M}$ has full rank k on $\mathcal{K}$, for every $x \in O$. □

Proof of 1.12 **(a)**.

In light of Theorem 2.13, parts **(a1)**, **(a2)** and **(a3)** are obtained as especial cases of Proposition 3.2, where $\mathcal{L} = \mathbb{R}$, $x_0 = 0$, and k is taken equal to $\mathsf{dim}(\mathcal{M})$, $\mathsf{dim}(\mathcal{K})$, and $\mathsf{dim}(\mathcal{K}) = \mathsf{dim}(\mathcal{M})$, respectively.

(a4) Let $h : \mathbb{R} \to \mathrm{C}^\infty(\mathcal{K}, \mathcal{M})$ be a smooth homotopy such that $h(0)$ is a local subduction. That is, $h(0)$ is a surjective diffeological submersion. By part **(a1)**, one can find a positive number $\epsilon > 0$ such that $h(t) : \mathcal{K} \to \mathcal{M}$ is a diffeological submersion, for all $t \in (-\epsilon, \epsilon)$. In particular, each $h(t)$ is a D-open map. By [**13**, Lemma 4.50(a)], each $h(t)$ is a D-closed map, too. Due to the fact that a diffeological space is locally connected by [**11**], its connected components are both D-open and D-closed. Thus we conclude that, if $C \subseteq \mathcal{K}$ is a connected component, so is $h(t)(C) \subseteq \mathcal{M}$, for all $t \in (-\epsilon, \epsilon)$. Now fix $t \in (-\epsilon, \epsilon)$ and take any $y \in \mathcal{M}$. By surjectivity of $h(0)$, one can find a point $x \in \mathcal{K}$ with $h(0)(x) = y$. But by Cartesian closedness, the map $\gamma_x : \mathbb{R} \to \mathcal{M}$ defined by $\gamma_x(s) = h(st)(x)$ is a smooth path connecting $h(0)(x) = y$ and $h(t)(x)$. Thus, $h(0)(x) = y$ and $h(t)(x)$ belong to the same component. Actually, $h(t)$ maps the component of x onto the component of y. Therefore, $h(t)$ is surjective and so a local subduction, for all $t \in (-\epsilon, \epsilon)$. □

Proof of 1.12 **(b)**.

(b1) Let $h : \mathbb{R} \to \mathrm{C}^\infty(\mathcal{K}, \mathcal{M})$ be a smooth homotopy such that $h(0)$ is an injective diffeological immersion. By part **(a2)**, we can find a positive number $\epsilon > 0$ such that $h(t) : \mathcal{K} \to \mathcal{M}$ is a diffeological immersion, for all $t \in (-\epsilon, \epsilon)$. Thus, we only need to show that there exists a positive number $0 < \delta < \epsilon$ such that $h(t)$ is injective, for all $t \in (-\delta, \delta)$. Otherwise, for every positive integer n, there are $t_n \in \mathbb{R}$ and distinct points $x_n, y_n \in \mathcal{K}$ such that $|t_n| < \dfrac{1}{n}$ and $h(t_n)(x_n) = h(t_n)(y_n)$. Since $\mathcal{K}$ is compact, up to suitable subsequences, we get $\lim_{n\to\infty} x_n = x_0$ and $\lim_{n\to\infty} y_n = y_0$, for some $x_0, y_0 \in \mathcal{K}$. Then

$$h(0)(x_0) = \lim_{n\to\infty} h(t_n)(x_n) = \lim_{n\to\infty} h(t_n)(y_n) = h(0)(y_0).$$

Injectivity of $h(0)$ implies that $x_0 = y_0$. On the other hand, by Lemma 2.16 and Theorem 2.13, we observe that

$$(\Pr_1, \hat{h}) : (-\epsilon, \epsilon) \times \mathcal{K} \to (-\epsilon, \epsilon) \times \mathcal{M}, \quad (t, x) \mapsto \big(t, h(t)(x)\big)$$

is a diffeological immersion. Because $\mathcal{M}$ is a usual manifold, by Proposition 2.4, $(\mathrm{Pr}_1, \hat{h})$ is actually an immersion of diffeological spaces and therefore, locally injective. In particular, there exists a neighborhood $(-\epsilon', \epsilon') \times O \subseteq (-\epsilon, \epsilon) \times \mathcal{K}$ of $(0, x_0)$ such that $(\mathrm{Pr}_1, \hat{h})|_{(-\epsilon',\epsilon')\times O}$ is injective. Due to convergence, for some sufficiently large m, we have $(t_m, x_m), (t_m, y_m) \in (-\epsilon', \epsilon') \times O$. Now from injectivity $(\mathrm{Pr}_1, \hat{h})|_{(-\epsilon',\epsilon')\times O}$ and the equality $\big(t_m, h(t_m)(x_m)\big) = \big(t_m, h(t_m)(y_m)\big)$ we get $x_m = y_m$ for some m, which is a contradiction with our hypothesis that x_m and y_m are distinct points.

(b2) Let $h : \mathbb{R} \to \mathrm{C}^\infty(\mathcal{K}, \mathcal{M})$ be a smooth homotopy such that $h(0)$ is a diffeological embedding. So $h(0)$ is an injective diffeological immersion. By part **(b1)**, there exists a positive number $\epsilon > 0$ such that $h(t) : \mathcal{K} \to \mathcal{M}$ is an injective diffeological immersion, for all $t \in (-\epsilon, \epsilon)$. Now, in view of [**13**, Lemma 4.50(c)], the result is obtained.

(b3) It is sufficient to consider the class of diffeomorphisms from $\mathcal{K}$ onto $\mathcal{M}$ as the intersection of the class of local subductions from $\mathcal{K}$ onto $\mathcal{M}$ with the class of injective diffeological immersions from $\mathcal{K}$ to $\mathcal{M}$. □

PROOF OF THEOREM 1.12 **(c)**.

(**c1**) By [**13**, Lemma 4.50(c)], all elements of $\mathsf{Inj}^\infty(\mathcal{K},\mathcal{M})$ are D-embeddings. So the class of diffeological embeddings from $\mathcal{K}$ to $\mathcal{M}$ is equal to the intersection of $\mathsf{Inj}^\infty(\mathcal{K},\mathcal{M})$ with the class of diffeological immersions from $\mathcal{K}$ to $\mathcal{M}$. Because the subspace topology is coarser than the D-topology of the subspace diffeology, the result is achieved by (**a2**).

(**c2**) Likewise, the class of diffeomorphisms from $\mathcal{K}$ onto $\mathcal{M}$ is equal to the intersection of $\mathsf{Inj}^\infty(\mathcal{K},\mathcal{M})$ with the class of local subductions from $\mathcal{K}$ onto $\mathcal{M}$, which is a D-open subset of $\mathsf{Inj}^\infty(\mathcal{K},\mathcal{M})$ by (**a4**). □

Acknowledgments

The author would like to thank Jean-Pierre Magnot and Jordan Watts, the organizers of Special Session on Recent Advances in Diffeology and their Applications, AMS-EMS-SMF Joint International Meeting, Grenoble, France, 2022. He also expresses his thanks to Patrick Iglesias-Zemmour, Katsuhiko Kuribayashi, and Enxin Wu for helpful discussions.

References

[1] A. Ahmadi, *Submersions, immersions, and étale maps in diffeology*, arXiv:2203.05994.

[2] J. D. Christensen, G. Sinnamon, and E. Wu, *The D-topology for diffeological spaces*, Pacific J. Math. **272** (2014), no. 1, 87–110, DOI 10.2140/pjm.2014.272.87. MR3270173

[3] J. D. Christensen and E. Wu, *Tangent spaces and tangent bundles for diffeological spaces* (English, with English and French summaries), Cah. Topol. Géom. Différ. Catég. **57** (2016), no. 1, 3–50. MR3467758

[4] M. Golubitsky and V. Guillemin, *Stable mappings and their singularities*, Graduate Texts in Mathematics, Vol. 14, Springer-Verlag, New York-Heidelberg, 1973. MR0341518

[5] V. Guillemin and A. Pollack, *Differential topology*, AMS Chelsea Publishing, Providence, RI, 2010. Reprint of the 1974 original, DOI 10.1090/chel/370. MR2680546

[6] S. Gürer and P. Iglesias-Zemmour, *Differential forms on manifolds with boundary and corners*, Indag. Math. (N.S.) **30** (2019), no. 5, 920–929, DOI 10.1016/j.indag.2019.07.004. MR3996773

[7] G. Hector, *Géométrie et topologie des espaces difféologiques* (French, with English summary), Analysis and geometry in foliated manifolds (Santiago de Compostela, 1994), World Sci. Publ., River Edge, NJ, 1995, pp. 55–80. MR1414196

[8] G. Hector and E. Macías-Virgós, *Diffeological groups*, Recent advances in Lie theory (Vigo, 2000), Res. Exp. Math., vol. 25, Heldermann, Lemgo, 2002, pp. 247–260. MR1937983

[9] M. W. Hirsch, *Differential topology*, Graduate Texts in Mathematics, No. 33, Springer-Verlag, New York-Heidelberg, 1976. MR0448362

[10] P. Iglesias-Zemmour, *Diffeology*, Mathematical Surveys and Monographs, vol. 185, American Mathematical Society, Providence, RI, 2013, DOI 10.1090/surv/185. MR3025051

[11] P. Iglesias-Zemmour, *Diffeological spaces are locally connected*, Available at `http://math.huji.ac.il/~piz/documents/DBlog-Rmk-DSALC.pdf`

[12] A. Kriegl and P. W. Michor, *The convenient setting of global analysis*, Mathematical Surveys and Monographs, vol. 53, American Mathematical Society, Providence, RI, 1997, DOI 10.1090/surv/053. MR1471480

[13] J. M. Lee, *Introduction to topological manifolds*, 2nd ed., Graduate Texts in Mathematics, vol. 202, Springer, New York, 2011, DOI 10.1007/978-1-4419-7940-7. MR2766102

[14] J. M. Lee, *Introduction to smooth manifolds*, 2nd ed., Graduate Texts in Mathematics, vol. 218, Springer, New York, 2013. MR2954043

[15] P. W. Michor, *Manifolds of differentiable mappings*, Shiva Mathematics Series, vol. 3, Shiva Publishing Ltd., Nantwich, 1980. MR583436

Department of Mathematical Sciences, Yazd University, Yazd, 89195–741, Iran
Email address: ahmadi@stu.yazd.ac.ir; alirezaahmadi13@yahoo.com

Contemporary Mathematics
Volume **794**, 2024
https://doi.org/10.1090/conm/794/15928

Smooth maps on convex sets

Yael Karshon and Jordan Watts

ABSTRACT. There are several notions of a smooth map from a convex set to a cartesian space. Some of these notions coincide, but not all of them do. We construct a real-valued function on a convex subset of the plane that does not extend to a smooth function on any open neighbourhood of the convex set, but that for each k extends to a C^k function on an open neighbourhood of the convex set. It follows that the diffeological and Sikorski notions of smoothness on convex sets do not coincide. We show that, for a convex set that is locally closed, these notions do coincide. With the diffeological notion of smoothness for convex sets, we then show that the category of diffeological spaces is isomorphic to the category of so-called exhaustive Chen spaces.

1. Introduction

There are many generalisations of smooth manifolds in the literature. Two of them are diffeological spaces and Chen spaces. A diffeology on a set X is given by a family of maps (called "plots") from open subsets of cartesian spaces to X; a Chen structure on a set X is given by a family of maps (called "Chen plots") from convex subsets of cartesian spaces to X. Every Chen space has an underlying diffeology, and one can ask whether the Chen structure can be recovered from this diffeology. It cannot. In fact, Stacey proves that the categories of diffeological spaces and of Chen spaces are not equivalent [**15**, Corollary 8.9].

The problem is that different Chen structures on the same set may have the same underlying diffeology. Even the standard diffeology on the closed interval $[0,1]$ can come from different Chen structures (see Example 6.2, which is due to Stacey). However, we show that among the Chen structures with the same underlying diffeology there is a maximal one, characterised by a property that we call "exhaustive", and after restricting to exhaustive Chen spaces, the natural functor from Chen spaces to diffeological spaces becomes an isomorphism of categories (Theorem 6.5).

The details behind the discussion above, and moreover the mere definition of a Chen space, rely on the notion of a smooth map from a convex set to a cartesian space. Chen does not define this notion. We are aware of several such notions in the literature, which all coincide when the convex set is closed or locally closed (e.g., see Remark 5.7 and Proposition 4.2). However, these notions do not all coincide

2020 *Mathematics Subject Classification.* Primary 57R55; Secondary 58-02.
Key words and phrases. convex, diffeology, Sikorski structure, Frölicher structure, Chen structure.

when the convex set is not locally closed. We demonstrate this by constructing a real-valued function on a convex subset X of $\mathbb{R}^2$ that, for each k, extends to a C^k function on an open neighbourhood of X, but that does not extend to a C^∞ function on any open neighbourhood of X (Proposition 3.1).

Thus, one needs to be careful when considering the definition of a Chen space. The applications that appear in Chen's papers [**5**–**8**], however, as well as other important works such as those of Stacey [**15**] and Baez-Hoffnung [**1**], either explicitly or implicitly take the diffeological notion of smoothness or one equivalent to it. Moreover, this is the notion of smoothness for which we have the isomorphism of categories between exhaustive Chen spaces and diffeological spaces.

Another generalisation of smooth manifolds is Sikorski spaces. A Sikorski structure on a set X is given by a family of real-valued functions on X. For a real-valued function on a convex set X, the Sikorski notion of smoothness requires local extensions to the ambient cartesian space, whereas the diffeological notion coincides with a more classical notion requiring the partial derivatives on the interior to extend continuously to the boundary (see Proposition 4.2, which relies on a result of Kriegl).

The best possible scenario is when there are both a diffeology and a Sikorski structure that determine each other. We refer to such structures as *reflexive.* Such structures form a category that is isomorphic to the category of Frölicher spaces; see [**2**] and [**17**, Chapter 2]. For more details on Frölicher spaces, see [**10**].

Arbitrary convex sets are not necessarily reflexive; this is our Corollary 3.2, which follows from our Proposition 3.1. In contrast, closed convex sets *are* reflexive; this is our Theorem 4.1, whose proof relies on a result of Kriegl. Moreover, in Section 4 we show that reflexivity is a local property (with a careful interpretation of "local"; see Propositions 5.2 and 5.4). As a result, we conclude that any Sikorski space that is locally Sikorski diffeomorphic to closed convex subsets of cartesian spaces is reflexive (Theorem 5.6).

And so, this paper has three purposes. The first is to demonstrate that various notions of smoothness (specifically, the diffeological and Sikorski notions) do not coincide on convex sets that are not locally closed. We achieve this in Section 3. The second is to show that Sikorski spaces that are locally diffeomorphic to closed convex sets, such as manifolds-with-corners, are reflexive. We achieve this in Sections 4 and 5, after establishing "locality of reflexivity", which is interesting in its own right. The third is to show the isomorphism of categories between diffeological spaces and exhaustive Chen spaces. We achieve this in Section 6.

2. Preliminaries

We recall the definitions of a Sikorski structure and of a diffeology. Good references are, respectively, Śniatycki's book [**14**] and Iglesias-Zemmour's book [**12**]. Note that in the literature, including Śniatcyki's book and our paper [**2**], Sikorski structures are often called "differential structures".

A *cartesian space* is $\mathbb{R}^d$ for some $d \in \mathbb{N}$, equipped with its standard smooth structure. A *convex set* is a convex subset of a cartesian space.

Let X be a set. A *Sikorski structure* on X is a non-empty collection $\mathcal{F}_X$ of real-valued functions on X that satisfies the following two axioms. We equip X with the *initial topology* induced by $\mathcal{F}_X$, which is the weakest topology for which all the functions in $\mathcal{F}_X$ are continuous.

(S1) If $f_1, \ldots, f_n \in \mathcal{F}_X$ and $g \in C^\infty(\mathbb{R}^n)$, then $g \circ (f_1, \ldots, f_n) \in \mathcal{F}_X$.
(S2) Given $f\colon X \to \mathbb{R}$, if for each point of X there exist a neighbourhood V and a function $g \in \mathcal{F}_X$ such that $g|_V = f|_V$, then $f \in \mathcal{F}_X$.

A *diffeology* on X is a collection $\mathcal{D}_X$ of maps from open subsets of cartesian spaces to X, called plots, that satisfies the following three axioms.

(D1) Constant maps to X are plots.
(D2) The precomposition of a plot with a C^∞ map between open subsets of cartesian spaces is a plot.
(D3) Given $p\colon U \to X$, if each point of U has a neighbourhood V such that $p|_V$ is a plot, then p is a plot.

Any manifold X has a natural Sikorski structure and a natural diffeology, consisting of those functions $X \to \mathbb{R}$ and those maps $U \to X$ that are C^∞ in the usual sense.

Given Sikorski spaces $(X, \mathcal{F}_X)$ and $(Y, \mathcal{F}_Y)$, a map $X \to Y$ is *Sikorski smooth* if its composition with every function in $\mathcal{F}_Y$ is in $\mathcal{F}_X$. Given diffeological spaces $(X, \mathcal{D}_X)$ and $(Y, \mathcal{D}_Y)$, a map $X \to Y$ is *diffeologically smooth* if its precomposition with every plot of X is a plot of Y. Both Sikorski spaces with Sikorski smooth maps between them, and diffeological spaces with diffeologically smooth maps between them, form categories, denoted **Sik** and **Diffeol**, resp. Both categories contain the category of smooth manifolds as full subcategories.

When considering diffeologies and Sikorski structures on the same set X, it is convenient to combine them into one structure. This approach was introduced in the introduction of Watts' thesis [**17**]; its topological aspects are studied in Virgin's paper [**16**]. Thus, we consider triples $(\mathcal{D}_X, X, \mathcal{F}_X)$, where $\mathcal{D}_X$ is a diffeology on X and $\mathcal{F}_X$ is a Sikorski structure on X, that are *compatible* in the following sense: for any $p\colon U \to X$ in $\mathcal{D}_X$ and $f\colon X \to \mathbb{R}$ in $\mathcal{F}_X$, the composition $f \circ p\colon U \to \mathbb{R}$ is smooth.

Given a Sikorski space $(X, \mathcal{F}_X)$, the set of Sikorski smooth maps from open subsets of cartesian spaces to X is a diffeology, which we denote $\Pi\mathcal{F}_X$. Given a diffeological space $(X, \mathcal{D}_X)$, the set of diffeologically smooth real-valued functions on X is a Sikorski structure, which we denote $\Phi\mathcal{D}_X$. A Sikorski structure $\mathcal{F}_X$ is *reflexive* if $\Phi\Pi\mathcal{F}_X = \mathcal{F}_X$. A diffeology $\mathcal{D}_X$ is *reflexive* if $\Pi\Phi\mathcal{D}_X = \mathcal{D}_X$. For any Sikorski structure $\mathcal{F}_X$, the diffeology $\Pi\mathcal{F}_X$ is reflexive. For any diffeology $\mathcal{D}_X$, the Sikorski structure $\Phi\mathcal{D}_X$ is reflexive. A triple is $(\mathcal{D}_X, X, \mathcal{F}_X)$ is *reflexive* if $\Phi\mathcal{D}_X = \mathcal{F}_X$ and $\Pi\mathcal{F}_X = \mathcal{D}_X$.

For any Sikorski space $(X, \mathcal{F}_X)$ and subset V of X, the *subspace Sikorski structure* is the set of those functions $f\colon V \to \mathbb{R}$ such that for each point of V there is a neighbourhood W in X and a function g in $\mathcal{F}_X$ such that $g|_{V\cap W} = f|_{V\cap W}$. For any diffeological space $(X, \mathcal{D}_X)$ and subset V of X, the *subset diffeology* is the set of those maps $p\colon U \to V$ from open subsets U of cartesian spaces to V whose composition with the inclusion map $V \to X$ is in $\mathcal{D}_X$.

For any subset X of any cartesian space, except where we say otherwise, we equip X with its subspace Sikorski structure $\mathcal{F}_X$ and with its subset diffeology $\mathcal{D}_X$ that are induced from the ambient cartesian space. When X is open in its ambient cartesian space, these structures coincide with its Sikorski and diffeological structures as a manifold.

The *D-topology* on a diffeological space is the strongest topology for which all the plots are continuous.

LEMMA 2.1. *For any subset X of any cartesian space $\mathbb{R}^d$,*

(1) $\mathcal{D}_X = \Pi\mathcal{F}_X$;

(2) *The initial topology on X coincides with its subspace topology;*

(3) *If X is convex, then the D-topology on X coincides with its subspace topology.*

PROOF. Item (1) is straightforward. Item (2) is [**17**, Lemma 2.28]. Item (3) is [**9**, Theorem 3.17]. □

3. An example

Let X be the union of the open upper half plane and the non-negative real axis in $\mathbb{R}^2$:

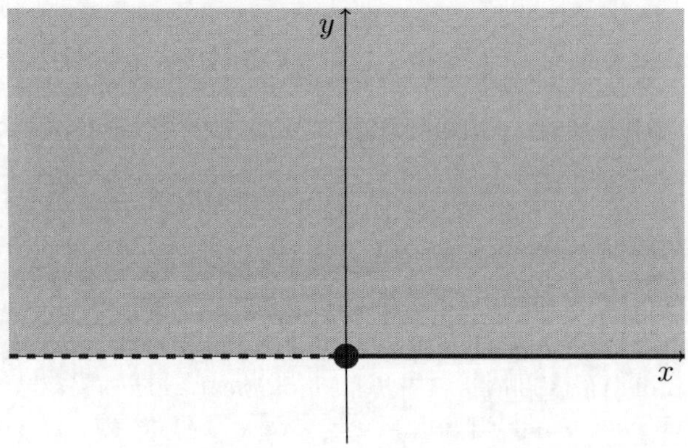

$$X := \{(x, y) \in \mathbb{R}^2 \mid y > 0\} \cup \{(x, 0) \mid x \geq 0\}.$$

PROPOSITION 3.1. *There exists a function*

$$f\colon X \to \mathbb{R}$$

that does not extend to a C^∞ function on any open neighbourhood of X in $\mathbb{R}^2$, but that, for each $k \in \mathbb{N}$, extends to a C^k function on some open neighbourhood of X in $\mathbb{R}^2$.

PROOF. For each $m \in \mathbb{N}$, let $\varphi_m\colon \mathbb{R} \to \mathbb{R}$ be a C^m function that vanishes on $(-\infty, 0]$, is smooth on $(0, \infty)$, whose derivatives of all orders less than or equal to m are bounded, and whose $m + 1^{\text{st}}$ derivative is not bounded on any punctured neighbourhood of 0. For example, we can take such a φ_m that vanishes outside of $(0, 2)$ and is equal to $x^{m+1/2}$ on $(0, 1)$.

Let $b\colon \mathbb{R} \to \mathbb{R}$ be a smooth function that is equal to 0 on $(-\infty, -1]$, is equal to -1 on $[0, \infty)$, and is strictly decreasing on $[-1, 0]$. For each $0 < \epsilon \leq 1$, let $b_\epsilon(x) := \epsilon b(x - 1 + \epsilon)$. Note that $\epsilon \mapsto b_\epsilon(x)$ is decreasing for all x and strictly decreasing for all $x \geq 0$.

For each $m \in \mathbb{N}$, define $h_m\colon \mathbb{R}^2 \to \mathbb{R}$ by

$$h_m(x, y) := y - b_{1/m}(x).$$

Then h_m is smooth, and every iterated derivative of h_m of every positive order is bounded on $\mathbb{R}^2$. Let

$$U_m := \{(x, y) \mid h_m(x, y) > 0\}.$$

Then $U_1 \supset U_2 \supset \dots$, and X is the intersection of the open sets U_m.

Because the derivatives of order less than or equal to m of φ_m and of h_m are bounded, so are the derivatives of order less than or equal to m of $\varphi_m \circ h_m$. Let $c_m > 0$ be a bound on the absolute values of these derivatives. Define

$$f := \sum_{m=1}^{\infty} \frac{1}{c_m 2^m} \varphi_m \circ h_m \Big|_X .$$

For each $k \in \mathbb{N}$, the series

$$g_{k+1} := \sum_{m=k+1}^{\infty} \frac{1}{c_m 2^m} \varphi_m \circ h_m$$

converges to a C^{k+1} function $\mathbb{R}^2 \to \mathbb{R}$. For $1 \leq m \leq k$, the functions $\varphi_m \circ h_m$ are smooth on the set U_k. Thus the series

$$\sum_{m=1}^{\infty} \frac{1}{c_m 2^m} \varphi_m \circ h_m \Big|_{U_k}$$

converges to a C^{k+1} function on U_k whose restriction to X is f.

We claim that f does not extend to a smooth function on any neighbourhood of the origin. Indeed, assuming otherwise, fix $k \in \mathbb{N}$ such that $(-1/k, 0)$ is in such a neighbourhood. For $1 \leq m < k$, we have $h_m(-1/k, 0) > 0$, and so the functions $\varphi_m \circ h_m$ are smooth on some neighbourhood of $(-1/k, 0)$. Since g_{k+1} is C^{k+1} on $\mathbb{R}^2$, the series

$$\sum_{\substack{m=1 \\ m \neq k}}^{\infty} \frac{1}{c_m 2^m} \varphi_m \circ h_m$$

converges to a C^{k+1} function on some neighbourhood of $(-1/k, 0)$. Subtracting it from f, we obtain that $\varphi_k \circ h_k|_X$ extends to a C^{k+1} function on some neighbourhood of $(-1/k, 0)$. This implies that the $k + 1^{\text{st}}$ derivatives of $\varphi_k \circ h_k$ are bounded on the intersection of the interior of X with some neighbourhood of $(-1/k, 0)$. But $\varphi_k(h_k(-1/k, y)) = \varphi_k(y)$, and so the $k + 1^{\text{st}}$ derivative of $\varphi_k \circ h_k$ with respect to y is not bounded on the intersection of $\{-1/k\} \times (0, \infty)$ with any neighbourhood of $(-1/k, 0)$. □

COROLLARY 3.2. *The Sikorski structure on X is not reflexive.*

PROOF. Let $f\colon X \to \mathbb{R}$ be a function as in Proposition 3.1. Then f is diffeologically smooth. Indeed, let $p\colon U \to X$ be a plot, *i.e.*, a map from an open subset U of some cartesian space $\mathbb{R}^\ell$ to X that is smooth as a map to $\mathbb{R}^2$. For each $k \in \mathbb{N}$, because f extends to a C^k function on some neighbourhood of X in $\mathbb{R}^2$, the composition $f \circ p\colon U \to \mathbb{R}$ is C^k. Because this holds for all $k \in \mathbb{N}$, the composition $f \circ p\colon U \to \mathbb{R}$ is smooth. Because the plot p is arbitrary, $f\colon X \to \mathbb{R}$ is diffeologically smooth.

However, f is not Sikorski smooth, because it does not extend to a C^∞ function on any open neighbourhood of X in $\mathbb{R}^2$. Because the subset diffeology on X in induced from the subspace Sikorski structure on X (Lemma 2.1), the Sikorski structure on X is not reflexive. □

4. Closed convex sets

The phenomenon exemplified in Corollary 3.2 does not occur with convex subsets of $\mathbb{R}^n$ that are closed:

THEOREM 4.1 (Closed Convex Sets are Reflexive). *Let $K \subseteq \mathbb{R}^n$ be closed and convex. Then the Sikorski structure on K is reflexive.*

In preparation for the proof of Theorem 4.1, we use results of Boman and Kriegl to show the following result.

PROPOSITION 4.2. *Let K be a convex subset of $\mathbb{R}^n$ with non-empty interior $\mathring{K}$. A map $f\colon K \to \mathbb{R}$ is diffeologically smooth if and only if $f|_{\mathring{K}}\colon \mathring{K} \to \mathbb{R}$ is smooth and all of its partial derivatives of all orders extend continuously to K.*

PROOF. We will use a result of Kriegl. Kriegl allows convex subsets in infinite dimensions. He works with the c^∞-topology, and with a notion of smoothness to which we will refer here as c^∞-smoothness. These notions rely on the notion of a smooth curve in K. When K is a subset of $\mathbb{R}^n$, a smooth curve in K is a map $c\colon \mathbb{R} \to K$ that is smooth (infinitely-differentiable) in the usual sense as a map to $\mathbb{R}^n$.

The c^∞-topology on K is defined to be the strongest topology on K for which all smooth curves $c\colon \mathbb{R} \to K$ are continuous. In our finite dimensional setup, the c^∞-topology coincides with the subspace topology. Indeed, the c^∞-topology coincides with the D-topology by [**9**, Theorem 3.7], and the D-topology coincides with the standard topology by [**9**, Theorem 3.17]. Thus, we may refer without ambiguity to the interior $\mathring{K}$, and to a function on $\mathring{K}$ having a continuous extension to K.

Suppose f is diffeologically smooth. In particular, for all smooth curves $c\colon \mathbb{R} \to K$, the composition $f \circ c$ is smooth. By definition, this means that f is c^∞-smooth. Conversely, suppose f is c^∞-smooth. Let $p\colon U \to K$ be a plot of K, that is, a smooth map in the usual sense. For any smooth curve $c\colon \mathbb{R} \to U$, the composition $p \circ c$ is a smooth curve in K. Since f is c^∞-smooth, the composition $f \circ p \circ c$ is smooth. By a result of Boman [**4**, Theorem 1], since c is arbitrary, $f \circ p$ is smooth. Since p is arbitrary, f is diffeologically smooth.

By a result of Kriegl [**13**, Theorem 1.5], c^∞-smoothness of f is equivalent to the restriction of f to the interior $\mathring{K}$ of K being smooth and it and all of its derivatives extending to functions on K continuously (with respect to the c^∞-topology on K, hence with respect to the standard topology on K).

We pause to explain Kriegl's term *derivatives*. Kriegl works with the setup of Frölicher and Kriegl [**11**]. Unravelling [**11**, Definitions 4.1.9, 4.3.9, and 4.3.26] for the special case of a smooth real-valued function on an open subset of $\mathbb{R}^n$, the derivatives of $f|_{\mathring{K}}\colon \mathring{K} \to \mathbb{R}$ are the same as its iterated differentials in the usual sense. These having continuous extensions to K is equivalent to all the partial derivatives of f of all orders (in the usual sense) having continuous extensions to K. □

We are now ready to prove Theorem 4.1.

PROOF OF THEOREM 4.1. Recall (Lemma 2.1) that the diffeology on K is induced from the Sikorski structure on K. So we need to show that, for every function $f\colon K \to \mathbb{R}$, if f is diffeologically smooth, then f is Sikorski smooth. By replacing $\mathbb{R}^n$ with the affine span of K, we may assume that the interior of K is non-empty.

Fix a function $f\colon K \to \mathbb{R}$ that is diffeologically smooth. By Proposition 4.2, the restriction of f to the interior $\mathring{K}$ of K is smooth and all its partial derivatives of all orders extend to continuous functions on K. Because K is closed and convex, it satisfies the hypothesis of the Whitney Extension Theorem (see, for instance, [**3**, Theorem 2.6 and Proposition 2.16]), and hence f extends to a smooth function $g\colon \mathbb{R}^n \to \mathbb{R}$. So f is Sikorski smooth. □

EXAMPLE 4.3. Let P be the solid square pyramid. By Theorem 4.1, P obtains a reflexive Sikorski structure and a reflexive diffeological structure from $\mathbb{R}^3$. One may interpret this to mean that P has an unambiguous smooth structure, although it is not a manifold-with-corners (*cf.* Example 5.8).

5. Locality of reflexivity

In this section, we show that compatibility and reflexivity of a triple $(\mathcal{D}, X, \mathcal{F})$ are local properties; see Propositions 5.2 and 5.4. We apply this to subcartesian spaces that are locally compact and locally convex; see Theorem 5.6. In particular, manifolds-with-corners are reflexive; see Example 5.8.

We begin with a lemma.

LEMMA 5.1. *Let $(X, \mathcal{F})$ be a Sikorski space, let $(V, \mathcal{F}_V)$ be a Sikorski subspace, and let $p \in \Pi\mathcal{F}$ with image in V. Then $p \in \Pi\mathcal{F}_V$.*

PROOF. As usual, we equip X with its initial topology that is induced by $\mathcal{F}$. Let $f_V \in \mathcal{F}_V$. For each $y \in V$ there exist an open neighbourhood W of y and a function $g \in \mathcal{F}$ such that $f_V|_{V\cap W} = g|_{V\cap W}$. Because $p \in \Pi\mathcal{F}$, the preimage $p^{-1}(W)$ is open and the composition $g \circ p$ is smooth. So $f_V \circ p|_{p^{-1}(W)} = g \circ p|_{p^{-1}(W)}$ is smooth. Because the sets $p^{-1}(W)$ form an open cover of the domain of p, the composition $f_V \circ p$ is smooth. It follows that $p \in \Pi\mathcal{F}_V$. □

For a diffeology $\mathcal{D}$ and Sikorski structure $\mathcal{F}$ on a set X and for a subset V of X, we denote by $\mathcal{D}_V$ and $\mathcal{F}_V$ the subset diffeology and the subspace Sikorski structure that are induced by X.

We now prove locality of "compatible" and "reflexive".

PROPOSITION 5.2 (Compatibility and Reflexivity are Local). *Let $\mathcal{D}$ and $\mathcal{F}$ be a diffeology and a Sikorski structure on a set X. Denote by $\tau_{\mathcal{D}}$ the D-topology on X induced by $\mathcal{D}$ and by $\tau_{\mathcal{F}}$ the initial topology on X induced by $\mathcal{F}$.*

1. *Suppose for every point in X there is a $\tau_{\mathcal{D}}$-open neighbourhood V such that $(\mathcal{D}_V, V, \mathcal{F}_V)$ is a compatible triple. Then $(\mathcal{D}, X, \mathcal{F})$ is a compatible triple. In particular, the topology $\tau_{\mathcal{D}}$ contains the topology $\tau_{\mathcal{F}}$.*
2. *Suppose that the topology $\tau_{\mathcal{D}}$ contains the topology $\tau_{\mathcal{F}}$ and that for every point in X there is a $\tau_{\mathcal{F}}$-open neighbourhood V such that $(\mathcal{D}_V, V, \mathcal{F}_V)$ is a reflexive triple. Then $(\mathcal{D}, X, \mathcal{F})$ is a reflexive triple.*

PROOF. We prove Claim (1). Fix $p \in \mathcal{D}$, $f \in \mathcal{F}$, and $x \in X$. By hypothesis, there is a $\tau_{\mathcal{D}}$-open neighbourhood V of x in X such that $(\mathcal{D}_V, V, \mathcal{F}_V)$ is a compatible triple. The restriction $p_V := p|_{p^{-1}(V)}$ takes values in V, and its domain is open in the domain of p, so $p_V \in \mathcal{D}_V$. Since $f_V := f|_V$ is in $\mathcal{F}_V$ and by compatibility, the composition $f \circ p|_{p^{-1}(V)} = f_V \circ p_V$ is smooth. Since smoothness of $f \circ p$ is a local condition and the sets $p^{-1}(V)$ form an open cover of the domain of p, this shows that $f \circ p$ is smooth. Claim (1) follows.

We prove Claim (2). By Claim (1), it is enough to show that $\Pi\mathcal{F} \subseteq \mathcal{D}$ and $\Phi\mathcal{D} \subseteq \mathcal{F}$. For the first inclusion, fix $p \in \Pi\mathcal{F}$. By the locality axiom of a diffeology, it suffices to show that for each $x \in X$ there is a $\tau_{\mathcal{D}}$-open neighbourhood V such that $p_V := p|_{p^{-1}(V)} \in \mathcal{D}_V$. Let $x \in X$. By hypothesis, there is a $\tau_{\mathcal{F}}$-open neighbourhood V of x such that $(\mathcal{D}_V, V, \mathcal{F}_V)$ is a reflexive triple. In particular, $\mathcal{D}_V = \Pi\mathcal{F}_V$. Since $\tau_{\mathcal{F}}$ is contained in $\tau_{\mathcal{D}}$, it suffices to show that $p_V \in \Pi\mathcal{F}_V$. This, in turn, follows from Lemma 5.1, since $p \in \Pi\mathcal{F}$.

For the second inclusion $\Phi\mathcal{D} \subseteq \mathcal{F}$, fix $f \in \Phi\mathcal{D}$. By the locality axiom of a Sikorski space, it suffices to show that for each $x \in X$ there is a $\tau_{\mathcal{F}}$-open neighbourhood W of x and a function $g \in \mathcal{F}$ such that $f|_W = g|_W$. Fix $x \in X$. By hypothesis, there is a $\tau_{\mathcal{F}}$-open neighbourhood V of x such that $(\mathcal{D}_V, V, \mathcal{F}_V)$ is a reflexive triple. In particular, $\mathcal{F}_V = \Phi\mathcal{D}_V$. Define $f_V := f|_V$. For any $p \in \mathcal{D}_V$, because p is in $\mathcal{D}$ and f is in $\Phi\mathcal{D}$, the composition $f_V \circ p = f \circ p$ is smooth. So $f_V \in \Phi\mathcal{D}_V = \mathcal{F}_V$. By definition of $\mathcal{F}_V$, there exist a $\tau_{\mathcal{F}}$-open neighbourhood $W \subseteq V$ of x and a function $g \in \mathcal{F}$ such that $g|_W = f_V|_W = f|_W$, proving what is required. □

In Part (2) of Proposition 5.2, the interaction between the two topologies is not superfluous. Indeed, let X be $\mathbb{R}$ and $\mathcal{D}$ be the standard diffeology on it. Take $\mathcal{F}$ to be *all* real-valued functions. Then the initial topology induced by $\mathcal{F}$ is discrete, and taking singleton open sets, we satisfy the hypotheses of both claims in the proposition. However, compatibility (let alone reflexivity) fails: no non-constant smooth curve in $\mathcal{D}$ is in $\Pi\mathcal{F}$.

We pause to recall that the initial topology admits bump functions.

LEMMA 5.3. *Let $(X, \mathcal{F})$ be a Sikorski space, let $\tau_{\mathcal{F}}$ be the initial topology on X induced by $\mathcal{F}$, let $x \in X$, and let V be a $\tau_{\mathcal{F}}$-open neighbourhood of x in X. Then there exists $\rho \in \mathcal{F}$ whose $\tau_{\mathcal{F}}$-support is contained in V and that is equal to 1 on a $\tau_{\mathcal{F}}$-neighbourhood of x in X.*

PROOF. Because V is a $\tau_{\mathcal{F}}$-open neighbourhood of x, there exist $h_1, \dots, h_N \in \mathcal{F}$ such that $x \in \cap_{i=1}^N h_j^{-1}((0,1)) \subseteq V$. Let $h := (h_1, \dots, h_N)$, and let $b\colon \mathbb{R}^N \to [0,1]$ be a smooth function that is identically 1 on an open neighbourhood of $h(x)$ and whose support is contained in $(0,1)^N$. Note that $b \circ h \in \mathcal{F}$, that the $\tau_{\mathcal{F}}$-support of $b \circ h$ is contained in V, and that $(b \circ h)^{-1}(\{1\})$ is a $\tau_{\mathcal{F}}$-neighbourhood of x in X. Define $\rho\colon X \to \mathbb{R}$ to be equal to $b \circ h$ on V and 0 elsewhere. Then the $\tau_{\mathcal{F}}$-support K of ρ coincides with that of $b \circ h$, and $\rho^{-1}(\{1\}) = (b \circ h)^{-1}(\{1\})$. Finally, because ρ coincides with an element of $\mathcal{F}$ on each of the two sets V and $X \smallsetminus K$ and these two sets form an open cover of X, we conclude that $\rho \in \mathcal{F}$. □

We now prove a converse statement to Proposition 5.2: global compatibility implies local compatibility, and global reflexivity implies local reflexivity.

PROPOSITION 5.4. *Let $\mathcal{D}$ and $\mathcal{F}$ be a diffeology and a Sikorski structure on a set X. If the triple $(\mathcal{D}, X, \mathcal{F})$ is compatible and $V \subset X$ is any subset, then the triple $(\mathcal{D}_V, V, \mathcal{F}_V)$ is compatible. If the triple $(\mathcal{D}, X, \mathcal{F})$ is reflexive and $V \subset X$ is an open subset in the initial topology, then the triple $(\mathcal{D}_V, V, \mathcal{F}_V)$ is reflexive.*

PROOF. Suppose that $(\mathcal{D}, X, \mathcal{F})$ is compatible, and let $V \subset X$. Fix $p \in \mathcal{D}_V$, $f \in \mathcal{F}_V$, and $x \in V$. There exist a $\tau_{\mathcal{F}}$-open neighbourhood W of x in X and a function $g \in \mathcal{F}$ such that $f|_{V \cap W} = g|_{V \cap W}$. Compatibility implies that $p^{-1}(W)$ is open in the domain of p and that $g \circ p$ is smooth. So $f \circ p|_{p^{-1}(W)} = g \circ p|_{p^{-1}(W)}$ is smooth. Because such sets $p^{-1}(W)$ are an open cover of V, the composition $f \circ p$ is smooth. Compatibility of $(\mathcal{D}_V, V, \mathcal{F}_V)$ follows.

Suppose now that $(\mathcal{D}, X, \mathcal{F})$ is reflexive and $V \subset X$ is $\tau_{\mathcal{F}}$-open. By the previous claim, it suffices to show that $\Pi\mathcal{F}_V \subset \mathcal{D}_V$ and $\Phi\mathcal{D}_V \subset \mathcal{F}_V$. For the first inclusion, fix $p \in \Pi\mathcal{F}_V$. For any $f \in \mathcal{F}$, the restriction $f|_V$ is in $\mathcal{F}_V$, so $f \circ p = f|_V \circ p$ is smooth. Thus, $p \in \Pi\mathcal{F} = \mathcal{D}$ with image in V, and so $p \in \mathcal{D}_V$.

For the second inclusion $\Phi\mathcal{D}_V \subseteq \mathcal{F}_V$, fix $f \in \Phi\mathcal{D}_V$. It suffices to show that for any $x \in V$ there exist a $\tau_{\mathcal{F}}$-open neighbourhood W of x in V and a function $g \in \mathcal{F}$ such that $f|_W = g|_W$. Fix $x \in V$. By Lemma 5.3, there exists a function $\rho \in \mathcal{F}$ whose $\tau_{\mathcal{F}}$-support $\operatorname{supp}(\rho)$ is contained in V and that is equal to 1 on a $\tau_{\mathcal{F}}$-neighbourhood of x in X. Define $g\colon X \to \mathbb{R}$ to be equal to the product $\rho \cdot f$ on V and 0 elsewhere. Fix $p \in \mathcal{D}$. Write

$$(g \circ p)|_{p^{-1}(V)} \;=\; \rho \circ p|_{p^{-1}(V)} \;\cdot\; f \circ p|_{p^{-1}(V)}.$$

By compatibility, $p^{-1}(V)$ is open in the domain of p, the first factor on the right hand side is smooth, and since $f \in \Phi\mathcal{D}_V$, the second factor on the right hand side is smooth. So the composition $g \circ p$ is smooth in the open subset $p^{-1}(V)$ of the domain of p. Since this composition vanishes outside $p^{-1}(\operatorname{supp}(\rho))$, which is a closed subset of the domain of p and is contained in $p^{-1}(V)$, this composition is smooth. We conclude that $g \in \Phi\mathcal{D} = \mathcal{F}$. Since f and g coincide on the $\tau_{\mathcal{F}}$-interior of $\rho^{-1}(1)$, which contains x, this proves what is required. □

We now prove that Sikorski spaces that are locally compact and locally convex are reflexive. We start with a technical lemma:

LEMMA 5.5. *Let C' be a subset of $\mathbb{R}^n$. Suppose that C' is locally compact and that $C' = C \cap U$ for some convex subset C of $\mathbb{R}^n$ and some open subset U of $\mathbb{R}^n$. Then the Sikorski structure of C' is reflexive.*

PROOF. Since C' is locally compact and is contained in the open set U, there is a collection $\{\overline{B}_i\}$ of closed balls in $\mathbb{R}^n$ that are contained in U, whose interiors $\mathring{B}_i$ cover C', and such that each $C' \cap \overline{B}_i$ is closed in $\mathbb{R}^n$. By Theorem 4.1, the Sikorski structure of each $C' \cap \overline{B}_i$ is reflexive. By Proposition 5.4, the Sikorski structure of each $C' \cap \mathring{B}_i$ is reflexive. By Proposition 5.2, the Sikorski structure of C' is reflexive. □

We continue to general Sikorski spaces that are locally compact and locally convex:

THEOREM 5.6. *Let X be a locally compact Sikorski space satisfying the following condition: for each point of X there is a neighbourhood that is Sikorski diffeomorphic to a convex subset of a cartesian space. Then the Sikorski structure on X is reflexive.*

PROOF. By hypothesis, there is a collection $\{V_k\}$ of subsets of X whose interiors $\mathring{V}_k$ cover X such that each V_k is Sikorski diffeomorphic to a convex subset C_k of a cartesian space. Because a Sikorski diffeomorphism is in particular a homeomorphism, each $\mathring{V}_k$ is diffeomorphic to an open subset C_k' of C_k. Because X is locally compact, so is each $\mathring{V}_k$, and hence so is each C_k'. By Lemma 5.5, each C_k' is reflexive, and hence so is each $\mathring{V}_k$. By Proposition 5.2, X is reflexive. □

REMARK 5.7. Theorem 5.6 implies the following: let K be a locally closed convex set, and $F\colon K \to \mathbb{R}^n$ a map. Then F is diffeologically smooth if and only if it is Sikorski smooth.

EXAMPLE 5.8. Let M be a manifold-with-corners, and let $\mathcal{F}$ and $\mathcal{D}$ be its natural Sikorski structure and diffeology, resp. Then $(\mathcal{D}, M, \mathcal{F})$ is a reflexive triple. Indeed, recall that a manifold-with-corners of dimension n is by definition locally diffeomorphic to open subsets of $[0, \infty)^n$. The result then follows from Theorem 5.6.

6. Chen spaces

Kuo-Tsai Chen introduced structures that are similar to diffeologies but in which the domains of plots are convex subsets of cartesian spaces rather than open subsets of cartesian spaces. Chen's papers [**5**–**8**], which range over the years 1973 to 1986, contain several different definitions of such spaces. They converge to a definition [**7**, **8**] that is now typically taken to be "the" definition of a Chen space. Baez and Hoffnung [**1**, page 5793] identify these Chen spaces with concrete sheaves over the site of convex subsets of cartesian spaces. Stacey's paper [**15**, Section 3] compares different categories of smooth structures, including Chen's different definitions (see [**15**, p. 80]) and diffeology, and describes functors between them.

Chen's smooth compatibility axiom relies on the notion of a "smooth map between convex sets", which Chen does not explicitly define. In practice, Chen seems to only need convex sets that are locally closed. For these, there is an unambiguous notion of "smooth map between convex sets" (see Remark 5.7 and Proposition 4.2). However, Chen does not assume "locally closed" in his latest definitions [**7**, **8**], and without this assumption, as we have shown in §3, the notion of "smooth map between convex sets" is ambiguous.

We will interpret "smooth map between convex sets" as a map between the convex sets that is smooth *in the diffeological sense.* See Remark 6.6 for the rationale behind this choice. For maps from a closed interval to $\mathbb{R}$, this is equivalent to smoothness in the interior of the interval and one-sided smoothness at the boundary points.

Recall that a *convex set* is a convex subset of a cartesian space, and that the D-topology on a convex set coincides with its subspace topology that is induced from its ambient cartesian space.

A *Chen structure* on a set X is a collection $\mathcal{C}_X$ of maps, called *Chen plots*, from convex sets to X, that satisfies the following axioms.

(C1) Constant maps to X are Chen plots.

(C2) The precomposition of a Chen plot with a (diffeologically) smooth map between convex sets is a Chen plot.

(C3) Given a convex set C and a map $p\colon C \to X$, if each point of C has a convex neighbourhood V in C such that $p|_V$ is a Chen plot, then p is a Chen plot.

A *Chen space* is a set equipped with a Chen structure. A map from a Chen space X to a Chen space Y is *Chen smooth* if its precomposition with every Chen plot of X is a Chen plot of Y, and it is a *Chen diffeomorphism* if it is bijective and both it and its inverse are Chen smooth. Chen spaces with Chen smooth maps form a category, which we denote **Chen**.

Except where we say otherwise, we equip any subset of a cartesian space with the *standard Chen structure*, consisting of the collection of all diffeologically smooth maps from convex sets to that subset.

EXAMPLES 6.1. A map between open subsets of cartesian spaces is Chen smooth if and only if it is smooth in the usual sense. As a second example, the function $f\colon X \to \mathbb{R}$ that we constructed in Section 3 is Chen smooth.

There is a natural functor $\mathfrak{Di}\colon$ **Chen** $\to$ **Diffeol** defined as follows:

- For any Chen space $(X, \mathcal{C}_X)$, define $\mathfrak{Di}(X, \mathcal{C}_X)$ to be the diffeological space $(X, \mathfrak{Di}(\mathcal{C}_X))$, in which a map $p\colon U \to X$ from an open subset of a Cartesian space is in $\mathfrak{Di}(\mathcal{C}_X)$ if it is Chen smooth.
- For any Chen smooth map $F\colon (X, \mathcal{C}_X) \to (Y, \mathcal{C}_Y)$, define $\mathfrak{Di}(F)$ to be F.

Similarly, there is a natural function $\mathfrak{Ch}\colon$ **Diffeol** $\to$ **Chen** defined as follows:

- For any diffeological space $(X, \mathcal{D}_X)$, define $\mathfrak{Ch}(X, \mathcal{D}_X)$ to be the Chen space $(X, \mathfrak{Ch}(\mathcal{D}_X))$, in which a map $p\colon C \to X$ from a convex set is in $\mathfrak{Ch}(\mathcal{D}_X)$ if it is diffeologically smooth.
- For any diffeologically smooth map $F\colon (X, \mathcal{D}_X) \to (Y, \mathcal{D}_Y)$, define $\mathfrak{Ch}(F)$ to be F.

The functors $\mathfrak{Di}$ and $\mathfrak{Ch}$ are exactly the functors $\mathfrak{So}$ and $\mathfrak{Ch}^\sharp$ of Stacey's paper [**15**]. We refer the reader to [**15**] for the proofs that $\mathfrak{Di}$ indeed takes Chen structures to diffeologies and Chen smooth maps to diffeologically smooth maps, and that $\mathfrak{Ch}$ indeed takes diffeologies to Chen structures and diffeologically smooth maps to Chen smooth maps. Moreover, Stacey shows that these functors are adjoint. A natural question is whether these functors form an equivalence of categories between **Chen** and **Diffeol**. They don't;[1] see Example 6.2. The main issue is that there can be "missing Chen plots". We will now show that, if we add these "missing Chen plots", then we do obtain equivalent (in fact, isomorphic) categories.

We define a Chen space $(X, \mathcal{C}_X)$ to be *exhaustive* if it has the following property:

> Let $p\colon C \to X$ be a map from a convex set C to X. Assume that, for every convex open subset U of a cartesian space and smooth map $f\colon U \to C$, the composition $p \circ f$ is in $\mathcal{C}_X$. Then $p \in \mathcal{C}_X$.

It is straightforward to show that the quality of being exhaustive is invariant under Chen diffeomorphism. The following example is due to Stacey [**15**, page 100].

EXAMPLE 6.2. Equip the interval $[0,1]$ with the set of those maps $p\colon C \to [0,1]$ from a convex set C such that p locally factors through a usual smooth map $U \to [0,1]$, where U is an open subset of a cartesian space. This is a non-standard Chen structure on $[0,1]$. It is not exhaustive; for example, the identity map on $[0,1]$ is not in this Chen structure, but its precomposition with any smooth map from an open convex subset U of a cartesian space to $[0,1]$ is in this non-standard Chen structure. In contrast, the standard Chen structure on $[0,1]$ is exhaustive. Thus, these two Chen structures on $[0,1]$ are not isomorphic. But the functor $\mathfrak{Di}$ takes

[1] In fact, according to Stacey, [**15**, Corollary 8.9], there does not exist *any* equivalence of categories between **Chen** and **Diffeol**.

both of these structures to the standard diffeology on $[0,1]$. So the functor $\mathfrak{Di}$ is not one-to-one on isomorphism classes, and so it is not an equivalence of categories.

Given a Chen space $(X, \mathcal{C}_X)$, we define the *exhaustion* of $\mathcal{C}_X$ to be the Chen structure $E(\mathcal{C}_X) := \mathfrak{Ch}(\mathfrak{Di}(\mathcal{C}_X))$ on X.

LEMMA 6.3 (Properties of Exhaustions). *Let $(X, \mathcal{C}_X)$ be a Chen space.*

(1) *$(X, E(\mathcal{C}_X))$ is an exhaustive Chen space.*
(2) *$\mathcal{C}_X \subseteq E(\mathcal{C}_X)$.*
(3) *If $(X, \mathcal{C}_X)$ is exhaustive, then $E(\mathcal{C}_X) = \mathcal{C}_X$.*

The proofs rely on some results that we defer to an appendix.

LEMMA 6.4. *Any object in the image of the functor $\mathfrak{Ch}$ is an exhaustive Chen space.*

PROOF. Fix a diffeological space $(X, \mathcal{D}_X)$, and consider $\mathcal{C}_X := \mathfrak{Ch}(\mathcal{D}_X)$. By Item (1) of Proposition A.3, $\mathfrak{Di}(\mathfrak{Ch}(\mathcal{D}_X)) = \mathcal{D}_X$. Applying $\mathfrak{Ch}$, we obtain $\mathfrak{Ch}(\mathfrak{Di}(\mathfrak{Ch}(\mathcal{D}_X))) = \mathfrak{Ch}(\mathcal{D}_X)$, that is, $\mathfrak{Ch}(\mathfrak{Di}(\mathcal{C}_X)) = \mathcal{C}_X$. By Item (2) of Proposition A.3, this implies that $\mathcal{C}_X$ is exhaustive. □

PROOF OF LEMMA 6.3. Item (1) is a consequence of Lemma 6.4.

Item (2) is Item (3) of Proposition A.3, but here is also a direct proof. Let $p\colon C \to X$ be a Chen plot in $\mathcal{C}_X$. Then p is Chen smooth with respect to $\mathcal{C}_X$ (Item (2) of Lemma A.1). Because $\mathfrak{Di}$ is a functor, p is diffeologically smooth with respect to $\mathfrak{Di}(\mathcal{C}_X)$. Because $\mathfrak{Ch}$ is a functor, p is Chen smooth with respect to $\mathfrak{Ch}(\mathfrak{Di}(\mathcal{C}_X))$. So p is a Chen plot in $\mathfrak{Ch}(\mathfrak{Di}(\mathcal{C}_X))$ (Item (2) of Lemma A.1).

To prove Item (3), suppose $(X, \mathcal{C}_X)$ is exhaustive. By Item (2), it suffices to show that $E(\mathcal{C}_X) \subseteq \mathcal{C}_X$. This, in turns, follows from Item (2) of Proposition A.3. □

THEOREM 6.5. *The restriction of $\mathfrak{Di}$ to exhaustive Chen spaces is an isomorphism of categories onto* **Diffeol**, *with inverse $\mathfrak{Ch}$.*

PROOF. By Lemma 6.4, $\mathfrak{Ch}$ takes diffeological spaces to exaustive Chen spaces. By Item (1) of Proposition A.3, the composition $\mathfrak{Di} \circ \mathfrak{Ch}$ is the identity functor on the category of diffeological spaces. By Part (3) of Lemma 6.3, the composition $\mathfrak{Ch} \circ \mathfrak{Di}$ restricts to the identity functor on the category of exhaustive Chen spaces. □

REMARK 6.6. One motive for taking the notion of a "smooth map between convex subsets" to be the diffeological notion is that an equivalent notion is used in Baez and Hoffnung [**1**]. A second motive is that this definition is equivalent to many other standard notions; see Proposition 4.2 and [**1**, page 5794]. We now have a third motive: the isomorphism of Theorem 6.5 no longer holds if we take "smooth map between convex subsets" to be the Sikorski notion. Indeed, the map $f\colon X \to \mathbb{R}$ of Section 3 is Chen smooth when we take smoothness between convex subsets to be the diffeological notion. However, if in the definition of "Chen space" we were to take smoothness between convex subsets to be the Sikorski notion, then the map $f\colon X \to \mathbb{R}$ of Section 3 would not be Chen smooth, and hence the functor $\mathfrak{Di}$ would no longer be full.

Appendix A. Back and forth between Diffeologies and Chen structures

We begin with a "sanity check" for our definitions.

LEMMA A.1.

(1) *Given a diffeological space* $(X, \mathcal{D}_X)$ *and a map* $p\colon U \to X$ *from an open subset* U *of a cartesian space, the map* p *is diffeologically smooth if and only if it is a plot.*

(2) *Given a Chen space* $(X, \mathcal{C}_X)$ *and a map* $p\colon C \to X$ *from a convex set* C, *the map* p *is Chen smooth if and only if it is a Chen plot in* $\mathcal{C}_X$.

(3) *Given a Sikorski differential space* $(X, \mathcal{F}_X)$ *and a function* $f\colon X \to \mathbb{R}$, *the function* f *is Sikorski smooth if and only if it is in* $\mathcal{F}_X$.

PROOF. We prove Item (1). Every plot in $\mathcal{D}_X$ is diffeologically smooth because $\mathcal{D}_X$ satisfies the smooth compatibility axiom in the definition of a diffeology. Every diffeologically smooth map from an open subset U of a cartesian space to X is a plot by the definition of "diffeologically smooth" and because the identity map of U is a plot.

We prove Item (2). Every plot in $\mathcal{C}_X$ is Chen smooth because $\mathcal{C}_X$ satisfies the smooth compatibility axiom of a Chen structure. Every Chen smooth map from a convex subset C to X is a Chen plot by the definition of "Chen smooth" and because the identity map on C is a Chen plot.

We prove Item (3). Every function in $\mathcal{F}_X$ is Sikorski smooth because $\mathcal{F}_X$ satisfies the smooth compatibility axiom in the definition of a Sikorski structure. Every Sikorski smooth function from X to $\mathbb{R}$ is in $\mathcal{F}_X$ by the definition of "Sikorski smooth" and because the identity map on $\mathbb{R}$ is in $\mathcal{F}_{\mathbb{R}}$. □

We continue with a simple characterisation of diffeological smoothness.

LEMMA A.2. *Let* C *and* X *be diffeological spaces, and let* $p\colon C \to X$ *be any map of their underlying sets. Suppose that, for every plot* $f\colon U \to C$ *of* C *whose domain* U *is convex, the composition* $p \circ f\colon U \to X$ *is a plot of* X. *Then* p *is diffeologically smooth.*

PROOF. Fix a plot $q\colon V \to C$ of C. We need to prove that the composition $p \circ q\colon V \to X$ is a plot of X. By the smooth compatibility axiom for the diffeology of C, for every convex open subset U of V, the restriction $q|_U\colon U \to C$ is also a plot of C. By the hypothesis on p, for every such U, the composition $p \circ q|_U\colon U \to X$ is a plot of X. Because V is covered by such convex open subsets U, and by the locality axiom for the diffeology of X, $p \circ q\colon V \to X$ is a plot of X, as required. □

We now give the main result of this appendix.

PROPOSITION A.3.

(1) *For any diffeological space* $(X, \mathcal{D}_X)$,
$$\mathfrak{Di}(\mathfrak{Ch}(\mathcal{D}_X)) = \mathcal{D}_X.$$

(2) *For any Chen space* $(X, \mathcal{C}_X)$, $\mathfrak{Ch}(\mathfrak{Di}(\mathcal{C}_X))$ *is the collection of those maps* $p\colon C \to X$ *from convex sets* C *to* X *that have the following property.*

> *For every convex open subset* V *of a cartesian space and every smooth map* $f\colon V \to C$, *the composition* $p \circ f\colon V \to X$ *is a Chen plot in* $\mathcal{C}_X$.

Consequently, a Chen space $(X, \mathcal{C}_X)$ is exhaustive if and only if

$$\mathfrak{Ch}(\mathfrak{Di}(\mathcal{C}_X)) \subset \mathcal{C}_X.$$

(3) *For any Chen space $(X, \mathcal{C}_X)$,*

$$\mathcal{C}_X \subset \mathfrak{Ch}(\mathfrak{Di}(\mathcal{C}_X)).$$

Proof of Item (1) of Proposition A.3. Fix a diffeology $\mathcal{D}_X$.

We will first prove that $\mathfrak{Di}(\mathfrak{Ch}(\mathcal{D}_X)) \subset \mathcal{D}_X$. Take a plot $p\colon U \to X$ in $\mathfrak{Di}(\mathfrak{Ch}(\mathcal{D}_X))$. Because $\mathfrak{Di}(\mathfrak{Ch}(\mathcal{D}_X))$ is a diffeology, it satisfies the smooth compatibility axiom in the definition of a diffeology, so for any convex open subset $V \subset U$, the restriction $p|_V\colon V \to X$ is also a plot in $\mathfrak{Di}(\mathfrak{Ch}(\mathcal{D}_X))$. By the definition of $\mathfrak{Di}(\mathfrak{Ch}(\mathcal{D}_X))$, every such $p|_V\colon V \to X$ is Chen smooth with respect to $\mathfrak{Ch}(\mathcal{D}_X)$. By Item (2) of Lemma A.1, every such $p|_V$ is a Chen plot in $\mathfrak{Ch}(\mathcal{D}_X)$. By the definition of $\mathfrak{Ch}(\mathcal{D}_X)$, every such $p|_V$ is diffeologically smooth with respect to $\mathcal{D}_X$. By Item (1) of Lemma A.1, Every such $p|_V$ is a plot in $\mathcal{D}_X$. Because U can be covered by convex open subsets V, and by the locality property of $\mathcal{D}_X$, p is a plot in $\mathcal{D}_X$, as required.

We will now prove that $\mathcal{D}_X \subset \mathfrak{Di}(\mathfrak{Ch}(\mathcal{D}_X))$. Take a plot $p\colon U \to X$ in $\mathcal{D}_X$. Because $\mathcal{D}_X$ satisfies the smooth compatibility axiom in the definition of a diffeology, for every convex open subset $V \subset U$, the restriction $p|_V\colon V \to X$ is a plot in $\mathcal{D}_X$. By Item (1) of Lemma A.1, every such $p|_V$ is diffeologically smooth with respect to $\mathcal{D}_X$. Because $\mathfrak{Ch}$ is a functor, every such $p|_V$ is Chen smooth with respect to $\mathfrak{Ch}(\mathcal{D}_X)$. Because $\mathfrak{Di}$ is a functor, every such $p|_V$ is diffeologically smooth with respect to $\mathfrak{Di}(\mathfrak{Ch}(\mathcal{D}_X))$. By Item (1) of Lemma A.1, every such $p|_V$ is a plot in $\mathfrak{Di}(\mathfrak{Ch}(\mathcal{D}_X))$. Because U can be covered by convex open subsets V, and because $\mathfrak{Di}(\mathfrak{Ch}(\mathcal{D}_X))$ satisfies the locality axiom in the definition of a diffeology, p is a plot in $\mathfrak{Di}(\mathfrak{Ch}(\mathcal{D}_X))$, as required. □

Proof of Item (2) of Proposition A.3. Fix a Chen structure $\mathcal{C}_X$ and a map $p\colon C \to X$ from a convex set C to X.

First, suppose that p is a Chen plot in $\mathfrak{Ch}(\mathfrak{Di}(\mathcal{C}_X))$, and fix a convex open subset V of a cartesian space and a smooth map $f\colon V \to C$. By the definition of $\mathfrak{Ch}(\mathfrak{Di}(\mathcal{C}_X))$, p is diffeologically smooth with respect to $\mathfrak{Di}(\mathcal{C}_X)$. Because the composition of diffeologically smooth maps is diffeologically smooth, the composition $p \circ f\colon V \to X$ is diffeologically smooth with respect to $\mathfrak{Di}(\mathcal{C}_X)$. By Item (1) of Lemma A.1, $p \circ f$ is a plot in $\mathfrak{Di}(\mathcal{C}_X)$. By the definition of $\mathfrak{Di}(\mathcal{C}_X)$, $p \circ f$ is Chen smooth with respect to $\mathcal{C}_X$. By Item (2) of Lemma A.1, $p \circ f$ is a Chen plot in $\mathcal{C}_X$.

Next, suppose that for every convex open subset V of a cartesian space and every smooth map $f\colon V \to C$, the composition $p \circ f\colon V \to X$ is a Chen plot in $\mathcal{C}_X$. By Item (2) of Lemma A.1, every such $p \circ f$ is Chen smooth with respect to $\mathcal{C}_X$. By the definition of $\mathfrak{Di}(\mathcal{C}_X)$, every such $p \circ f$ is a plot in $\mathfrak{Di}(\mathcal{C}_X)$. By Lemma A.2, p is diffeologically smooth with respect to $\mathfrak{Di}(\mathcal{C}_X)$. By the definition of $\mathfrak{Ch}(\mathfrak{Di}(\mathcal{C}_X))$, p is in $\mathfrak{Ch}(\mathfrak{Di}(\mathcal{C}_X))$. □

Proof of Item (3) of Proposition A.3. Fix a Chen space $(X, \mathcal{C}_X)$. Let $p\colon C \to X$ be a Chen plot in $\mathcal{C}_X$. Because $\mathcal{C}_X$ satisfies the smooth compatibility axiom in the definition of a Chen structure, p has the property listed in Item (2). By Item (2), p is in $\mathfrak{Ch}(\mathfrak{Di}(\mathcal{C}_X))$. □

Acknowledgements. This paper is motivated by correspondence between the authors and Patrick Iglesias-Zemmour many years ago on the question of whether the category of Chen spaces is equivalent to the category of diffeological spaces.

Yael Karshon acknowledges the support of the Natural Sciences and Engineering Research Council of Canada (NSERC).

References

[1] J. C. Baez and A. E. Hoffnung, *Convenient categories of smooth spaces*, Trans. Amer. Math. Soc. **363** (2011), no. 11, 5789–5825, DOI 10.1090/S0002-9947-2011-05107-X. MR2817410

[2] Augustin Batubenge, Yael Karshon, and Jordan Watts, *Diffeological, Frölicher, and differential spaces*, Rocky Mountain J. Math. (to appear).

[3] E. Bierstone, *Differentiable functions*, Bol. Soc. Brasil. Mat. **11** (1980), no. 2, 139–189, DOI 10.1007/BF02584636. MR671464

[4] J. Boman, *Differentiability of a function and of its compositions with functions of one variable*, Math. Scand. **20** (1967), 249–268, DOI 10.7146/math.scand.a-10835. MR237728

[5] K.-t. Chen, *Iterated integrals of differential forms and loop space homology*, Ann. of Math. (2) **97** (1973), 217–246, DOI 10.2307/1970846. MR380859

[6] K. T. Chen, *Iterated integrals, fundamental groups and covering spaces*, Trans. Amer. Math. Soc. **206** (1975), 83–98, DOI 10.2307/1997148. MR0377960

[7] K. T. Chen, *Iterated path integrals*, Bull. Amer. Math. Soc. **83** (1977), no. 5, 831–879, DOI 10.1090/S0002-9904-1977-14320-6. MR454968

[8] K. T. Chen, *On differentiable spaces*, Categories in continuum physics (Buffalo, N.Y., 1982), Lecture Notes in Math., vol. 1174, Springer, Berlin, 1986, pp. 38–42, DOI 10.1007/BFb0076932. MR842915

[9] J. D. Christensen, G. Sinnamon, and E. Wu, *The D-topology for diffeological spaces*, Pacific J. Math. **272** (2014), no. 1, 87–110, DOI 10.2140/pjm.2014.272.87. MR3270173

[10] A. Frölicher, *Smooth structures*, Category theory (Gummersbach, 1981), Lecture Notes in Math., vol. 962, Springer, Berlin-New York, 1982, pp. 69–81. MR682945

[11] A. Frölicher and A. Kriegl, *Linear spaces and differentiation theory*, Pure and Applied Mathematics (New York), John Wiley & Sons, Ltd., Chichester, 1988. A Wiley-Interscience Publication. MR961256

[12] P. Iglesias-Zemmour, *Diffeology*, Mathematical Surveys and Monographs, vol. 185, American Mathematical Society, Providence, RI, 2013, DOI 10.1090/surv/185. MR3025051

[13] A. Kriegl, *Remarks on germs in infinite dimensions*, Acta Math. Univ. Comenian. (N.S.) **66** (1997), no. 1, 117–134. MR1474553

[14] J. Śniatycki, *Differential geometry of singular spaces and reduction of symmetry*, New Mathematical Monographs, vol. 23, Cambridge University Press, Cambridge, 2013, DOI 10.1017/CBO9781139136990. MR3112888

[15] A. Stacey, *Comparative smootheology*, Theory Appl. Categ. **25** (2011), No. 4, 64–117. MR2805746

[16] B. Virgin, *Watts spaces and smooth maps*, PUMP J. Undergrad. Res. **4** (2021), 63–85. MR4214544

[17] J. Watts, *Diffeologies, Differential Spaces, and Symplectic Geometry*, ProQuest LLC, Ann Arbor, MI, 2012. Thesis (Ph.D.)–University of Toronto (Canada). MR3153238

School of Mathematical Sciences, Tel-Aviv University, Tel-Aviv, Israel; and: University of Toronto, Toronto, Ontario, Canada

Email address: yaelkarshon@tauex.tau.ac.il, karshon@math.toronto.edu

Department of Mathematics, Central Michigan University, Mount Pleasant, Michigan, USA 48859

Email address: jordan.watts@cmich.edu

Contemporary Mathematics
Volume **794**, 2024
https://doi.org/10.1090/conm/794/15930

A survey on diffeological vector spaces and applications

Enxin Wu

ABSTRACT. Diffeological vector spaces arise naturally in the study of tangent bundles in geometry and operators in functional analysis. We give a concise survey on the current state of the art for the development of the theory of diffeological vector spaces, the relationship with topological vector spaces and fibre bundle theory, and some applications. We also present several new results, and propose many open questions.

1. Introduction

Diffeology provides a uniform framework for smooth manifolds, singular spaces and infinite-dimensional spaces. Diffeological vector spaces appear naturally for tangent bundles in geometry and operators in functional analysis, etc. In this short survey, we start from a concrete example of a non-splitting short exact sequence of diffeological vector spaces from analysis, develop the basic smooth homological algebra, discuss the relationship with topological vector spaces and further examples of projective diffeological vector spaces by pushforward of vector bundles in classical differential geometry, compare various properties of diffeological vector spaces, and eventually reach some applications. This is essentially a summary of my recent work related to diffeological vector spaces with several coauthors, together with some new results and many open questions for future work.

We use the following conventions throughout the paper. Every smooth manifold is assumed to be finite-dimensional, Hausdorff, second-countable and without boundary. Every vector space is over the field $\mathbb{R}$ of real numbers, and every linear map is $\mathbb{R}$-linear. By dimension, we always mean the vector space dimension over $\mathbb{R}$.

2. Background

The theory of diffeological spaces was first introduced by Souriau in the 1980's ([**S80**, **S84**]). [**I13**] is currently the standard textbook on the subject.

DEFINITION 2.1. A *diffeological space* is a set X together with a collection of maps $U \to X$ (called *plots*) from all open subsets $U \subseteq \mathbb{R}^n$ for all $n \in \mathbb{N}$, such that

(1) every constant map is a plot;

2020 *Mathematics Subject Classification.* Primary 18F99; Secondary 53A45.
Key words and phrases. Diffeological vector space, fibre bundle, topological vector space.

(2) The composite $V \to U \to X$ is a plot if the first map is smooth between Euclidean open domains and the second one is a plot;
(3) $U \to X$ is a plot if there is an open cover $\{U_i\}$ of U such that each restriction $U_i \to X$ is a plot.

A *smooth map* $X \to Y$ between diffeological spaces is a map which sends plots of X to plots of Y. Diffeological spaces with smooth maps form a category denoted Diff.

Here are the basic properties of the category Diff:

THEOREM 2.2.
(1) *The category of smooth manifolds and smooth maps is a full subcategory of* Diff.
(2) [**IKZ**] *Orbifolds in the sense of Satake are diffeological spaces. This approach solves the problem of how to define smooth maps between orbifolds.*[1]
(3) *Fix a set* X. *The collection of all diffeologies on* X *forms a complete lattice under inclusion. The smallest one is called the* discrete *diffeology, and the largest one is called the* indiscrete *diffeology.*
(4) Diff *is bicomplete and cartesian closed.*

In other words, every smooth manifold can be viewed as a diffeological space in a natural way, and a smooth map between smooth manifolds in the usual sense is the same as a smooth map in the diffeological sense. From now on, we always equip a smooth manifold with this standard diffeology. Moreover, one has subspaces, quotient spaces, products, pullbacks, function spaces, etc. For more detailed constructions, see [**I13**] or [**CSW**, Section 2]. From now on, we always equip such constructions with the corresponding diffeologies whenever no confusion will occur.

DEFINITION 2.3. Let $f : X \to Y$ be a smooth map between diffeological spaces.
(1) f is called an *induction* if it is equivalent to an inclusion of a subspace;
(2) f is called a *subduction* if it is equivalent to a quotient map.

The concept of D-topology connects diffeological spaces to topological spaces, first introduced in [**I85**]:

DEFINITION 2.4. The *D-topology* on a diffeological space X is the finest topology on the underlying set X such that every plot (with domain equipped with the Euclidean topology) is continuous. We use $D(X)$ to denote the topological space of X with the D-topology.

The merit is, every smooth map between diffeological spaces is automatically continuous when the domain and the codomain are equipped with the D-topology.

Now we introduce our main character; see [**W15**, Section 3] for all the details of the subsequent results.

DEFINITION 2.5. A *diffeological vector space* is both a diffeological space and a vector space, such that scalar multiplication and addition are both smooth.

The category of diffeological vector spaces with smooth linear maps is denoted DVect. A *linear induction* (resp. *linear subduction*) is a linear map between diffeological vector spaces which is also an induction (resp. a subduction).

[1] Alternatively, Moerdijk and Pronk resolved this same issue using Lie groupoids earlier in [**MP**].

PROPOSITION 2.6. DVect *is a bicomplete additive category, but not an abelian category.*

The following technique is extremely useful, which was also discussed in [**V**]:

PROPOSITION 2.7. *Given a vector space V and a collection $\mathcal{I}$ of maps $\{U_i \to V\}$ from Euclidean open domains U_i, there is a smallest vector space diffeology $\mathcal{D}$ on V containing $\mathcal{I}$. We call $\mathcal{D}$ the* dvsification *of $\mathcal{I}$.*

One can extend this as follows: Given a collection $\mathcal{J}$ of maps $\{X_j \to V\}$ from diffeological spaces X_j to a fixed vector space V, we obtain a collection $\mathcal{J}'$ of maps $\{U_{jk} \to X_j \to V\}$ with $U_{jk} \to X_j$ going through all plots of X_j for all j. The dvsification of $\mathcal{J}'$ is simply called the dvsification of $\mathcal{J}$.

Fix a vector space V. The collection of all vector space diffeologies on V forms a complete lattice under inclusion. The smallest one is the dvsification of the discrete diffeology (called the *fine* diffeology), and the largest one is the indiscrete diffeology.

DEFINITION 2.8. Let V_1, V_2 and V_3 be diffeological vector spaces. Let $i : V_1 \to V_2$ and $p : V_2 \to V_3$ be smooth linear maps. We say that

$$0 \longrightarrow V_1 \xrightarrow{i} V_2 \xrightarrow{p} V_3 \longrightarrow 0$$

is a *short exact sequence* in DVect, if

(1) it is a short exact sequence in the category of vector spaces and linear maps;
(2) i is an induction and p is a subduction.

As expected, we have the following result:

PROPOSITION 2.9. *Let*

$$0 \longrightarrow V_1 \xrightarrow{i} V_2 \xrightarrow{p} V_3 \longrightarrow 0$$

be a short exact sequence in DVect. *The following are equivalent:*

(1) *There exists a smooth linear map $r : V_2 \to V_1$ such that $r \circ i = 1_{V_1}$.*
(2) *There exists a smooth linear map $q : V_3 \to V_2$ such that $p \circ q = 1_{V_3}$.*
(3) *The short exact sequence is isomorphic to*

$$0 \longrightarrow V_1 \xrightarrow{i_1} V_1 \times V_3 \xrightarrow{p_2} V_3 \longrightarrow 0$$

in DVect, *with identity maps on V_1 and V_3. In particular, V_2 is isomorphic to $V_1 \times V_3$ in* DVect.

In this case, we say that the short exact sequence

$$0 \longrightarrow V_1 \xrightarrow{i} V_2 \xrightarrow{p} V_3 \longrightarrow 0$$

in DVect *splits smoothly*, and we call V_1 (and V_3) a *smooth direct summand* of V_2.

Now we present some new results:

DEFINITION 2.10.

(1) A map $f : V \to W$ between diffeological vector spaces is called *suitable* if it is smooth linear, and the induced map $V/\ker(f) \to W$ is an induction.

(2) We say that a sequence of diffeological vector spaces

$$V_1 \xrightarrow{f} V_2 \xrightarrow{g} V_3$$

is *smoothly exact at* V_2 if both f and g are suitable, and it is an exact sequence for the underlying vector spaces. This generalizes Definition 2.8. In this way, one can define a long(er) exact sequence to be smoothly exact at every middle spot.

Note that composite of two suitable maps may not be suitable.

We have the following Five Lemma:

LEMMA 2.11. *Given a commutative diagram in* DVect

$$\begin{array}{ccccccccc} V_1 & \longrightarrow & V_2 & \xrightarrow{f_2} & V_3 & \xrightarrow{f_3} & V_4 & \xrightarrow{f_4} & V_5 \\ \downarrow{\scriptstyle h_1} & & \downarrow{\scriptstyle h_2} & & \downarrow{\scriptstyle h_3} & & \downarrow{\scriptstyle h_4} & & \downarrow{\scriptstyle h_5} \\ W_1 & \longrightarrow & W_2 & \xrightarrow[g_2]{} & W_3 & \xrightarrow[g_3]{} & W_4 & \longrightarrow & W_5 \end{array}$$

with both rows smoothly exact.

(1) *If* h_2 *and* h_4 *are linear subductions, and* h_5 *is an injection, then* h_3 *is also a linear subduction.*

(2) *If* h_1, h_2, h_4, h_5 *are isomorphisms in* DVect, *then so is* h_3.

Note that, comparing with the classical Five Lemma, we miss a corresponding result which concludes h_3 to be an induction. The proof below uses Proposition 3.2(2) to avoid locality for simplicity.

PROOF. (2) follows from (1). To prove (1), the classical Five Lemma already shows that h_3 is surjective. Given a plot $p : U \to W_3$, $g_3 \circ p$ lifts to a plot $q : U \to V_4$ since h_4 is a linear subduction. Then $f_4 \circ q = 0$ since h_5 is injective. By smooth exactness at V_4, we get a plot $r : U \to V_3$ with $q = f_3 \circ r$. One checks that $g_3 \circ (p - h_3 \circ r) = 0$. By smooth exactness at W_3, we get a plot $s : U \to W_2$ with $p - h_3 \circ r = g_2 \circ s$. Now since h_2 is a linear subduction, we get a plot $t : U \to V_2$ lifting s. Therefore, $p = h_3 \circ (r + f_2 \circ t)$, which implies that h_3 is a linear subduction. □

3. A motivating example

Consider the map $d : C^\infty(\mathbb{R}, \mathbb{R}) \to \prod_{\mathbb{N}} \mathbb{R}$ given by

$$f \mapsto (f(0), f'(0), f''(0), f'''(0), \cdots).$$

Borel's lemma claims that this linear map is surjective. On the first glance, one might think that this result is trivial, since one can write down a formula for a function from its prescribed Taylor's expansion at a point. However, this infinite sum might not converge to a well-defined function. Borel's trick is to encode a suitable compactly supported function inside each term of the infinite sum, so that the sum converges uniformly to a smooth function by the Weierstrass M-test.

Note that $C^\infty(\mathbb{R}, \mathbb{R})$ with the functional diffeology and $\prod_{\mathbb{N}} \mathbb{R}$ with the product diffeology are diffeological spaces. Indeed, they are diffeological vector spaces under their usual linear structures. It is straightforward to verify that d is a smooth linear map. As in [**W15**, Claim 1 of Example 4.3], one can extend Borel's trick a bit to show that d is in fact a subduction. In other words, $\prod_{\mathbb{N}} \mathbb{R}$ is the quotient space of $C^\infty(\mathbb{R}, \mathbb{R})$, although their diffeologies look quite different.

Write K for the kernel of d. It consists of all smooth functions $f : \mathbb{R} \to \mathbb{R}$ which are flat at $0 \in \mathbb{R}$, i.e., $f^{(n)}(0) = 0$ for all $n \in \mathbb{N}$.

THEOREM 3.1. *We have a short exact sequence of diffeological vector spaces*

$$0 \longrightarrow K \longrightarrow C^\infty(\mathbb{R}, \mathbb{R}) \xrightarrow{d} \prod_{\mathbb{N}} \mathbb{R} \longrightarrow 0 \tag{3.1}$$

which does not split smoothly.

Compare the statements in the following proposition:

PROPOSITION 3.2. *Let $f : X \to Y$ be a smooth map between diffeological spaces.*

(1) *If f is a subduction, then for every plot $p : U \to Y$, there is an open cover $\{U_i\}$ of U such that each $p|_{U_i}$ lifts to a plot of X.*
(2) *If f is a linear subduction between diffeological vector spaces, then every plot $p : U \to Y$ has a global lift to a plot of X.*

The second statement is an application of the smooth paracompactness of U to the first statement. In the current case, we obtain the following generalization of Borel's lemma:

COROLLARY 3.3. *For any Euclidean open domain U and any countable collection of smooth functions $\{f_i : U \to \mathbb{R}\}_{i \in \mathbb{N}}$, there exists a smooth function $F : U \times \mathbb{R} \to \mathbb{R}$ such that $\frac{\partial^k F}{\partial y^k}(x, 0) = f_k(x)$ for all $k \in \mathbb{N}$ and all $x \in U$.*

The non-splitting of the short exact sequence (3.1) in DVect was proved in [**W15**] as follows. One first shows that the D-topology on $\prod_{\mathbb{N}} \mathbb{R}$ is the product topology, and the D-topology on $C^\infty(\mathbb{R}, \mathbb{R})$ contains the compact-open topology. Using these results, one can then show that for any smooth linear map $q : \prod_{\mathbb{N}} \mathbb{R} \to C^\infty(\mathbb{R}, \mathbb{R})$, $d \circ q$ cannot be injective.

Open Question 1. . Can you find an alternative proof of the non-splitting part without reference to the D-topology?

4. Homological algebra

Since not every short exact sequence in DVect splits smoothly, we shall use homological algebra to classify them. There are several ways to build up homological algebra: relative homological algebra, model structures, etc. We explain the relative homological algebra approach here. The interested reader might consult [**W15**, Section 8] for a cofibrantly generated model structure on the category of diffeological chain complexes and diffeological chain maps.

It is known from [**BH**] that the category of diffeological spaces is equivalent to the category of concrete sheaves over the concrete site consisting of Euclidean open domains, smooth maps and the usual open covers. Experience from sheaf theory tells us that we shall look for relative injective objects in DVect.

DEFINITION 4.1. A diffeological vector space V is called *injective* if for every linear induction $i : V_1 \to V_2$ and every smooth linear map $f : V_1 \to V$, there exists a smooth linear map $g : V_2 \to V$ making the following triangle commutative:

$$\begin{array}{ccc} V_1 & \xrightarrow{i} & V_2 \\ {\scriptstyle f}\big\downarrow & \swarrow{\scriptstyle g} & \\ V & & \end{array}$$

One can easily check that every vector space with the indiscrete diffeology is injective. Also see Proposition 7.4 for a new result related to injective diffeological vector spaces.

Open Questions 2. :

(1) Can you find any other injective diffeological vector space?
(2) Do there exist enough injective diffeological vector spaces? In other words, for any fixed diffeological vector space V, does there always exist an injective diffeological vector space V' together with a linear induction $V \to V'$?

Surprisingly, there are solutions to the dual problem.

DEFINITION 4.2. A diffeological vector space V is called *projective* if for every linear subduction $p : V_2 \to V_3$ and every smooth linear map $f : V \to V_3$, there exists a smooth linear map $g : V \to V_2$ making the following triangle commutative:

$$\begin{array}{ccc} & & V \\ & {}^{g}\swarrow & \downarrow{\scriptstyle f} \\ V_2 & \xrightarrow[p]{} & V_3 \end{array}$$

THEOREM 4.3. *Write $\mathcal{S}$ for the collection of all linear subductions, and write $\mathcal{P}$ for the collection of all projective diffeological vector spaces. Then $(\mathcal{S}, \mathcal{P})$ forms a relative homological algebra for* DVect, *in the sense that*

(1) *$V \in \mathcal{P}$ iff for each $f : V_2 \to V_3 \in \mathcal{S}$, $f_* : \mathrm{DVect}(V, V_2) \to \mathrm{DVect}(V, V_3)$ defined by $g \mapsto f \circ g$ is surjective;*
(2) *$f : V_2 \to V_3 \in \mathcal{S}$ iff for every $V \in \mathcal{P}$, $f_* : \mathrm{DVect}(V, V_2) \to \mathrm{DVect}(V, V_3)$ is surjective;*
(3) *for any diffeological vector space V, there exists $V' \in \mathcal{P}$ together with a smooth linear map $V' \to V \in \mathcal{S}$.*

The proof goes as follows, which is tightly connected to the definition of a diffeological space:

Notice that (1) is the definition of a projective diffeological vector space.

Step 1: Given any diffeological space X, define the *free diffeological vector space* $F(X)$ generated by X in the following way: As a vector space, it is $\bigoplus_{x \in X} \mathbb{R}$. So we have a canonical map $i_X : X \to F(X)$ by $x \mapsto [x]$, where $[x]$ denotes the element in $F(X)$ which is 1 in the copy of $\mathbb{R}$ corresponding to $x \in X$ and 0 elsewhere. Equip $F(X)$ with the dvsification of $\{i_X\}$.

These free diffeological vector spaces have the following universal property: Given any smooth map $f : X \to V$ to a diffeological vector space V, there exists a unique smooth linear map $g : F(X) \to V$ making the following triangle commutative:

$$\begin{array}{ccc} X & \xrightarrow{i_X} & F(X) \\ {\scriptstyle f}\downarrow & \swarrow{\scriptstyle g} & \\ V & & \end{array}$$

Step 2: When U is a Euclidean open domain, one can show that the free diffeological vector space $F(U)$ generated by U is projective, again by the smooth paracompactness of U.[2] (2) is then implied by the universal property of $F(U)$.

[2]This also proves Proposition 3.2(2).

Step 3: For any diffeological vector space V, consider $V' := \bigoplus_p F(U_p)$ where the direct sum is over all plots $p : U_p \to V$. The universal property of free diffeological vector space and the universal property of direct sum imply a smooth linear map $V' \to V$, which is a subduction by construction. This proves (3).

Moreover, we have the following structural theorem for projective diffeological vector spaces:

THEOREM 4.4. *Every projective diffeological vector space is a smooth direct summand of a direct sum of $F(U)$'s with each U a Euclidean open domain.*

Using projective diffeological vector spaces, the Ext groups can be defined, and Ext^1 classifies all isomorphism classes of short exact sequences in DVect as expected.

For more details for this section, one should consult [**W15**, Sections 6 and 7].

5. Relationship with topological vector spaces

The D-topology transforms every diffeological space into a topological space. Indeed it gives rise to a left adjoint $D : \mathrm{Diff} \to \mathrm{Top}$, and hence it preserves all colimits. However, it does not preserve limits (in particular, product) in general. Moreover, the D-topology does not necessarily make a diffeological vector space into a topological vector space.

To state the known result, we adopt the following terminology of cardinals from set theory: a denotes the infinite countable cardinal and c denotes the continuum cardinal. Given a cardinal w, we say that a diffeological vector space V is *at most w-generated* if there is a linear subduction from a direct sum indexed over w of $F(U)$'s (as in the structural theorem of projective diffeological vector spaces) to V. Similarly one can define *at least w-generatedness* for a diffeological vector space.

THEOREM 5.1.

(1) *A projective diffeological vector space is a topological vector space under the D-topology, if it is at most a-generated.*
(2) *A projective diffeological vector space is not a topological vector space under the D-topology, if it is at least c-generated.*
(3) *Assuming the continuum hypothesis, a projective diffeological vector space is a topological vector space under the D-topology iff it is at most a-generated.*
(4) *For a diffeological space X such that $D(X)$ is discrete, the free diffeological vector space $F(X)$ is a topological vector space under the D-topology.*

The key for the proof of (1) and (2) are from two sides.

The positive side is the use of k_ω-space theory from general topology. Recall that a k-space is a Hausdorff space such that the topology of the space is determined by all compact subsets, and a k_ω-space is a Hausdorff space such that the topology is determined by a countable covering family of compact subsets. The essential difference is, the product of two k_ω-spaces with the product topology is again a k_ω-space, while the product of two k-spaces with the product topology may not be a k-space in general.

The negative side is an example of Dowker ([**D**]) showing the non-commutativity of product topology and weak topology, adapted to the diffeological case. In this example, the index set with cardinality of the continuum is indispensable.

As an application of Theorem 5.1, we have:

PROPOSITION 5.2. *Neither $\prod_{\mathbb{N}} \mathbb{R}$ nor $C^\infty(\mathbb{R}, \mathbb{R})$ is projective, assuming continuum hypothesis.*

In fact, Theorem 3.1 already showed that $\prod_{\mathbb{N}} \mathbb{R}$ is not projective, without assuming continuum hypothesis. Now we give an alternative proof using Theorem 5.1. One first shows that the D-topology of $\prod_{\mathbb{N}} \mathbb{R}$ is the product topology, and the D-topology of $C^\infty(\mathbb{R}, \mathbb{R})$ is the Whitney weak topology. Therefore, they are topological vector spaces under the D-topology. On the other hand, one can show that $\prod_{\mathbb{N}} \mathbb{R}$ is not a-generated by using Sard's theorem in differential geometry. As $d : C^\infty(\mathbb{R}, \mathbb{R}) \to \prod_{\mathbb{N}} \mathbb{R}$ in Borel's lemma is a linear subduction, $C^\infty(\mathbb{R}, \mathbb{R})$ is not a-generated as well. So the result follows from Theorem 5.1(3).

See [**WY**] for all details of the results in this section.

Open Question 3. : Can you find a proof of the non-projectivity of $C^\infty(\mathbb{R}, \mathbb{R})$ without assuming continuum hypothesis?

Open Questions 4. : Are the following statements true for a diffeological space X?

(1) If $D(X \times X) = D(X) \times D(X)$, then the free diffeological vector space $F(X)$ generated by X is a topological vector space under the D-topology.
(2) If $D(X \times X) \neq D(X) \times D(X)$, then $F(X)$ is not a topological vector space under the D-topology.

Open Question 5. : Is it true that for every projective diffeological vector space V, $D(V \times V)$ and $D(V) \times D(V)$ have the same compact subsets, or in other words, $D(V \times V)$ is the Kellification of $D(V) \times D(V)$?

Open Question 6. : Is it true that if a single point of a diffeological vector space V is D-closed, then $D(V)$ is Hausdorff?

6. Relationship with fibre bundle theory

We already know from Theorem 4.3 that there are enough projective diffeological vector spaces in DVect. Besides the free diffeological vector space $F(M)$ generated by a smooth manifold M, can we get more concrete examples?

One way is to relate this to the homotopy theory of [**CW14**]. More precisely, let A and B be diffeological spaces. If either B is cofibrant, or $F(A)$ is a projective diffeological vector space and there is a cofibration $A \to B$ in the sense of [**CW14**], then $F(B)$ is also a projective diffeological vector space.

A typical example is the following:

EXAMPLE 6.1. Let I be an arbitrary index set, and write X_I for the gluing of I-copies of $\mathbb{R}$ at the origins. Then $F(X_I)$ is a projective diffeological vector space.

Since we do not know of too many cofibrant objects or fibrations in Diff, (e.g., we do not even know whether every smooth manifold is cofibrant), we cannot get too many extra examples in this way.

Here is an alternative approach. We introduce some background first.

On every diffeological space, we can still define tangent bundle; see [**CW16**, **CW17**]. Such a bundle is no longer locally trivial (with respect to the D-topology). So we introduce the following concept to encode tangent bundles:

DEFINITION 6.2. A *diffeological vector pseudo-bundle* $\pi : E \to B$ is a smooth surjective map between diffeological spaces, such that each fibre $E_b := \pi^{-1}(b)$ for $b \in B$ is a vector space. Moreover, the following conditions hold:

(1) Fibrewise addition $E \times_B E \to E$ is smooth.
(2) Fibrewise scalar multiplication $\mathbb{R} \times E \to E$ is smooth.
(3) The zero section $\sigma : B \to E$ is smooth.

From this, we get two categories: The category DVPB of all diffeological vector pseudo-bundles and bundle maps, and the category DVPB_B of all diffeological vector pseudo-bundles over the fixed base space B and bundle maps over B.

THEOREM 6.3. *The category* DVPB *is bicomplete, and the category* DVPB_B *is bicomplete and additive.*

In particular, we have fibrewise operations on DVPB_B, such as product, direct sum, free bundle, tensor product, exterior power, Hom and dual, etc. Note that when $B = \mathbb{R}^0$, we get back DVect. Moreover, a smooth map $f : B \to B'$ gives rise to the pullback functor $f^* : \mathrm{DVPB}_{B'} \to \mathrm{DVPB}_B$. It has both a left adjoint and a right adjoint. The left adjoint f_* is called the *pushforward.*

Since DVPB_B is an additive category, we can again consider relative homological algebra. As in the case of DVect, we have a similar projective class for every DVPB_B. Here is a key observation:

PROPOSITION 6.4.

(1) *Pushforward preserves projectives.*
(2) *Vector bundles $E \to B$ in classical differential geometry are projectives in* DVPB_B.

Therefore, given any vector bundle $p : E \to B$ in classical differential geometry, consider the pushforward induced by the map $f : B \to \mathbb{R}^0$, we get a projective diffeological vector space $f_*(p)$, simply called the *pushforward projective diffeological vector space* by p.

For more details on operations in DVPB and DVPB_B, see [**CW23**]; for more details on pushforward, see [**W23**].

Open Question 7. : Is it true that for non-isomorphic vector bundles of the same rank[3] in classical differential geometry, the corresponding pushforward projective diffeological vector spaces are not isomorphic in DVect?

A related but more fundamental question can be formulated as follows:

We know from [**W23**] that the canonical map $i_X : X \to F(X)$ may not be an induction. We say that a diffeological space $\tilde{X}$ is a *D-basis* of the free diffeological vector space $F(X)$, if the underlying set of $\tilde{X}$ is a subset of $F(X)$, $F(X) = F(\tilde{X})$ and $i_{\tilde{X}} : \tilde{X} \to F(\tilde{X})$ is an induction.

A D-basis always exists, since for any free diffeological vector space $F(X)$, let $\tilde{X}$ be the underlying set X equipped with the subset diffeology of $F(X)$ from the

[3]This condition comes from Example 8.2, since the pushforwards of the projections $S^1 \times \mathbb{R} \to S^1$ and $S^1 \times \mathbb{R}^3 \to S^1$ along $S^1 \to \mathbb{R}^0$ are isomorphic as diffeological vector spaces.

canonical map i_X. The universal property of free diffeological vector space implies that $F(X) = F(\tilde{X})$, and hence $i_{\tilde{X}}$ is an induction.

Open Question 8. : Are any two D-bases of a free diffeological vector space diffeomorphic?

7. Hierachies of diffeological vector spaces

Up to now, we have the following classes of diffeological vector spaces:

Fine diffeological vector spaces are vector spaces with the smallest vector space diffeology. The concept of a fine diffeological vector space was first introduced in [**I07**]. Here are some further results:

PROPOSITION 7.1.

(1) *The fineness of a diffeological vector space can be tested by curves.*
(2) *A diffeological vector space V is fine iff all linear maps $V \to \mathbb{R}$ are smooth.*

Projective diffeological vector spaces are those with the lifting property in DVect for all linear subductions.

We distinguish the following classes of diffeological vector spaces:

DEFINITION 7.2.

(1) Write $\mathcal{FV}$ for the class of all fine diffeological vector spaces.
(2) Write $\mathcal{PV}$ for the class of all projective diffeological vector spaces.
(3) Write $\mathcal{SV}$ for the class of all diffeological vector spaces such that the smooth linear functionals separate points.
(4) Write $\mathcal{SD}$ for the class of all diffeological vector spaces such that the smooth functionals separate points.
(5) Write $\mathcal{FFV}$ for the class of all diffeological vector spaces such that all finite-dimensional linear subspaces have the fine diffeology.
(6) Write $\mathcal{HT}$ for the class of all diffeological vector spaces such that the D-topology is Hausdorff.

THEOREM 7.3. *We have the following chain of containments for these classes of diffeological vector spaces:*

$$\mathcal{FV} \subset \mathcal{PV} \subset \mathcal{SV} \subseteq \mathcal{SD} \subseteq \mathcal{FFV} \cap \mathcal{HT}$$

When restricted to finite-dimensional diffeological vector spaces, we have

$$\mathcal{FV} = \mathcal{PV} = \mathcal{SV} = \mathcal{SD} = \mathcal{FFV}$$

The first statement follows from the second one. The conclusions in the second statement are more or less straightforward except for $\mathcal{SD} = \mathcal{FFV}$. The interested reader shall consult [**CW19**, Section 5] for the proof of this harder part.

Open Question 9. : Do the proper inclusions $\mathcal{SV} \subset \mathcal{SD}$ and $\mathcal{SD} \subset \mathcal{FFV} \cap \mathcal{HT}$ hold?

It is known that there is no containment relation between $\mathcal{FFV}$ and $\mathcal{HT}$. The example for $\mathcal{FFV} \not\subseteq \mathcal{HT}$ in [**CW19**, Proposition 3.23] has a new conclusion in this paper. Recall that this example $V \in \mathcal{FFV} \setminus \mathcal{HT}$ is constructed as follows. It is a vector space with basis $\mathbb{R}$. Fix a bijection $f : \mathbb{R} \to \mathbb{R}$ such that $f^{-1}(U)$ is dense in $\mathbb{R}$ for every open neighbourhood U of $0 \in \mathbb{R}$. Define two maps $p : \mathbb{R} \to V$ with

$x \mapsto [f(x)]$ and $\bar{p} : \mathbb{R} \to V$ with $x \mapsto [x]$,[4] and equip V with the dvsification of $\{p, \bar{p}\}$.

Here is a new result:

PROPOSITION 7.4. *$\mathbb{R}$ is not injective in* DVect. *More precisely, there is no theorem of Hahn-Banach type in $\mathcal{FFV}$.*

PROOF. Let V be the diffeological vector space as in the example $\mathcal{FFV} \not\subseteq \mathcal{HT}$ above, and let W be an arbitrary finite-dimensional linear subspace of V of dimension ≥ 1. Since $V \in \mathcal{FFV}$, W has the fine diffeology, and hence every linear map $W \to \mathbb{R}$ is smooth. However, the only smooth linear maps $V \to \mathbb{R}$ are the constant maps by the definition of the plots p and $\bar{p}$. The result hence follows. □

Compare the above proposition with the following:

PROPOSITION 7.5. *If $V \in \mathcal{SV}$, then every finite-dimensional linear subspace of V is a smooth direct summand.*

The example used in the proof of Proposition 7.4 implies that $\mathcal{SV}$ cannot be replaced by $\mathcal{FFV}$ for Proposition 7.5. Here is another new result for $\mathcal{SV}$:

PROPOSITION 7.6. *For $V \in \mathcal{SV}$ and any two finite-dimensional linear subspaces V_1 and V_2 of V, any isomorphism $V_1 \to V_2$ can be extended to an automorphism $V \to V$ in* DVect.

PROOF. Write V' for the linear subspace of V spanned by V_1 and V_2. So V' is also finite-dimensional, and hence it has the fine diffeology. Proposition 7.5 implies that $V \cong V' \oplus V''$. There is an automorphism $V' \to V'$ extending the isomorphism $V_1 \to V_2$. Together with the identity map $1_{V''}$, we get an automorphism $V \to V$ extending the original isomorphism $V_1 \to V_2$. □

We have another class $\mathcal{F}$ of diffeological vector spaces, namely the free diffeological vector spaces generated by diffeological spaces.

PROPOSITION 7.7.

(1) $\mathcal{F} \not\subseteq \mathcal{PV}$. *For example, if $D(X)$ is not Hausdorff, then $F(X)$ is not projective.*
(2) *The free diffeological vector space generated by a smooth manifold is projective.*
(3) *The free diffeological vector spaces generated by the orbifolds $\mathbb{R}^2/C_n$ and $\mathbb{R}^n/\{\pm 1\}$ of the linear actions of cyclic groups are both projective.*

Open Question 10. : Is every free diffeological vector space generated by an arbitrary orbifold projective?

Open Question 11. : Is every free diffeological vector space generated by an arbitrary subset of a Euclidean space projective?[5]

Open Question 12. : Can you find necessary and sufficient conditions for a free diffeological vector space to be projective?

For more details of the results in this section, see [**CW19**].

---

[4]See Step 1 after Theorem 4.3 for the meaning of the notation $[\cdot]$.

[5]This question is closely related to the description of the projectives in DVPB_B.

8. Some applications

Now we discuss some applications.

8.1. Construction of non-splitting short exact sequences.

Observe that if V is a diffeological vector space such that $D(V)$ is T_1 and $V = V_1 \oplus V_2$, then both V_1 and V_2 are D-closed in V. Therefore, for a linear subspace V' of V which is not D-closed in V, we get a short exact sequence

$$0 \to V' \to V \to V/V' \to 0$$

which does not split in DVect. This applies e.g. to the case when $V = F(M)$ with M a smooth manifold of dimension ≥ 1, and V' is the linear subspace of V whose basis is M minus finitely many points. Of course, the non-splitting in this case can also be verified by showing that the quotient diffeological vector space V/V' is not fine.

This approach does not apply to the example in Section 3, since it is not hard to show that K is D-closed in $C^\infty(\mathbb{R}, \mathbb{R})$. Therefore, this is a truely non-trivial example.

8.2. Testing non-filteredness.

It is well-known that diffeology provides a uniform framework for smooth manifolds, singular spaces (such as orbifolds and irrational tori) and infinite-dimensional spaces (e.g., function spaces and diffeomorphism groups of smooth manifolds). We do not define filteredness and weakly filteredness here – they are useful categorical properties for the pointed germ categories of a diffeological space; the interested reader may consult [**CW17**] for the concepts.

One can show that

(1) Smooth manifolds and irrational tori are filtered.
(2) Orbifolds and infinite-dimensional spaces with diffeological vector spaces as local models are weakly filtered.

Filtered spaces behave nicely, e.g.:

THEOREM 8.1.

(1) *For any filtered space, the Hector tangent bundle coincides with the dvs tangent bundle. Moreover, every tangent space with the subset diffeology from the dvs tangent bundle is a fine diffeological vector space.*
(2) *For a filtered space, several natural ways of defining differential forms coincide.*

Note that for a general diffeological space, its tangent bundle with Hector's diffeology may not be a diffeological vector pseudo-bundle. For all corresponding concepts and precise statements, see [**CW16, CW17, CW23**].

Here is an alternative way to see the importance of filteredness:

We see that the spirit of a fixed diffeological space is really in its plots. There are so many plots, and most of them are in fact redundant for doing calculations – e.g., in the case of a smooth manifold, an atlas of charts or its refinement suffices. There is a similar concept called a generating set of plots for a diffeological space.

Note that smooth manifolds are the gluing of Euclidean open domains via diffeomorphisms, and diffeological spaces are the gluing of Euclidean open domains via smooth maps. This then implies that infinite-dimensional spaces are the gluing of finite-dimensional pieces. However, we have to pay the price that such gluings are somehow "folded", in the sense that you cannot find a generating set of plots such that each plot is injective. We will make this clear via a concrete example of an infinite-dimensional diffeological vector space below, which can also be translated into a concrete constructive analysis problem. We solve this problem by examing the diffeology of the tangent space as a structural invariant, which bypasses the concrete construction.

We call a diffeological space *injectively generated* if it has a generating set of plots each of which is injective as a function. It is easy to see from definitions that (1) every diffeological group is weakly filtered; (2) if a diffeological space is both weakly filtered and injectively generated, then it is filtered.

We consider the diffeological vector space $\prod_{\mathbb{N}} \mathbb{R}$, and we will show that it is not injectively generated. This can be translated into the following question:

Question. : A function $U \to \prod_{\mathbb{N}} \mathbb{R}$ from a Euclidean open domain is called smooth if its composition with each projection to $\mathbb{R}$ is smooth. Write $\mathcal{I}$ for the collection of all functions $p : U \to \prod_{\mathbb{N}} \mathbb{R}$ which are both smooth and injective for all open $U \subseteq \mathbb{R}^n$ and all $n \in \mathbb{N}$. Do there exist a smooth function $q : U' \to \prod_{\mathbb{N}} \mathbb{R}$ and a point $u' \in U'$ such that for any open neighbourhood U'' of u' in U', one cannot find $p : U \to \prod_{\mathbb{N}} \mathbb{R} \in \mathcal{I}$ and a smooth map $U'' \to U$ making the following square commutative:

$$\begin{array}{ccc} U'' & \longrightarrow & U \\ \big\downarrow & & \big\downarrow{\scriptstyle p} \\ U' & \underset{q}{\longrightarrow} & \prod_{\mathbb{N}} \mathbb{R}? \end{array}$$

Solution. : The answer to this question is yes, but the proof below is non-constructive: One can show that the tangent space $T_0(\prod_{\mathbb{N}} \mathbb{R})$ is isomorphic to $\prod_{\mathbb{N}} \mathbb{R}$ as diffeological vector spaces, and the diffeology on $\prod_{\mathbb{N}} \mathbb{R}$ is not fine. The diffeological vector space $\prod_{\mathbb{N}} \mathbb{R}$ is weakly filtered, so it cannot be injectively generated.

Open Question 13. : Can you find a constructive proof to this question?

Note that the function space $C^\infty(M, N)$ is also weakly filtered but not filtered, where M is a compact smooth manifold with dimension ≥ 1 and N is a smooth manifold. One can also translate this into a concrete analysis result as above.

8.3. Splitting of projective diffeological vector spaces via classical vector bundle theory.

As a notation, given a vector bundle $\pi : X \to Y$ (or ξ) in classical differential geometry, we write V_π or V_X (or V_ξ) for the corresponding pushforward diffeological vector space along the map $Y \to \mathbb{R}^0$.

Now let $p : E \to B$ be a principal G-bundle in classical differential geometry. Then it induces a linear subduction $F(E) \to F(B)$, and we are able to identify its kernel geometrically, and hence we reach the splitting of $F(E)$ in DVect as $F(B)$ is projective. Here are some simple examples:

EXAMPLE 8.2.

(1) From the covering $S^n \to \mathbb{R}P^n$, we get

$$F(S^n) \cong F(\mathbb{R}P^n) \oplus V_{\gamma_n^1}$$

where γ_n^1 denotes the tautological line bundle γ_n^1 on $\mathbb{R}P^n$. Taking $n = 1$, we therefore get

$$F(S^1) \cong F(S^1) \oplus (V_{\gamma_1^1})^m$$

for every $m \in \mathbb{N}$.

(2) From the tangent bundle $TS^n \to S^n$, we get

$$F(S^n)^{n+1} \cong F(S^n) \oplus V_{TS^n}$$

From the tangent bundle $T\mathbb{R}P^n \to \mathbb{R}P^n$, we get

$$(V_{\gamma_n^1})^{n+1} \cong F(\mathbb{R}P^n) \oplus V_{T\mathbb{R}P^n}$$

From these and taking $n = 1$, we therefore get

$$F(S^1)^3 \cong F(S^1) \oplus (V_{\gamma_1^1})^3 \cong F(S^1)$$

In fact, the last isomorphism $F(S^1)^3 \cong F(S^1)$ can also be derived directly by considering the covering space $S^1 \to S^1$ with $z \mapsto z^3$.

For all details in this part, see [**W23**, Section 5].

Open Questions 14. :

(1) Is $F(S^1)$ isomorphic to $F(S^1)^2$ as diffeological vector spaces?
(2) Are there any physical explanation for some of these curious isomorphisms in DVect?
(3) When the tangent bundle $TS^n \to S^n$ has a direct summand of a trivial bundle on S^n of rank k, V_{TS^n} has a smooth direct summand of $F(S^n)^k$. Can you use this approach to give an alternative solution to the maximal number of linearly independent vector fields on S^n?

9. A note on other related works

Here is a skim of some works by other researchers related to diffeological vector spaces and diffeological vector pseudo-bundles.

Kriegl and Michor introduced convenient vector spaces, and further studied manifolds modelled on them; see [**KM**]. Convenient vector spaces are locally convex vector spaces with any one of the equivalent completeness conditions ([**KM**, Theorem 2.14]). They can be viewed naturally as diffeological vector spaces via the continuous linear functionals.

Iglesias-Zemmour was the first to introduce and study fine diffeological vector spaces; see [**I07**]. We have encoded this important class of diffeological vector spaces in the context.

Pervova studied finite-dimensional diffeological vector spaces in [**P18**] and [**Pp**]. She showed in [**P18**, Section 3] that every such diffeological vector space V has a smooth direct summand V^* which is the smooth dual of V and also a fine diffeological vector space.

Diffeological vector pseudo-bundles were also studied by various researchers under different names. They were called "regular vector bundles" in [**V**], and "diffeological vector spaces over the base space" in [**CW16**]. We adopt the notation from [**P16**]. For further development, see [**M**].

References

[BH] J. C. Baez and A. E. Hoffnung, *Convenient categories of smooth spaces*, Trans. Amer. Math. Soc. **363** (2011), no. 11, 5789–5825, DOI 10.1090/S0002-9947-2011-05107-X. MR2817410

[CSW] J. D. Christensen, G. Sinnamon, and E. Wu, *The D-topology for diffeological spaces*, Pacific J. Math. **272** (2014), no. 1, 87–110, DOI 10.2140/pjm.2014.272.87. MR3270173

[CW14] J. D. Christensen and E. Wu, *The homotopy theory of diffeological spaces*, New York J. Math. **20** (2014), 1269–1303. MR3312059

[CW16] J. D. Christensen and E. Wu, *Tangent spaces and tangent bundles for diffeological spaces* (English, with English and French summaries), Cah. Topol. Géom. Différ. Catég. **57** (2016), no. 1, 3–50. MR3467758

[CW17] J. D. Christensen and E. Wu, *Tangent spaces of bundles and of filtered diffeological spaces*, Proc. Amer. Math. Soc. **145** (2017), no. 5, 2255–2270, DOI 10.1090/proc/13334. MR3611335

[CW19] J. D. Christensen and E. Wu, *Diffeological vector spaces*, Pacific J. Math. **303** (2019), no. 1, 73–92, DOI 10.2140/pjm.2019.303.73. MR4044857

[CW23] J.D. Christensen and E. Wu, *Exterior bundles in diffeology*, Israel J. Math. **253** (2023), no. 2, 673–713.

[D] C. H. Dowker, *Topology of metric complexes*, Amer. J. Math. **74** (1952), 555–577, DOI 10.2307/2372262. MR48020

[I85] P. Iglesias-Zemmour, *Fibrations difféologiques et Homotopie*, Thèse de Doctorat Es-sciences, L'Université de Provence, 1985.

[I07] P. Iglesias-Zemmour, *Diffeology of the infinite Hopf fibration*, Geometry and topology of manifolds, Banach Center Publ., vol. 76, Polish Acad. Sci. Inst. Math., Warsaw, 2007, pp. 349–393, DOI 10.4064/bc76-0-17. MR2346968

[I13] P. Iglesias-Zemmour, *Diffeology*, Mathematical Surveys and Monographs, vol. 185, American Mathematical Society, Providence, RI, 2013, DOI 10.1090/surv/185. MR3025051

[IKZ] P. Iglesias, Y. Karshon, and M. Zadka, *Orbifolds as diffeologies*, Trans. Amer. Math. Soc. **362** (2010), no. 6, 2811–2831, DOI 10.1090/S0002-9947-10-05006-3. MR2592936

[KM] A. Kriegl and P. W. Michor, *The convenient setting of global analysis*, Mathematical Surveys and Monographs, vol. 53, American Mathematical Society, Providence, RI, 1997, DOI 10.1090/surv/053. MR1471480

[M] J.-P. Magnot, *On the geometry of diffeological vector pseudo-bundles and infinite dimensional vector bundles: automorphisms, connections and covariant derivatives*, Preprint, arXiv:abs/2207.06824

[MP] I. Moerdijk and D. A. Pronk, *Orbifolds, sheaves and groupoids*, K-Theory **12** (1997), no. 1, 3–21, DOI 10.1023/A:1007767628271. MR1466622

[P16] E. Pervova, *Diffeological vector pseudo-bundles*, Topology Appl. **202** (2016), 269–300, DOI 10.1016/j.topol.2016.01.014. MR3464165

[P18] E. Pervova, *On the notion of scalar product for finite-dimensional diffeological vector spaces*, Electron. J. Linear Algebra **34** (2018), 18–27, DOI 10.13001/1081-3810.3121. MR3780575

[Pp] E. Pervova, *Finite-dimensional diffeological vector spaces being and not being coproducts*, Preprint, arXiv:abs/2212.02297.

[S80] J.-M. Souriau, *Groupes différentiels* (French), Differential geometrical methods in mathematical physics (Proc. Conf., Aix-en-Provence/Salamanca, 1979), Lecture Notes in Math., vol. 836, Springer, Berlin, 1980, pp. 91–128. MR607688

[S84] J.-M. Souriau, *Groupes différentiels de physique mathématique* (French), South Rhone seminar on geometry, II (Lyon, 1983), Travaux en Cours, Hermann, Paris, 1984, pp. 73–119. MR753860

[V] M. Vincent, *Diffeological differential geometry*, Master's Thesis, University of Copenhagen, 2008, available at http://www.math.ku.dk/english/research/tfa/top/paststudents/martinvincent.msthesis.pdf.

[W15] E. Wu, *Homological algebra for diffeological vector spaces*, Homology Homotopy Appl. **17** (2015), no. 1, 339–376, DOI 10.4310/HHA.2015.v17.n1.a17. MR3350086

[W23] E. Wu, *Pushforward and smooth vector pseudo-bundles*, Pacific J. Math. **322** (2023), no. 1, 195–219, DOI 10.2140/pjm.2023.322.195. MR4582959

[WY] E. Wu and Z. Yang, *Topology on diffeological vector spaces*, Preprint, arXiv:abs/2205.09562.

Email address: exwu@stu.edu.cn

DEPARTMENT OF MATHEMATICS, SHANTOU UNIVERSITY, GUANGDONG, PEOPLE'S REPUBLIC OF CHINA

Contemporary Mathematics
Volume **794**, 2024
https://doi.org/10.1090/conm/794/15941

Finite-dimensional diffeological vector spaces being and not being coproducts

Ekaterina Pervova

Abstract. It is known that for a topological vector space it is possible to be the coproduct of two of its subspaces in the category of vector spaces while not being the coproduct of the same subspaces in the category of topological vector spaces. There are however wide classes of spaces where this cannot occur, notably finite-dimensional spaces (but also some infinite-dimensional ones, for instance, Banach spaces). In contrast, this kind of phenomen occurs easily (and frequently, as we here show) for finite-dimensional diffeological vector spaces, where its numerous instances are readily obtained in any dimension starting from 2. After briefly reviewing what is known on this question in some classical categories, we provide an overview of this phenomenon and some of its implications for finite-dimensional diffeological vector spaces, indicating briefly its connections with some other subjects.

1. Vector spaces in general as coproducts or not

There is a simple mathematical curiosity that arises when a vector space is endowed with an extra structure, such as that of a topological vector space, a smooth structure of sorts, and so on. In the most general terms, this curiosity can be described as follows: let C be a category endowed with a faithful functor Ξ to the category **Vect** of vector spaces, and let c, a, b be objects of C such that $\Xi(c)$ is the coproduct in the category **Vect** of $\Xi(a)$ and $\Xi(b)$; then one may wonder whether c itself is the coproduct of a and b in the initial category C.

1.1. Topological vector spaces. The most obvious instance of course is that of the category C being the category of topological vector spaces and Ξ being the forgetful functor into **Vect**; given a topological vector space V whose underlying vector space decomposes into a direct sum of two of its subspaces, $\Xi(V) = \Xi(V_1) \oplus \Xi(V_2)$, where V_1 and V_2 are vector subspaces of V endowed with the subspace topology, one may wonder whether the product topology on $V_1 \oplus V_2$ (relative to their subspace topologies inherited from V) coincides with the topology of V itself (this is usually expressed as V being an *algebraic direct sum* of V_1 and V_2 vs. it being their *topological direct sum*).

Now, in this case it is rather well-known that the answer is positive for any finite-dimensional V, while it may turn out to be negative in the case of infinite

2020 *Mathematics Subject Classification.* Primary 53C15; Secondary 46A19, 53-02.

I would like to thank the organizers of the *Special session on recent advances in diffeology and their applications* for their kind invitation.

dimension. Indeed, any finite-dimensional Hausdorff topological vector space is homeomorphic to $\mathbb{R}^n$ (if V is over $\mathbb{R}$, to $\mathbb{C}^n$ if it is over $\mathbb{C}$), while in the case of non-Hausdorff topological vector spaces it suffices to recall that every such space is isomorphic as a topological vector space to the product of a Hausdorff topological vector space and an indiscrete vector space (see, for instance, [**19**], p. 63).

Thus, for topological vector spaces the question of an algebraic direct sum of vector subspaces being also a topological direct sum is a matter limited to infinite dimension. There indeed the two concepts may relatively easily turn out to be distinct as can be illustrated by the space V defined as the subspace of $C[0,1]$ (the space of all continuous functions with the uniform convergence topology) given by $V = \mathrm{Span}(M, N)$ with M the space of all polynomials and N the span of some non-polynomial continuous function, and endowed with the subspace topology relative to its inclusion in $C[0,1]$: indeed, $V = M \oplus N$ algebraically, however its topology is larger than the direct sum topology relative to the inherited topologies on M and N.[1] On the other hand, for some specific classes of infinite-dimensional vector spaces (such as, for instance, Banach spaces) it may turn out that for every member of that class any its decomposition into an algebraic direct sum (of closed subspaces) is also a decomposition into a topological direct sum (see [**18**], Theorem 3.6).

1.2. Vector spaces with a "smooth" structure. Similarly to the case of topological vector spaces, if V is a vector space endowed with some type of a smooth structure such as that of a finite-dimensional manifold (relative to which the vector space operations on V are smooth), a Hilbert space, a Fréchet space, etc, and the subspaces V_1, V_2 of V are such that they inherit from V an appropriate smooth structure (as is automatic in many cases), then again one may wonder whether the corresponding product smooth structure on $V_1 \oplus V_2$ coincides with that of V. Here again, if the dimension of V is finite, one observes that V is in particular a connected Abelian Lie group with respect to the addition operation; all such groups are of form $\mathbb{R}^k \times (\mathbb{S}^1)^l$ ([**17**]) and so any such vector space must again be some $\mathbb{R}^n$ with its usual smooth structure.

1.3. Finite dimension appearing. Contrary to the above-listed cases, it can be noted that in categories different from **Vect** a phenomenon in some ways similar may occur even in finite dimension. For instance, as follows from [**8**], [**5**], [**9**], there exist smooth manifolds, such as the K3 surface, that admit topological decompositions (into a connected sum) but not smooth decompositions. The analogy is not wholly complete, however, since in general a connected sum is not the coproduct of its factors.

1.4. Diffeological vector spaces. Now, what happens when C is the category of diffeological vector spaces, is that a finite-dimensional diffeological vector space (and a rather simple one at that) may decompose as an algebraic direct sum of two of its subspaces without being their (diffeological) smooth direct sum[2]

---

[1] It must be said that I encountered this example in an online discussion at `https://math.stackexchange.com/questions/78917/example-of-a-topological-vector-space-such-that-e-m-oplus-n-algebraically`.

[2] Let V be a diffeological vector space, and let V_1, V_2 be two of its vector subspaces such that the vector space underlying V coincides with $V_1 \oplus V_2$. The diffeological vector space V is said to decompose as a smooth sum of V_1 and V_2 if the diffeology on V coincides with the product diffeology on $V_1 \oplus V_2$ relative to the subset diffeologies on V_1 and V_2 inherited from V.

[**16**], Example 3.5. Thus, the just-mentioned simple example indicates that a number of standard trivial properties involving direct sum decompositions of vector spaces cannot be guaranteed in the diffeological context, even for very simple finite-dimensional examples: a given direct sum decomposition may not be smooth (and checking whether one is or is not so may not be wholly trivial, as the example in the next section indicates), a given subspace may not — *a priori* — be complemented[3], and, also *a priori*, a given finite-dimensional diffeological vector space may not admit any smooth decomposition into a direct sum at all. We are not yet aware of specific examples as to the latter possibilities (we just give some indications as to how such examples could be obtained, possibly exploiting a certain conjectural property of so-called non-Baire functions, *i.e.* functions not belonging to any Baire class, [**13**]), although it is known ([**16**], Proposition 3.8 and Theorem 3.12) that their consideration should be limited to the vector spaces with trivial diffeological dual: indeed, every finite-dimensional diffeological vector space V admits a (non-unique) smooth decomposition into a direct sum of a subspace diffeomorphic to its diffeological dual (hence whose subset diffeology is standard) and the so-called *maximal isotropic subspace*, which is defined as the intersection of kernels of all smooth linear $\mathbb{R}$-valued functions.

Still another host of issues regards the classic direct sum decomposition of a vector space V relative to a given linear map $f: V \to W$, that as $V = \mathrm{Ker}(f) \oplus \mathrm{Im}(f)$. Unlike the standard context where the meaning of such equality is immediately understood, in the diffeological context and to our purposes much specification is needed as to what exactly should be meant by the subspace $\mathrm{Im}(f)$. Indeed, by the standard definition $\mathrm{Im}(f)$ is a subspace of the target space W (and the above equality indicates in fact an isomorphism), and so its most natural diffeology is the subset diffeology relative to the diffeology on W. However, this diffeology, while guaranteed to contain the pushforward of the diffeology of V by f, may easily turn out to be larger (which is an internal property of W). If this is the case, V will certainly not be diffeomorphic to $\mathrm{Ker}(f) \oplus \mathrm{Im}(f)$ — but for reasons fundamentally different from direct sums being or not being smooth.

Thus, the most natural way to pose the question appears to be that to ask whether V is always diffeomorphic to $\mathrm{Ker}(f) \oplus \mathrm{Im}(f)$, where $\mathrm{Im}(f)$ is endowed with the pushforward of the diffeology of V by f. Still another angle on the same question relates it to the question of smoothness of direct sums. Indeed, $\mathrm{Ker}(f)$ admits in general a multitude of algebraic complements in V; one may, on one hand, wonder whether any or even all of these complements, considered with the corresponding subset diffeologies, are diffeomorphic to $\mathrm{Im}(f)$ with its pushforward diffeology — and on the other hand whether any of these subspaces complements $\mathrm{Ker}(f)$ smoothly, the two questions being *a priori* distinct.

In what follows we first illustrate how even establishing a smoothness of a given direct sum decomposition of a diffeological vector space could be tricky, and then, after recalling the known example of a non-smooth direct sum decomposition, we provide some considerations regarding the existence of non-complemented subspaces, and then that of finite-dimensional diffeological vector spaces that do not admit any smooth decomposition into a direct sum.

[3]Meaning that it may not be part of any smooth direct sum decomposition of the ambient space.

2. When a maximal isotropic subspace is a (non-obvious) coproduct

In this section we consider in detail the following example.

EXAMPLE 2.1. Let V be $\mathbb{R}^2$ endowed with the vector space diffeology generated by the following two maps:

$$p : \mathbb{R} \to V,\ p(x) = (|x|, |x|),$$
$$q : \mathbb{R} \to V,\ q(x) = (0, \Delta_{\mathbb{Q}}(x)),$$

where $\Delta_{\mathbb{Q}} : \mathbb{R} \to \mathbb{R}$ acts by

$$\Delta_{\mathbb{Q}}(x) = \begin{cases} 0, & \text{if } x \in \mathbb{Q}, \\ 1, & \text{otherwise.} \end{cases}$$ [4]

We first observe that the space V possesses the following (rather evident) property.

LEMMA 2.2. *The maximal isotropic subspace of V coincides with V.*

PROOF. Let $f : V \to \mathbb{R}$ be a smooth linear function, and let $f(e_i) = a_i$ for $i = 1, 2$. Then $(f \circ p)(x) = (a_1 + a_2)|x|$, which is an ordinary smooth map if and only if $a_2 = -a_1$. Since $f \circ q = a_2\Delta_{\mathbb{Q}}$, we immediately obtain $a_1 = a_2 = 0$. □

The property of V that we are most interested in is the following one.

THEOREM 2.3. *The space V decomposes as a smooth direct sum of* $\text{Span}(e_1)$ *and* $\text{Span}(e_2)$*, with the subset diffeology on* $\text{Span}(e_1)$ *being generated by the map* $x \mapsto (|x|, 0)$ *and that on* $\text{Span}(e_2)$ *by the above-defined map* $q : x \mapsto (0, \Delta_{\mathbb{Q}}(x))$.

PROOF. The essence of the proof consists in establishing the following equality:

$$|x| = 2x\Delta_{\mathbb{Q}}(H_1(x)) - 2x\Delta_{\mathbb{Q}}(H_2(x)) + x$$

for all $x \in \mathbb{R}$ and two specific functions $H_1, H_2 : \mathbb{R} \to \mathbb{R}$ that we now define. The function H_1 is given by

$$H_1(x) = \begin{cases} e^{-\frac{1}{x^2}} & x > 0 \\ 0 & x \leqslant 0. \end{cases}$$

To define H_2 we need certain auxiliary maps. Specifically, we define $w : (0,1) \to (\frac{1}{\sqrt{2}}, 1)$ by setting $w(x) = \frac{\sqrt{2}-1}{\sqrt{2}}x + \frac{1}{\sqrt{2}}$. Next, consider the countable sets $\{a_i\} = (0,1) \cap \mathbb{Q}$ and $\{b_j\} = w^{-1}\left((\frac{1}{\sqrt{2}}, 1) \cap \mathbb{Q}\right)$; now, it is established in [**7**] (see Theorem I and its proof, p.94) that there exists a smooth (actually analytic) monotone map $f : (0,1) \to (0,1)$ such that f maps the set $\{a_i\}$ onto the set $\{b_j\}$ (that is, $w \circ f$ is a map $(0,1) \to (\frac{1}{\sqrt{2}}, 1)$ that sends rational points to rational points, and irrational ones, to irrational).

Let now $\gamma = w \circ f$, and define $\bar{\gamma} : [0,1) \to \mathbb{R}$ by $\bar{\gamma}(x) = \begin{cases} \gamma(x) & x \in (0,1) \\ \frac{1}{\sqrt{2}} & x = 0 \end{cases}$; observe that, since γ is analytic at 0, so is $\bar{\gamma}$. Finally, define

$$H_2 = \bar{\gamma} \circ H_1.$$

Observe that, since f is analytic at 0 ([**7**], p. 94), H_2 is smooth.

---

[4] Obviously, $\Delta_{\mathbb{Q}}$ is related to the well-known Dirichlet function D by $\Delta_{\mathbb{Q}} = 1 - D$. In fact, our choice to use $\Delta_{\mathbb{Q}}$ rather than the Dirichlet function turned out to be rather frivolous, since the same arguments as below, with very slight adjustments, work for both.

The equality $|x| = 2x\Delta_{\mathbb{Q}}(H_1(x)) - 2x\Delta_{\mathbb{Q}}(H_2(x)) + x$ is now a simple consequence of the choice of the functions H_1 and H_2. Indeed, since $H_1(x) \in \mathbb{Q}$ if and only if $H_2(x) \in \mathbb{Q}$, we have $2x\Delta_{\mathbb{Q}}(H_1(x)) - 2x\Delta_{\mathbb{Q}}(H_2(x)) = 0$ for all $x > 0$, while, since $H_1(x) \equiv 0$ and $H_2(x) \equiv \frac{1}{\sqrt{2}}$ for $x \leqslant 0$, it equals $-2x$ for $x \leqslant 0$. Hence the entire sum is x for postive x and $-x$ for all other x, that is, is the absolute value function.

To conclude the proof of the theorem, observe that it follows from the equality just proven that the subset diffeology on $\mathrm{Span}(e_2)$ contains the map $x \mapsto (0, |x|)$; indeed, this map is equal to $2xq(H_1(x)) - 2xq(H_2(x)) + F(x)$, where $F(x) = (0, x)$, which is a plot of the subset diffeology of $\mathrm{Span}(e_2)$. It then follows that the plot p of V is a sum of the map $x \mapsto (|x|, 0)$ and a plot of the subset diffeology of $\mathrm{Span}(e_2)$, and so the diffeology of V is generated by the maps $x \mapsto (|x|, 0)$ and q. In particular, it is the product diffeology corresponding to the subset diffeologies on $\mathrm{Span}(e_1)$ and $\mathrm{Span}(e_2)$, whence the conclusion. □

REMARK 2.4. As a mere curiosity, we observe that substituting in the expression of $|x|$ through $\Delta_{\mathbb{Q}}$'s, in place of x and $2x$, any other smooth functions f and g, we obtain a map equaling f on $(0, \infty)$ and $f - g$ on $(-\infty, 0]$, therefore, utilizing if necessary appropriate translations, any piecewise-smooth function with one singularity can be expressed in an analogous way through $\Delta_{\mathbb{Q}}$.

In addition, we can also conclude the following.

COROLLARY 2.5. *The subset diffeology on any non-zero subspace of V is non-standard.*

PROOF. Let W be a (proper) subspace of V, and let (a, b) be a generator of it. It follows directly from the proof of the above theorem that the diffeology of V contains plots of form $x \mapsto (|x|, 0)$ and $x \mapsto (0, |x|)$; being a vector space diffeology, it therefore contains a plot of form $x \mapsto (a|x|, b|x|)$, which is a non-standard plot for the subset diffeology of V. □

3. Diffeological vector spaces not being coproducts

We now turn to nonsmooth direct sums.

3.1. A nonsmooth direct sum decomposition. Let V be $\mathbb{R}^3$ endowed with the vector space diffeology generated by the plot $p : \mathbb{R} \to V$ given by $p(x) = |x|(e_2 + e_3)$. It was shown in [**16**] that the subset diffeologies on its subspaces $V_0 = \mathrm{Span}(e_1, e_2)$ and $V_1 = \mathrm{Span}(e_3)$ are both standard, hence, while V is an algebraic direct sum of V_0 and V_1, it is not their smooth direct sum.

In fact, for diffeological vector spaces non-smooth direct sum decompositions are a rather frequent phenomenon. To make this claim more precise, recall that, as was shown in [**16**] (Proposition 3.8), every finite-dimensional diffeological vector space V contains at least one subspace that is maximal for the folowing two properties: its subset diffeology is standard, and it splits off as a smooth direct summand (however, contrary to what was erroneously claimed in [**16**], there is in general more than one such subspace). Any such subspace is called a *characteristic subspace* of V, and it possesses the following property.

LEMMA 3.1. *Let V_0 be a characteristic subspace of V, and let $V = V_0 \oplus V_1$ be a smooth direct sum decomposition of V. Then V_1 coincides with the maximal isotropic subspace of V.*

PROOF. Let $f : V \to \mathbb{R}$ be a smooth linear function, and suppose that $f|_{V_1} \neq 0$. Since the decomposition $V = V_0 \oplus V_1$ is smooth, the map $f' = 0 \oplus f|_{V_1}$ is a smooth linear function, which implies that the dimension of the diffeological dual of V is strictly greater than that of V_0, which contradicts [**16**], Theorem 3.12. □

Thus, there are numerous diffeological vector spaces admitting non-smooth decompositions.

COROLLARY 3.2. *If the dimension and the codimension of the maximal isotropic subspace of a finite-dimensional diffeological vector space V are both positive, then V admits at least one smooth decomposition.*

PROOF. As a (by assumption) proper subspace of V, any chosen characteristic subspace has a multitude of algebraic direct complements, only one of which coincides with the maximal isotropic subspace, all the others giving rise to non-smooth direct sum decompositions of V. □

3.2. A non-complemented subspace. The discussion in the subsequent sections is contingent on the following assumption: *suppose that there exists a non-smooth function $\gamma : \mathbb{R} \to \mathbb{R}$ such that, if D_γ is the vector space diffeology on $\mathbb{R}$ generated by γ and $D_{|\cdot|}$ is the vector space diffeology on $\mathbb{R}$ generated by the absolute value function, then $D_\gamma \cap D_{|\cdot|}$ is the standard diffeology on $\mathbb{R}$.*

Assuming this, we can easily conclude that there do exist finite-dimensional diffeological vector spaces containing non-complemented subspaces and, later, that there exist ones that do not admit any smooth direct sum decomposition.

EXAMPLE 3.3. Let γ be a function as in the assumption, and let V be $\mathbb{R}^2$ endowed with the vector space diffeology generated by the following two plots: $p : x \mapsto (\gamma(x), \gamma(x))$ and $q : x \mapsto (0, |x|)$. Observe first that V coincides with its maximal isotropic subspace; indeed, let $f : V \to \mathbb{R}$ defined by $f(e_1) = a$, $f(e_2) = b$, be a smooth linear function. Then the functions $x \mapsto (a+b)\gamma(x)$, $x \mapsto b|x|$ are ordinary smooth functions, which readily implies that $a = b = 0$.

Observe now that the subspace $\mathrm{Span}(e_1)$ of V has standard subspace diffeology and therefore it follows from the above that it is not complemented. Indeed, a generic plot of V locally has form

$$\left(\sum_{i=1}^{k} h_i \cdot (\gamma \circ H_i) + \alpha_1, \sum_{i=1}^{k} h_i \cdot (\gamma \circ H_i) + \sum_{j=1}^{l} f_j|F_j| + \alpha_2\right),$$

where $\alpha_1, \alpha_2, h_i, H_i, f_j, F_j$ are some smooth functions $U \to \mathbb{R}$ for some domain $U \subseteq \mathbb{R}^n$. For this to be a plot of the subset diffeology of $\mathrm{Span}(e_1)$ we must have $\sum_{i=1}^{k} h_i \cdot (\gamma \circ H_i) + \sum_{j=1}^{l} f_j|F_j| + \alpha_2 \equiv 0$, that is, $\sum_{i=1}^{k} h_i \cdot (\gamma \circ H_i) = -\sum_{j=1}^{l} f_j|F_j| - \alpha_2$. Now, the function on the right belongs to $D_{|\cdot|}$, while the one on the left belongs to D_γ, hence by assumption they are both smooth, which implies the desired conclusion.

Notice however that V, although (presumably) containing a non-complemented subspace, does admit an obvious decomposition into a smooth direct sum, that as $\mathrm{Span}(e_1 + e_2) \oplus \mathrm{Span}(e_2)$.

Regarding the plausibility of the assumption stated in the beginning of the section and on which our conjectural example depends, we suggest that γ could be a non-Baire (not belonging to any Baire class) function (for instance, if one assumes the axiom of choice, it could be the indicator function of a non-measirable set. Even without this axiom, there do exist Lebesgue-measurable functions that are not Borel-measurable — again, indicator functions of sets with this property —, which therefore again do not belong to any Baire class, see for instance [**1**], Lemma 2). It should on the other hand be noted that our assumption is not as trivial as one might perhaps at first glance expect: for instance, it is known, for Baire 1 functions, that the class of a Baire function is not necessarily preserved by the operations of pre-composition with smooth functions, multiplication by such, and summation,[5] and we ourselves give in the previous section an example of a Baire 2 function being rendered a Baire 0 function by precisely these operations. Yet, it can be noticed that, if it were true that any subspace of a finite-dimensional diffeological vector space is complemented, our example would imply that any function $\mathbb{R} \to \mathbb{R}$ whatsoever could be made continuous via these operations.

Still independently of such considerations, an attempt to construct an explicit example of a non-complemented subspace could be made as follows. Let V be $\mathbb{R}^2$ endowed with the vector space diffeology generated by the plot $x \mapsto \left(\Delta_{\mathbb{Q}}(x), \Delta_{\mathbb{Q}}(\sqrt{|x|})\right)$. Observe first that the maximal isotropic subspace of V coincides with V. Indeed, let f, $f(e_1) = a$, $f(e_2) = b$, be a smooth linear function on V, then $x \mapsto a\Delta_{\mathbb{Q}} + b\Delta_{\mathbb{Q}}(\sqrt{|x|})$ is an ordinary smooth function $\mathbb{R} \to \mathbb{R}$. Since the set of irrational numbers is dense in $\mathbb{R}$, it must be a constant function with value $a+b$. On the other hand, since $\mathbb{R}$ contains both rational numbers with irrational root and those with rational root, this value must be equal to both b and 0, respectivly, hence $a = b = 0$.

Observe next that the result just obtained implies that if V contains a subspace with the subset diffeology that is standard, that subspace cannot split off as a smooth direct summand, and consider the subspace $\mathrm{Span}(e_1)$. It is trivial to observe that this subspace having standard subset diffeology depends on the following, at the moment conjectural, implication holding for all smooth functions h_i, H_i:

$$\sum_{i=1}^{k} h_i(\Delta_{\mathbb{Q}} \circ \sqrt{|H_i|}) \text{ is smooth } \Rightarrow \sum_{i=1}^{k} h_i(\Delta_{\mathbb{Q}} \circ H_i) \text{ is smooth }.$$

Whether this implication does actually hold is at the moment work in progress.[6]

3.3. Non-decomposable vector spaces. Let γ be a function as in the previous section, *i.e.*, satisfying the assumption that $D_\gamma \cap D_{|\cdot|}$ is the standard diffeology, whose existence we again assume. Then this yields the following example of a diffeological vector space that does not admit any non-trivial decompositions into a smooth direct sum.

EXAMPLE 3.4. Let W be $\mathbb{R}^2$ endowed with the vector space diffeology generated by the map $x \mapsto (|x|, \gamma(x))$. Observe first that the maximal isotropic subspace of

---

[5]It is preserved by the latter two operations [**6**], Propositions 7 and 8, but not necessarily by that of the pre-composition, there being a certain field of study of the relevant issue, see for instance [**11**], [**20**].

[6]The hope of it being true is based on potential use of arguments akin to those involved in the study of recoverability of Baire fuctions (see for instance [**14**], [**4**], although these regard Baire 1 functions): the potential recoverability of $\sum_{i=1}^{k} h_i(\Delta_{\mathbb{Q}} \circ \sqrt{|H_i|})$ from its values on certain specific sets and the assumption of it being smooth might impose certain conditions (local nullity conditions, in fact) on the coefficient functions h_i sufficient to ensure the smoothness of $\sum_{i=1}^{k} h_i(\Delta_{\mathbb{Q}} \circ H_i)$.

W coincides with W; indeed, if $f : W \to \mathbb{R}$ is a smooth linear map defined by $f(e_1) = a$, $f(e_2) = b$ then $x \mapsto a|x|$, $x \mapsto b\gamma(x)$ are ordinary smooth maps, which immediately implies $a = b = 0$.

We claim that every proper (nontrivial) subspace of W has standard diffeology. Indeed, let $ce_1 + de_2$ be a generator of such a subspace. A generic plot of W locally is a function of form $\left(\sum_{i=1}^k h_i|H_i| + \alpha_1, \sum_{i=1}^k h_i(\gamma \circ H_i) + \alpha_2\right)$ for some smooth maps $\alpha_1, \alpha_2, h_i, H_i : U \to \mathbb{R}$. For this to be a plot of the subset diffeology on $\mathrm{Span}(ce_1 + de_2)$ we must have $d\sum_{i=1}^k h_i|H_i| + d\alpha_1 = c\sum_{i=1}^k h_i(\gamma \circ H_i) + c\alpha_2$. However, the function on the left belongs to $D_{|\cdot|}$, while that on the right, to D_γ, therefore by the assumption both of them are smooth. Thus, every one-dimensional subspace of W has standard diffeology, and since the maximal isotropic subspace of W is W itself, no non-trivial subspace of W with standard subbset diffeology splits off as a smooth direct summand. Therefore W does not admit nontrivial decompositions into a smooth direct sum.

It is quite obvious that the choice of $|\cdot|$ is relatively arbitrary, and the same procedure as in the example above and that in the preceding section would work for any pair of non-smooth functions γ_1, γ_2 such that $D_{\gamma_1} \cap D_{\gamma_2}$ is the standard diffeology of $\mathbb{R}$. Since we are not aware of a formal proof of existence or non-existence of such functions, we again notice that a different kind of space, which in the previous section was conjectured to contain a non-complemented subspace, should this latter conjecture turn out to be correct, would turn out to be non-decomposable as well. Indeed, it is easy to see that the conjectural implication stated in the previous section would also imply that the subset diffeology on any proper subspace of the space in question be standard, and by the same reasoning the space would be nondecomposable.

REMARK 3.5. As a side remark, the question of a given (finite-dimensional) diffeological space admitting at least one decomposition into a smooth direct sum of its subspaces can be also stated in the following terms. Let V be the space in question, and let $\mathcal{D}_p$ be the category of its plots (*i.e.* the category whose objects are plots of V and whose arrows are commutative triangles of form

$$\begin{array}{ccc} U & \xrightarrow{\quad f \quad} & U' \\ & {\scriptstyle p}\searrow \quad \swarrow{\scriptstyle q} & \\ & V & \end{array},$$

where p and q are plots of V, and $f : U \to U'$ is an ordinary smooth map; this category is amply used in diffeology, for instance in studying the D-topology on diffeological spaces [**2**], defining (a version of) their tangent spaces and tangent bundles [**3**], defining diffeology on the Milnor classifying space of a diffeological group [**15**], defining sheaves for diffeological spaces [**12**], and so on).

Let V_0, V_1 be a pair of subspaces of V, and let $\mathcal{D}_p^0$ and $\mathcal{D}_p^1$ be respectively the categories of plots of the subset diffeologies on V_0 and V_1. Consider the subcategory $\mathcal{D}_p^{0,1}$ of the product category $\mathcal{D}_p^0 \times \mathcal{D}_p^1$ whose objects are pairs (p_0, p_1) such that p_0 and p_1 have the same domain.[7] Then it is obvious that V decomposes as a smooth

[7] Obviously, $\mathcal{D}_p^{0,1}$ is just the equalizer in the category **Cat** of all small categories of the two functors $F_0, F_1 : \mathcal{D}_p^0 \times \mathcal{D}_p^1 \to \mathbf{Cart}$ into the category of all domains in Cartesian spaces, that are

direct sum of V_0 and V_1 if and only if the assignment $(p_0, p_1) \mapsto p_0 + p_1$ defines an isomorphism of categories $\mathcal{D}_p^{0,1} \to \mathcal{D}_p$. In particular, V is decomposable if its category of plots $\mathcal{D}_p$ can be identified with some equalizer category $\mathcal{D}_p^{0,1}$.

3.4. On decompositions of form $\mathrm{Ker}(f) \oplus \mathrm{Im}(f)$**.** Finally, we observe that if nondecomposable diffeological vector spaces do exist, for any such space a diffeomorphism $V \cong \mathrm{Ker}(f) \oplus \mathrm{Im}(f)$ ($\mathrm{Im}(f)$ being considered, recall, with the pushforward diffeology) automatically cannot hold for any nontrivial f with nontrivial kernel (this by nature of what it means to be a nondecomposable space). On the other hand, the subspace $\mathrm{Ker}(f)$ being a summand in some non-smooth decomposition does not, of course, exclude that there may be such a diffeomorphism, as can be illustrated by the space V of [**16**], Example 3.5 ($\mathbb{R}^3$ with the vector space diffeology generated by $x \mapsto |x|(e_2 + e_3)$) and the map f defined on V by $e_1 \mapsto e_1$, $e_2 \mapsto e_2$, $e_3 \mapsto 0$, and taking values in $\mathbb{R}^2$ endowed with pushforward diffeology.

References

[1] L. G. Brown, *Baire functions and extreme points*, Amer. Math. Monthly **79** (1972), 1016–1020, DOI 10.2307/2318075. MR306880

[2] J. D. Christensen, G. Sinnamon, and E. Wu, *The D-topology for diffeological spaces*, Pacific J. Math. **272** (2014), no. 1, 87–110, DOI 10.2140/pjm.2014.272.87. MR3270173

[3] J. D. Christensen and E. Wu, *Tangent spaces and tangent bundles for diffeological spaces* (English, with English and French summaries), Cah. Topol. Géom. Différ. Catég. **57** (2016), no. 1, 3–50. MR3467758

[4] U. B. Darji, M. J. Evans, and R. J. O'Malley, *A first return characterization for Baire-one functions*, Real Anal. Exchange **19** (1993/94), no. 2, 510–515. MR1282666

[5] S. K. Donaldson, *An application of gauge theory to four-dimensional topology*, J. Differential Geom. **18** (1983), no. 2, 279–315. MR710056

[6] J. P. Fenecios and E. A. Cabral, *On some properties of Baire-1 functions*, Int. J. Math. Anal. (Ruse) **7** (2013), no. 5-8, 393–402, DOI 10.12988/ijma.2013.13035. MR3001729

[7] P. Franklin, *Analytic transformations of everywhere dense point sets*, Trans. Amer. Math. Soc. **27** (1925), no. 1, 91–100, DOI 10.2307/1989166. MR1501300

[8] M. H. Freedman, *The topology of four-dimensional manifolds*, J. Differential Geometry **17** (1982), no. 3, 357–453. MR679066

[9] M. H. Freedman, *There is no room to spare in four-dimensional space*, Notices Amer. Math. Soc. **31** (1984), no. 1, 3–6. MR728340

[10] P. Iglesias-Zemmour, *Diffeology*, Mathematical Surveys and Monographs, vol. 185, American Mathematical Society, Providence, RI, 2013, DOI 10.1090/surv/185. MR3025051

[11] O. Karlova and V. Mykhaylyuk, *On composition of Baire functions*, Topology Appl. **216** (2017), 8–24, DOI 10.1016/j.topol.2016.11.002. MR3584119

[12] D. Krepski, J. Watts, S. Wolbert, *Sheaves, principal bundles, and Cech cohomology for diffeological spaces*, arxiv:2111.01032 (2021).

[13] H. Lebesgue, *Sur les fonctions représentables analytiquement*, Journal de mathématiques pures et appliquées 6e série, 1 (1905), 139-216.

[14] D. Lecomte, *How can we recover Baire class one functions?*, Mathematika **50** (2003), no. 1-2, 171–198 (2005), DOI 10.1112/S0025579300014881. MR2136360

[15] J.-P. Magnot and J. Watts, *The diffeology of Milnor's classifying space*, Topology Appl. **232** (2017), 189–213, DOI 10.1016/j.topol.2017.10.011. MR3720892

[16] E. Pervova, *On the notion of scalar product for finite-dimensional diffeological vector spaces*, Electron. J. Linear Algebra **34** (2018), 18–27, DOI 10.13001/1081-3810.3121. MR3780575

[17] C. Procesi, *Lie groups: An approach through invariants and representations*, Universitext, Springer, New York, 2007. MR2265844

[18] D. S. Rakić and D. S. Djordjević, *A note on topological direct sum of subspaces*, Funct. Anal. Approx. Comput. **10** (2018), no. 1, 9–20. MR3782826

---

given by the compositions of the functor $\mathcal{D}_p^0 \times \mathcal{D}_p^1 \to \mathbf{Cart} \times \mathbf{Cart}$ that assigns to a pair of plots the pair of their domains, with the projections on factors.

[19] A. Wilansky, *Modern Methods in Topological Vector Spaces*. Mineola, New York: Dover Publications, Inc., 2013.

[20] D. Zhao, *Functions whose composition with Baire class one functions are Baire class one*, Soochow J. Math. **33** (2007), no. 4, 543–551. MR2404581

University of Pisa, Department of Mathematics, Via F. Buonarroti 1C, 56127 Pisa, Italy

Current address: University of Pisa, Department of Mathematics, Via F. Buonarroti 1C, 56127 Pisa, Italy

Email address: ekaterina.pervova@unipi.it

Contemporary Mathematics
Volume **794**, 2024
https://doi.org/10.1090/conm/794/15940

Singular foliations through diffeology

David Miyamoto

Abstract. A singular foliation is a partition of a manifold into leaves of perhaps varying dimension. Stefan and Sussmann carried out fundamental work on singular foliations in the 1970s. We survey their contributions, show how diffeological objects and ideas arise naturally in this setting, and highlight some consequences within diffeology. We then introduce a definition of transverse equivalence of singular foliations, following Molino's definition for regular foliations. We show that, whereas transverse equivalent singular foliations always have diffeologically diffeomorphic leaf spaces, the converse holds only for certain classes of singular foliations. We finish by showing that the basic cohomology of a singular foliation is invariant under transverse equivalence.

1. Introduction

A singular foliation is a partition of a manifold into connected submanifolds, called leaves, of perhaps varying dimension, which fit together smoothly in an appropriate sense. Singular foliations lie beneath many structures one frequently encounters in geometry: the partition of a manifold into orbits of a connected smooth Lie group action, the partition of a Poisson manifold into symplectic leaves, and the partition of a manifold into the maximal integral submanifolds of an involutive distribution are all singular foliations. In the last example, the leaves of the resulting singular foliation all have the same dimension, and thus we obtain a regular foliation. A global version of the Frobenius theorem asserts that every regular foliation arises in this way.

Theorem 1.1 (Global Frobenius). *Let M be a manifold. An involutive distribution Δ has integral submanifolds, and the partition of a manifold M into the maximal integral submanifolds is a regular foliation $\mathcal{F}_\Delta$. Conversely, given a regular foliation $\mathcal{F}$, the associated distribution $T\mathcal{F}$ is involutive. The assignments*

$$\mathcal{F} \mapsto T\mathcal{F}, \quad \Delta \mapsto \mathcal{F}_\Delta$$

are inverses of each other.

We employ the adjective "global" because in his original paper [**8**], Frobenius worked locally, and did not consider foliations; in modern terms (as opposed to

2020 *Mathematics Subject Classification.* Primary 53C12.

Key words and phrases. singular foliations, diffeology, transverse equivalence.

This work was partly funded by a Graduate Student Travel Grant provided by the Department of Mathematical and Computational Sciences at the University of Toronto Mississauga.

the language of Pfaffian systems used at the time), Frobenius proved that a regular distribution Δ has integral submanifolds if and only if it is involutive. Furthermore, Frobenius was not - and did not claim to be - the first to prove the result that now bears his name. He was building from the work of Jacobi [**15**], who gave sufficient integrability conditions in a special case, and of Clebsch [**3**], who expanded Jacobi's conditions the general case. Frobenius also singled out the contributions of Deahna [**5**], who gave necessary and sufficient conditions for integrability (c.f. Remark 2.21). For more on the history of Frobenius's theorem, see the treatments by Hawkins [**10**] and Samelson [**31**].

For singular foliations the situation remained unclear for some time. Hermann [**12**] was the first to give some sufficient conditions for integrability in terms of Lie subalgebras of vector fields (stated in Example 2.22). Shortly after, Nagano [**29**] established Frobenius's theorem holds in the analytic case. Matsuda [**25**], and Lobry [**24**] worked to extended the sufficient conditions of Hermann, but like Hermann did not propose integrability conditions on the singular distributions themselves. Then, in 1973-1974, Stefan [**32**] and Sussmann [**33**] independently published results that, taken together, successfully generalize the global Frobenius theorem as stated above; the integrability condition on Δ is that it is invariant under the flow of any vector field tangent to Δ. In Section 2 of this paper, we proceed carefully toward Stefan and Sussmann's results, highlighting the independent contributions of both authors. For a broader perspective on the history of integrability theorems for singular distributions, see Lavau's survey [**21**].

Diffeology was introduced in the 1980s by Souriau, and was therefore not available to Stefan and Sussmann as they wrote in the early 1970s. However, Stefan's work in particular uses many tools that arise naturally in diffeology, and he draws conclusions that are best interpreted diffeologically. We emphasize these connections to diffeology throughout Section 2, but highlight here Corollary 2.25 and Proposition 2.26, which imply that it is equivalent to consider singular foliations, and orbits of D-connected diffeological groups acting smoothly on M.

In Section 3, we contrast two approaches to the transverse geometry of a singular foliation: one equips the leaf space $M/\mathcal{F}$ with the quotient diffeology, and the other defines a notion of transverse equivalence directly. We call this Molino transverse equivalence, after Molino, whose notion of transverse equivalence ([**28**, Definition 2.1]) for regular foliations this generalizes. By Proposition 3.10, a Molino transverse equivalence always induces a diffeomorphism between the leaf spaces, but the converse does not hold in general; in Example 3.18, we give two regular foliations that are not Molino transverse equivalent, but which have diffeomorphic leaf spaces. It is thus a substantive question whether there is a class of singular foliations for which the diffeology of its leaf space determines its Molino transverse equivalence class. We give one affirmative answer in Proposition 3.20 in terms of quasifold groupoids, which were introduced in [**17**]. This specializes to orbifolds.

Our Molino transverse equivalence is intimately related to Garmendia and Zambon's Hausdorff Morita equivalence of singular foliations, defined in [**9**]. We note that Garmendia and Zambon begin with a substantially different notion of singular foliation, which we call an Androulidakis-Skandalis singular foliation, c.f. Definition 2.30. They therefore have a different, generally finer, notion of equivalence. Our Molino transverse equivalence may be viewed as an extension of Garmendia and Zambon's Hausdorff Morita equivalence to Stefan singular foliations (Definition 2.5).

Finally, in Section 4, we apply our Molino transverse equivalence to the basic cohomology associated of a singular foliation. This cohomology was first introduced by Reinhart **[30]** in the regular case. Our main result of this section, Corollary 4.4, shows that the basic cohomology is invariant under Molino transverse equivalence. This is consistent with the expectation that the basic cohomology depends only on the transverse geometry of the singular foliation. As a consequence, we see by Proposition 4.6 that if the quotient map $\pi : M \to M/\mathcal{F}$ induces an isomorphism $\pi^* : H^\bullet(M/\mathcal{F}) \to H_b^\bullet(M, \mathcal{F})$, where $H^\bullet(M/\mathcal{F})$ is the cohomology associated to the diffeological differential forms on $M/\mathcal{F}$, the same is true for all singular foliations in the Molino transverse equivalence class of $(M, \mathcal{F})$. Whether π always induces an isomorphism remains an open question.

We will assume familiarity with diffeology, and not review it here. For an excellent introduction, see the book by Iglesias-Zemmour **[13]**. For a history of the development of singular foliations, see the survey by Lavau **[21]**. If the reader is unfamiliar with Lie groupoids, we have provided a very brief overview in Appendix A, but refer to the book by Moerdijk and Mrcun **[27]**, and the survey by Lerman **[23]**, for more details.

We will reiterate the structure of this article. In Section 2, we give a careful description of the definition of a singular foliation, and an overview of the Stefan-Sussmann theorem. We pay special attention to diffeological considerations. In Section 3, we introduce Molino transverse equivalence of singular foliations, and relate this to the diffeology of the leaf space, and Morita equivalence of Lie groupoids. Finally, in Section 4, we show that the basic cohomology of a singular foliation is invariant under Molino transverse equivalence. Appendix A establishes the notation and language we need when dealing with Lie groupoids and algebroids.

2. The basics of singular foliations

A singular foliation is a partition of a manifold[1] M into connected submanifolds, called *leaves*, fitting together smoothly in an appropriate sense. Before we give a definition of singular foliation, however, we must establish what we mean by "submanifold." First candidates are immersed submanifolds.

DEFINITION 2.1. A subset L of M, together with a manifold structure σ_L, is an *immersed submanifold* of M if the inclusion $\iota : (L, \sigma_L) \hookrightarrow M$ is an injective immersion.

REMARK 2.2. The smooth structure σ_L is part of the data of the immersed submanifold (L, σ_L). There may be other smooth structures σ'_L on L for which (L, σ'_L) is immersed, but is not diffeomorphic to (L, σ_L). For example, a figure-eight may be immersed in $\mathbb{R}^2$ in two different ways. Note the topology of (L, σ_L) is generally finer than the subspace topology.

Diffeology provides other candidates for submanifolds, namely diffeological and weakly-embedded submanifolds.

[1]by "manifold," we mean set equipped with a maximal smooth atlas (smooth structure) whose induced topology is Hausdorff and second-countable. When required, we may denote the smooth structure by a letter such as σ, and refer to the manifold as (M, σ).

DEFINITION 2.3. A subset L of M is a *diffeological submanifold* of M if its subspace diffeology is a manifold diffeology. Diffeological submanifolds always carry the manifold structure given by their subspace diffeology. We call such L a *weakly-embedded submanifold* if the inclusion $i : L \hookrightarrow M$ is an immersion.

REMARK 2.4. Weakly-embedded submanifolds have unique smooth structures for which the inclusion is an immersion; they are unambiguously immersed submanifolds. The topology of a diffeological submanifold is finer than the subspace topology. Due to a theorem of Joris [**16**], the cusp $x^2 = y^3$ is a diffeological submanifold of $\mathbb{R}^2$ that is not weakly-embedded. For a detailed discussion, see [**18**].

Stefan required the leaves of a singular foliation to be weakly-embedded in his definition from [**32**].

DEFINITION 2.5. A *(Stefan) singular foliation* of a smooth manifold M is a partition $\mathcal{F}$ of M into connected, weakly-embedded submanifolds L, called *leaves* such that about each $x \in M$, there exists a chart ψ of M for which

(a) ψ is a diffeomorphism $V \to U \times W$, where U and W are open neighbourhoods of the origin in $\mathbb{R}^k$ and $\mathbb{R}^{n-k}$, respectively, and k is the dimension of the leaf through x;
(b) $\psi(x) = (0, 0)$;
(c) If L is any leaf of $\mathcal{F}$, then $\psi(L) = U \times \ell$, for some $\ell \subseteq W$.

REMARK 2.6. If every leaf has the same dimension, we call the foliation *regular*. In this case, Stefan's definition coincides with the usual chart-based definition of a foliation.

REMARK 2.7. In the next section, we will consider the assignment to each $x \in M$ the subspace $T_xL \subseteq T_xM$. This is an example of a singular distribution, and the leaves L are its maximal integral submanifolds (c.f. Definitions 2.13, 2.16, and 2.18).

An attractive feature of this definition is that it is intrinsic to the manifold M and the partition into leaves L (whose smooth structures are determined by the smooth structure on M, see Remark 2.4). This is the first intrinsic definition of singular foliation to appear in the literature. Prior to Stefan, singular foliations were known only through examples appearing, for instance, in control theory.

EXAMPLE 2.8. Given a Lie group G acting smoothly on a manifold M, the connected components of the G-orbits assemble into a singular foliation. More generally, the connected components of the orbits of a Lie groupoid $\mathcal{G} \rightrightarrows M$ form a singular foliation of M. Yet more generally, every Lie algebroid $A \implies M$ induces a singular foliation on M. We treat these examples later in Example 2.22. See Appendix A for basic information on Lie groupoids and algebroids.

The condition that the leaves of $\mathcal{F}$ are weakly-embedded submanifolds seems at first unnecessarily restrictive. *A priori*, we might gain generality by only requiring that the leaves are connected immersed submanifolds of M. Kubarski [**20**] in 1990 showed that in fact, we gain no generality from this weaker assumption.

THEOREM 2.9 (Kubarski). *Suppose M is a smooth manifold, and $\mathcal{F}$ is a partition of M such that*

- *Each $L \in \mathcal{F}$ comes equipped with a smooth structure σ_L for which (L, σ_L) is a connected immersed submanifold of M, and*

- *About each $x \in M$, there is a chart ψ of M satisfying (a) through (c) in Definition 2.5, where the k in condition (a) is the dimension of the leaf (L, σ_L) about x.*

Then each L is a weakly-embedded submanifold of M, and consequently, $\mathcal{F}$ is a Stefan singular foliation.

On the other hand, the notion of a diffeological submanifold provides another potential generalization to Stefan's definition. We show that again, we gain no generality under this weaker assumption.

PROPOSITION 2.10. *Suppose M is a smooth manifold, and $\mathcal{F}$ is a partition of M into connected, diffeological submanifolds such that about each $x \in M$, there exists a chart ψ of M satisfying (a) through (c) in Definition 2.5. Then each $L \in \mathcal{F}$ is a weakly-embedded submanifold of M, and consequently, $\mathcal{F}$ is a Stefan singular foliation.*

PROOF. It suffices to show the inclusion $\iota : L \hookrightarrow M$ is an immersion. Take $x \in L$, and fix a chart ψ satisfying (a) through (c) of Definition 2.5. The intersection $L \cap V$ is an open subset of L.[2] Consider the diagram below (slightly abusing notation)

$$\begin{array}{ccc} L \cap V & \overset{\iota}{\hookrightarrow} & M \\ {\scriptstyle \psi^{-1}}\uparrow & \nearrow_{\psi^{-1}} & \\ U \times \{0\}. & & \end{array}$$

Because the manifold structure on L is given by the subset diffeology inherited from M, all maps in this diagram are smooth. The map on the right, $\psi^{-1} : U \times \{0\} \to M$, is an immersion, and by the chain rule, so is the map on the left, $\psi^{-1} : U \times \{0\} \to L \cap V$. Because both $U \times \{0\}$ and $L \cap V$ are manifolds of dimension $\dim L$, the map on the left, $\psi^{-1} : U \times \{0\} \to L \cap V$, is a diffeomorphism onto its image $L_0 := \psi^{-1}(U \times \{0\})$. Therefore it is a chart of L. In this chart, the inclusion satisfies

$$\begin{array}{ccc} L_0 & \overset{\iota}{\hookrightarrow} & V \\ {\scriptstyle \psi^{-1}}\uparrow & & \downarrow{\scriptstyle \psi} \\ U \times \{0\} & \hookrightarrow & U \times W. \end{array}$$

Therefore ι is locally an inclusion, hence is an immersion. □

2.1. The Stefan-Sussmann Theorem. Stefan's definition is ultimately difficult to verify, even when the proposed foliation should be regular. In the regular case, we instead identify foliations with their associated distributions, and apply the Frobenius theorem. Stefan [**32**] and Sussmann [**33**] extended the Frobenius theorem simultaneously but independently in 1973-1974 to the singular setting, in what is now called the Stefan-Sussmann theorem. However, while similar, their contributions were not identical, and we will take care here to highlight their separate work.

Stefan worked with singular foliations generated by arrows. These are simply certain 1-plots of $\mathrm{Diff}_{\mathrm{loc}}(M)$, the space of diffeomorphisms between open subsets of M equipped with its functional diffeology (for a description of this diffeology, see [**14**]).

[2]recall the topology on L is finer than the subspace topology.

DEFINITION 2.11. An *arrow* on a manifold M is a 1-plot $a : U \to \mathrm{Diff}_{\mathrm{loc}}(M)$, such that $a(0) = \mathrm{id}$ and $\mathrm{dom}(a(t)) \subseteq \mathrm{dom}(a(s))$, whenever $0 \leq |s| \leq |t|$. We allow for $a(t)$ to be the empty map.

Given a collection $\mathcal{A}$ of arrows, the union of their images, $\bigcup_{a \in \mathcal{A}} a(U)$ generates a pseudogroup, which we denote $\Psi\mathcal{A}$. We recall here the definition of pseudogroup.

DEFINITION 2.12. A *pseudogroup* P on a manifold M is a set of diffeomorphisms between open subsets of M that contains the identity, is closed under composition, inversion, and restriction to open subsets, and satisfies: if f is a diffeomorphism between open subsets, and f is locally in P, then f is in P.

Also associated to $\mathcal{A}$ are two smooth singular distributions, by which we mean:

DEFINITION 2.13. A *(smooth) singular distribution* on a manifold M is a subset $\Delta \subseteq TM$ such that, for every $x \in M$, the set $\Delta_x := \{v \in \Delta \mid \pi(v) = x\}$ is a vector subspace of $T_x M$, and for every $v \in \Delta$, there is a locally defined vector field X such that $X_{\pi(v)} = v$ and $X_y \in \Delta_y$ for all $y \in \mathrm{dom}\, X$.

Namely, set

$$(\Delta\mathcal{A})_x := \left\{ \frac{d}{dt}\bigg|_{t=t_0} a(t)(y) \mid a(t)(y) = x \right\}$$

$$\overline{(\Delta\mathcal{A})}_x := \{d\varphi_y v \mid \varphi \in \Psi\mathcal{A} \text{ and } v \in (\Delta\mathcal{A})_x \text{ and } \varphi(y) = x\}.$$

Stefan [**32**, Theorem 1] proved:

THEOREM 2.14 (Stefan). *Given a collection of arrows $\mathcal{A}$ on M, the partition $\mathcal{F}$ of M into the orbits of $\Psi\mathcal{A}$ is a singular foliation, and $T_x L = \overline{(\Delta\mathcal{A})}_x$ for each $x \in M$, where L is the leaf about x.*

EXAMPLE 2.15. Consider two vector fields on $\mathbb{R}^2$,

$$X := \frac{\partial}{\partial x} \quad Y := \varphi(x)\frac{\partial}{\partial y},$$

where φ is a smooth, bounded, non-negative, increasing function which vanishes if and only if $x \leq 0$, and $\varphi(x) = 1$ if $x \geq 1$. Let $\mathcal{A}$ consist of the flows of X and Y. Then $\Psi\mathcal{A}$ acts transitively on $\mathbb{R}^2$: it clearly acts transitively on $x > 0$; if $x \leq 0$, first flow along X until we enter the region $x > 0$, flow along $\pm Y$ until we arrive at the required y-coordinate, and then flow along $-X$. It follows that $\overline{(\Delta\mathcal{A})}_{(x,y)} = \mathbb{R}^2$ at all (x, y), and the only orbit of $\Psi\mathcal{A}$ is $L = \mathbb{R}^2$.

In the setting of Stefan's Theorem 2.14, the fact that for each $x \in M$, we have $\overline{(\Delta\mathcal{A})}_x = T_x L$ (for the leaf L about x) means that the singular distribution $\overline{\Delta\mathcal{A}}$ is integrable, in the sense of the next definition below.

DEFINITION 2.16. A singular distribution Δ is *integrable* if, at every $x \in M$, it admits an *integral submanifold*, which is an immersed submanifold (L, σ_L) containing x such that $T_y(L, \sigma_L) = \Delta_y$ for all $y \in L$.

Suppose now that D is a collection of locally defined vector fields on M inducing a singular distribution Δ, and $\mathcal{A}_D$ is the collection of plots of the form $t \mapsto \Phi^X(t, \cdot)$, where Φ^X is the flow of an element of D. We have $\Delta\mathcal{A}_D = \Delta$. Stefan's theorem implies

COROLLARY 2.17. *The orbits of $\Psi\mathcal{A}_D$ are integral submanifolds of Δ if and only if $\Delta = \overline{\Delta}$.*

This corollary characterizes all integral singular distributions Δ whose integral submanifolds are the orbits of $\Psi\mathcal{A}_D$, for some (hence any) collection of locally defined vector fields D spanning Δ. However, assuming only that Δ is integrable, it is not obvious from the above statements that $\Delta = \overline{\Delta}$. We do not know *a priori* that the integral submanifolds of Δ are the orbits of $\Psi\mathcal{A}_D$. Sussmann [**33**, Theorem 4.2] showed that nevertheless, for an integrable singular distribution Δ, we do in fact have $\Delta = \overline{\Delta}$. He achieves this as a consequence of his Orbit Theorem [**33**, Theorem 4.1]), which we state for completeness.

DEFINITION 2.18. An integral submanifold (L, σ_L) of Δ is *maximal* if it is connected, and every connected integral submanifold (S, σ_S) of Δ which intersects L is an open subset of L (in its manifold topology).[3]

THEOREM 2.19 (Sussmann's orbit theorem). *Given a singular distribution Δ and set of locally-defined vector fields D as above, the orbits of $\Psi\mathcal{A}_D$ are maximal integral submanifolds of $\overline{\Delta}$.*

Combining Stefan and Sussmann's results yields what is today called the Stefan-Sussmann theorem.

THEOREM 2.20 (Stefan-Sussmann). *A singular distribution Δ is integrable if and only if $\Delta = \overline{\Delta}$, in which case its maximal integral submanifolds are leaves of a singular foliation $\mathcal{F}$ such that $T_xL = \Delta_x$, where L is the leaf through x.*

REMARK 2.21. Stefan called the condition $\Delta\mathcal{A} = \overline{\Delta\mathcal{A}}$ *homogeneity* of the set of arrows $\mathcal{A}$. Sussmann worked with a set D of partially-defined vector fields spanning a singular distribution Δ, and used the term *D-invariance* for the condition $\Delta = \overline{\Delta}$. We will adopt Stefan's term "homogeniety." This condition is implicit in earlier work by Lobry [**24**], c.f. [**21**, page 53].

When $\dim \Delta_x$ is constant over M, homogeneity of Δ is equivalent to involutivity, and we recover the global Frobenius theorem 1.1 for regular foliations. In fact, Deahna [**5**], who provided necessary and sufficient conditions for integrability of a regular distribution before Frobenius, gave homogeniety of Δ, and not involutivity, as the sufficient condition, c.f. [**31**, page 525].

We already remarked that only Sussmann proved maximality of the integral submanifolds of an integrable Δ. Only Stefan proved the orbits of $\Psi\mathcal{A}$ were weakly-embedded submanifolds, and not merely immersed, and only Stefan showed the orbits (i.e. the maximal integral submanifolds) assemble into a singular foliation.

EXAMPLE 2.22. Let $A \implies M$ be a Lie algebroid, with anchor $\rho : A \to TM$. The image of the anchor, $\rho(A)$, is an integrable singular distribution. In this case, a maximal integral submanifold of Δ through $x \in M$ coincides with the set of all points reachable from x via A-paths, which are paths γ in M for which there exists a path $\tilde{\gamma}$ in A such that $\rho \circ \tilde{\gamma} = \dot{\gamma}$.

By passing to its associated Lie algebroid, one finds that the connected components of the orbits of a Lie groupoid $\mathcal{G} \rightrightarrows M$, and the connected components of the orbits of a Lie group G acting smoothly on M, assemble into a singular foliation.

[3]Equivalently, (L, σ_L) if, at each point in L, it is maximal if it is maximal with respect to the inclusion.

Historically, integrability of $\rho(A)$ follows from a result of Hermann [**12**] that predates Stefan and Sussmann's work. Hermann showed that any *locally finitely generated Lie sub-algebra*[4] of $\mathfrak{X}(M)$ induces an integrable singular distribution, and we can show that $\rho(A)$ satisfies this hypothesis. That the integral submanifolds are maximal, and assemble into a singular foliation, still requires Stefan's contribution to the Stefan-Sussmann theorem.

As a corollary, we can completely describe singular foliations by their associated singular distributions. This generalizes the Frobenius theorem as stated in the Introduction.

COROLLARY 2.23. *Given a singular foliation $\mathcal{F}$, the collection $T\mathcal{F} := \bigcup_{x\in M} T_xL$ is an integrable singular distribution. Given an integrable singular distribution Δ, the partition of M into maximal integral submanifolds is a singular foliation, denoted $\mathcal{F}_\Delta$. The assignments*

$$\mathcal{F} \mapsto T\mathcal{F}, \quad \Delta \mapsto \mathcal{F}_\Delta$$

are inverses of each other.

REMARK 2.24. We make two comparisons to the regular case. First, homogeneity (see Remark 2.21) implies involutivity, but for non-regular singular distributions, homogeneity is stronger than involutivity. Indeed, the singular distribution spanned by X and Y in Example 2.15 is involutive, but it is not integrable. For regular distributions, involutivity and homogeneity are equivalent.

Second, the maximal integral submanifolds of an integrable regular distribution may also be described as equivalence classes of the relation: $x \sim y$ if there is a path γ from x to y tangent to Δ, i.e. $\dot\gamma(t) \in \Delta_{\gamma(t)}$ for all t. In general, this is not the case for non-regular singular distributions. For example, consider the singular distribution spanned by the Euler vector field $X := x\frac{\partial}{\partial x} + y\frac{\partial}{\partial y}$ on $\mathbb{R}^2$. This has maximal integral submanifolds the origin $\{(0,0)\}$, and the rays from the origin. However, every two points can be joined by a path tangent to Δ.

2.2. Stefan's results and diffeology. While Stefan's Theorem 2.14 is the most cited part of his paper [**32**], in the same work he also proved a result about orbits of subgroups of $\mathrm{Diff}(M)$, which has applications to diffeology. In this section, the *D-topology*, and in particular *D-connectedness* of a diffeological space X is central.[5] The D-topology on X is the finest topology for which all the plots are continuous, and its D-connected components coincide with its (smooth) path-connected components. See Chapters 2 and 5 of [**13**] for details.

Stefan proved [**32**, Theorem 3]:

COROLLARY 2.25. *Let G be a D-connected diffeological subgroup of* $\mathrm{Diff}(M)$. *Then the G-orbits assemble into a singular foliation of M.*

PROOF. Let $\mathcal{A}$ consist of all plots (paths) $a : \mathbb{R} \to G$ with $a(0) = \mathrm{id}$. Then $\mathcal{A}$ is a set of arrows. Because G is D-connected, it is path-connected ([**13**, Article 5.7]). Thus $\bigcup_{a\in\mathcal{A}} a(\mathbb{R}) = G$, and $\Psi\mathcal{A}$ is the pseudogroup generated by G, so the result follows from Theorem 2.14. □

[4] meaning a Lie subalgebra D of $\mathfrak{X}(M)$ such that, for every open $U \subseteq M$, there exist $X^1, \ldots, X^k \in D$ such that $D|_U \subseteq C^\infty(U)\mathrm{span}(X^1|_U, \ldots, X^k|_U)$.

[5] Here the "D" stands for "Diffeology." In this context, it does not represent some family of vector fields.

Conversely, we have:

PROPOSITION 2.26. *Every singular foliation of M is a partition of M into orbits of some D-connected subgroup of* $\mathrm{Diff}(M)$.

PROOF. Fix a singular foliation $\mathcal{F}$. For each $x \in M$, take a collection $\{X^x\}$ of compactly supported vector fields whose values at x span $T_x\mathcal{F}$. This is possible because $T\mathcal{F}$ is smooth (Definition 2.13, Corollary 2.23), and M supports bump functions. Let $\mathcal{A}$ denote the collection of all plots of the form $t \mapsto \Phi^{X^x}(t, \cdot)$. Then $\bigcup_{a\in\mathcal{A}} a(\mathbb{R})$ generates some D-connected subgroup of $\mathrm{Diff}(M)$, and its orbits coincide with the orbits of $\Psi\mathcal{A}$, which are precisely the leaves of $\mathcal{F}$ by the Stefan-Sussmann Theorem 2.20. □

REMARK 2.27. In the proof above, we required a collection of vector fields which spanned $T\mathcal{F}$. We were not concerned with how many vector fields were necessary. In 2008 and 2012, respectively, Sussmann [**34**] and Drager, Lee, Park, and Richardson [**7**] independently showed that only finitely many are required. Precisely, they showed that given a singular distribution Δ, there exist finitely many vector fields $X^1, \dots, X^n$ such that $\Delta_x = \mathrm{span}\{X^1_x, \dots, X^n_x\}$ for every $x \in M$. Drager et al. provided some bounds for n. Note, however, that the collection $X^1, \dots, X^n$ does not necessarily generate the sheaf of sections of Δ. In fact, Drager et. al. [**7**, Section 5] give an example of a singular distribution over $\mathbb{R}$ whose sheaf of sections is not finitely generated.

As an immediate of Proposition 2.26, we have:

COROLLARY 2.28. *For a singular foliation $\mathcal{F}$, the group of diffeomorphisms of M which fix the leaves of $\mathcal{F}$ acts transitively on each leaf.*

It follows that, for a manifold M, it is equivalent to consider singular foliations, to consider integrable distributions, and to consider orbits of D-connected subgroups of $\mathrm{Diff}(M)$. Here is one consequence of this fact.

PROPOSITION 2.29. *The partition given by a singular foliation $\mathcal{F}$ of M satisfies the frontier condition: if S and L are leaves, and $S \cap \overline{L} \neq \emptyset$, then $S \subseteq \overline{L}$ (here the closure is taken relative to M).*

PROOF. Fix a D-connected subgroup G of $\mathrm{Diff}(M)$ whose orbits are the leaves of $\mathcal{F}$. It suffices to show $\overline{L}$ is G-invariant. Suppose $x \in \overline{L}$, and let (x_n) be a sequence of points in L which converge (in M) to x. Then for any $g \in G$, we have $g(x_n) \to g(x)$, and all the $g(x_n)$ are in L, hence $g(x) \in \overline{L}$. Since g was arbitrary, this completes the proof. □

2.3. Androulidakis-Skandalis singular foliations. We view singular foliations as partitions of a manifold into leaves, and have seen that by the Stefan-Sussmann theorem, this is equivalent to considering integrable singular distributions. Some authors take a different approach, and view singular foliations as choices of submodules of vector fields. We use the definition given by Androulidakis and Skandalis [**1**].

DEFINITION 2.30. An *Androulidakis-Skandalis singular foliation* is a $C^\infty(M)$-submodule $\mathscr{D}$ of $\mathfrak{X}_c(M)$ that is locally finitely generated[6] and involutive.

[6]this means that, about every $x \in M$, there is a neighbourhood U and $Y^1, \dots, Y^k \in \mathscr{D}|_U$ such that $\mathscr{D}|_U = C^\infty_c(U)\mathrm{span}(Y^1, \dots, Y^k)$, where $\mathscr{D}|_U$ is the $C^\infty(U)$-submodule of $\mathfrak{X}_c(U)$ generated by all the $Y = fX|_U$, where $f \in C^\infty_c(U)$ and $X \in \mathscr{D}$.

REMARK 2.31. As noted, for example, by Garmendia and Zambon [**9**, Remark 1.8], and by Wang [**35**, Remark 2.1.13], it is equivalent to define an Androulidakis-Skandalis singular foliation as an involutive, locally finitely generated subsheaf of the sheaf of vector fields on M. This approach could allow a definition of singular foliations on possibly non-Hausdorff manifolds, where the submodule and sheaf theoretic notions no longer coincide, c.f. Remark 3.13.

The Stefan-Sussmann Theorem 2.20 implies that the singular distribution associated to an Androulidakis-Skandalis singular foliation $\mathscr{D}$ is integrable (or, one can apply an earlier result of Hermann, c.f. Example 2.22). Therefore $\mathscr{D}$ induces a partition of a manifold M into leaves, and this partition is a Stefan singular foliation. However, many different choices of $\mathscr{D}$ may induce the same singular foliation, as the example below shows.

EXAMPLE 2.32. Let $\mathscr{D}_k$ be generated by $x^k \frac{\partial}{\partial x}$. Each $\mathscr{D}_k$ induces the Stefan singular foliation of $\mathbb{R}$ with three leaves: $x < 0$, $\{0\}$, and $x > 0$. However, no two of the modules $\mathscr{D}_k$ are isomorphic.

Every Lie algebroid, hence every Lie groupoid and action of a Lie group on a manifold, induces an Androulidakis-Skandalis singular foliation. Conversely, Androulidakis and Zambon [**2**, Proposition 1.3] have described an Androulidakis-Skandalis singular foliation $\mathscr{D}_{\text{counter}}$ of $\mathbb{R}^2$ that is not induced by any Lie algebroid. We do not give the details here, but note that the underlying Stefan singular foliation is the partition of $\mathbb{R}^2$ into point leaves $\{(k,0)\}$ for natural k, and the complement $\mathbb{R}^2 \smallsetminus \bigcup_{k\geq 1}\{(k,0)\}$. This underlying Stefan singular foliation is induced by a smooth Lie group action on $\mathbb{R}^2$, namely that generated by vector fields $f(x,y)\frac{\partial}{\partial x}$ and $f(x,y)\frac{\partial}{\partial y}$, where f is a bounded, non-negative function that vanishes precisely on the point leaves $\{(k,0)\}$. But this Lie group action does not induce the Androulidakis-Skandalis singular foliation $\mathscr{D}_{\text{counter}}$.

At time of writing, the following question is open:

QUESTION. Is every Stefan singular foliation induced by some Androulidakis-Skandalis singular foliation?

As we see in Example 2.32 above, even if the answer is affirmative, the choice of Androulidakis-Skandalis singular foliation is not unique. It is also not natural, in general: consider the singular foliation $\mathcal{F}$ of $\mathbb{R}$ whose leaves are $\{x\}$ with $x \leq 0$, and $x > 0$. By [**7**, Proposition 5.3], the space of all vector fields tangent to $\mathcal{F}$ is not locally finitely generated, and it is not clear how to choose a "best" vector field tangent to $\mathcal{F}$ to use as a generator of an Androulidakis-Skandalis singular foliation.

3. Transverse equivalence of singular foliations

Given a singular foliation $\mathcal{F}$ of M, the leaf space $M/\mathcal{F}$ is naturally a diffeological space. For convenience, we recall the plots of $M/\mathcal{F}$.

DEFINITION 3.1. A map $p : U \to M/\mathcal{F}$, where U is an open subset of Cartesian space, is a *plot* of the quotient diffeology of $M/\mathcal{F}$ if, about each $r \in U$, there is an open neighbourhood V or r, and a smooth map $q : V \to M$, such that $p|_V = \pi \circ q$, where π is the quotient map.

The D-topology coincides with the quotient topology [**13**, Article 2.12]. However, whereas the quotient topology on $M/\mathcal{F}$ may be trivial, or lose information about the singular foliation $\mathcal{F}$, its diffeology is often richer, as the following examples show.

EXAMPLE 3.2. Let $\pi : \mathbb{R}^2 \to T^2 := \mathbb{R}^2/\mathbb{Z}^2$ denote the quotient map for the $\mathbb{Z}^2$ action on $\mathbb{R}^2$. For irrational α, let S_α be the line in $\mathbb{R}^2$ with slope α. The conjugacy classes of the subgroup $\pi(S_\alpha)$ of T^2 assemble into a regular foliation of T^2. We call the leaf space $T_\alpha := T^2/\pi(S_\alpha)$ an *irrational torus.*

The quotient topology on T_α is always trivial, but the quotient diffeology is not: Donato and Iglesias-Zemmour [**6**] proved that $T_\alpha \cong T_\beta$ if and only if $\beta = \frac{a+b\alpha}{c+d\alpha}$, where $a, b, c, d \in \mathbb{Z}$ and $ad - bc = \pm 1$.

EXAMPLE 3.3. Consider the action of the orthogonal group $O(n)$ on $\mathbb{R}^n$. The leaves of the induced singular foliation are the orbits of the action, namely the origin $\{0\}$ and the concentric spheres. Topologically, the quotient spaces $\mathbb{R}^n/O(n)$ are all homeomorphic to the half-line $[0, \infty)$ with its subspace topology. But by [**13**, Exercise 50,51] they are all diffeologically distinct, and none are diffeologically diffeomorphic to $[0, \infty)$ with its subset diffeology. For $n = 1$, we get a diffeological orbifold $\mathbb{R}/\mathbb{Z}_2$.

We may, therefore, propose the diffeological space $M/\mathcal{F}$ as a model for the transverse geometry of $\mathcal{F}$. In the next sections, we compare this notion with another model.

3.1. Molino transverse equivalence. In [**28**, Definition 2.1], Molino defines a notion of transverse equivalence for regular foliations. In [**9**], Garmendia and Zambon extended this notion to Androulidakis-Skandalis singular foliations, and our definition below is similar to theirs and extends Molino's. However, due to the differences between Androulidakis-Skandalis and Stefan singular foliations outlined in Section 2.3, our definition does not coincide with Garmendia and Zambon's.

We need two technical Lemmas first.

LEMMA 3.4. *Suppose $p : M \to N$ is a surjective submersion with connected fibers, and assume N is connected. Then M is connected.*

PROOF. Take any continuous function $g : M \to \{0, 1\}$. Because the fibers of p are connected, g is constant on the fibers. Therefore, as p is a submersion, there is a smooth function $h : N \to \{0, 1\}$ such that $h \circ p = g$. Since N is connected, $h(N)$ is a single point; since p is surjective, $g(M) = h(N)$ is a single point. Therefore every continuous function $g : M \to \{0, 1\}$ is constant, and we conclude that M is connected. □

LEMMA 3.5. *Suppose $p : M \to N$ is a submersion. For every locally defined vector field Y on N, and every $v \in T_xM$ such that $dp_x(v) = Y_{p(x)}$, there is a locally defined vector field X on M such that $X_x = v$ and X is p-related to Y.*

PROOF. Because p is a submersion, without loss of generality we may assume that $p : \mathbb{R}^m \to \mathbb{R}^n$ is the projection of the first n coordinates. For $v = (v^1, \ldots, v^m) \in T_xM$, define

$$X_{x'} := (Y_{p(x')}, v^{n+1}, \ldots, v^m).$$

Then X is p-related to Y, because $dp_{x'}$ is the projection of the first n coordinates, and the condition $dp_x(v) = Y_{p(x)}$ writes

$$(v^1, \dots, v^n) = Y_{p(x)},$$

so $X_x = v$. □

PROPOSITION 3.6. *Let $p : M \to N$ be a surjective submersion with connected fibers. If Δ is an integrable singular distribution on N, then $(dp)^{-1}(\Delta)$ is an integrable singular distribution on M. If L is a leaf of the singular foliation of N induced by Δ, then $p^{-1}(L)$ is a leaf of the singular foliation of M induced by $(dp)^{-1}(\Delta)$.*

PROOF. To see that $(dp)^{-1}(\Delta)$ is a smooth singular distribution, take $v \in ((dp)^{-1}(\Delta))_x$. Then $dp_x(v) \in \Delta_{p(x)}$, and since Δ is smooth, there is some partially defined vector field Y of N tangent to Δ with $dp_x(v) = Y_{p(x)}$. By Lemma 3.5, we get a partially defined vector field X of M such that $X_x = v$, and $dp(X) = Y$. This vector field is tangent to $(dp)^{-1}(\Delta)$, and passes through v. Hence $(dp)^{-1}(\Delta)$ is smooth.

To see that $(dp)^{-1}(\Delta)$ is integrable, let $x \in M$ and let L be the maximal integral submanifold of Δ through $p(x)$. Because p is a surjective submersion, $p^{-1}(L)$ is a weakly-embedded submanifold of M, and for each $x' \in p^{-1}(L)$,

$$T_{x'}(p^{-1}(L)) = (dp_{x'})^{-1}(T_{p(x')}L) = ((dp)^{-1}(\Delta))_{x'}.$$

The Stefan-Sussmann Theorem 2.20 implies that the maximal integral submanifolds of $(dp)^{-1}(\Delta)$ assemble into a singular foliation of M. But the $p^{-1}(L)$ already partition M into connected (by Lemma 3.4) integral submanifolds of $(dp)^{-1}(\Delta)$. Corollary 2.23 lets us conclude that the $p^{-1}(L)$ are the maximal integral submanifolds of $(dp)^{-1}(\Delta)$. □

In light of Proposition 3.6, we can use the following notation.

DEFINITION 3.7. Given a surjective submersion with connected fibers $p : M \to N$, and an integrable singular distribution Δ on N with associated singular foliation $\mathcal{F}$, let $p^{-1}\Delta := (dp)^{-1}(\Delta)$, and let $p^{-1}\mathcal{F}$ denote the singular foliation associated to $p^{-1}\Delta$; its leaves are the sets $p^{-1}(L)$, where the L are leaves of $\mathcal{F}$. We call $p^{-1}\Delta$ and $p^{-1}\mathcal{F}$ the *pullbacks* of Δ and $\mathcal{F}$.

Using the pullback foliation, we now define Molino transverse equivalence of singular foliations.

DEFINITION 3.8. Two singular foliations, $(N_0, \mathcal{F}_0)$ and $(N_1, \mathcal{F}_1)$, are *Molino transverse equivalent* if there exists a singular foliation $(M, \mathcal{F})$ and surjective submersions with connected fibers $p_i : M \to N_i$ such that $p_i^{-1}(\mathcal{F}_i) = \mathcal{F}$. We will write $\mathcal{F}_0 \cong \mathcal{F}_1$.

Naturally, we must show:

PROPOSITION 3.9. *Molino transverse equivalence is an equivalence relation on singular foliations.*

PROOF. Reflexivity is witnessed by the identity. Symmetry is clear because we can reverse the roles of the N_i. For transitivity, build the following diagram:

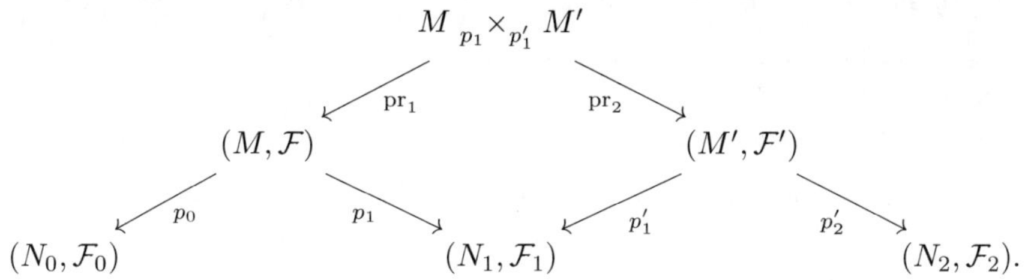

The bottom two rows indicate the assumed Molino transverse equivalences. Denote the fiber product at the top by M''. This is a manifold because p_1 is a submersion. The projections are surjective submersions because the p_i and p_i' are surjective submersions, thus so too are the compositions $p_0 \circ \mathrm{pr}_1$ and $p_2' \circ \mathrm{pr}_2$. We claim that, moreover, these have connected fibers, and pull back $\mathcal{F}_0$ and $\mathcal{F}_2$, respectively, to the same singular foliation of M''.

First, in light of Lemma 3.4, it suffices to show, without loss of generality, that pr_1 has connected fibers. Let $x \in M$, and take (x, z) and (x, z') in M''. Then z and z' are in the same fiber of p_1', which is connected, hence there is a path γ' joining them in this fiber. Then $(x, \gamma'(t))$ is a path in the fiber of pr_1 joining (x, z) and (x, z'), as required.

The second assertion follows from the fact that any leaf L of $\mathcal{F}$ has the form $p_0^{-1}(L_0) = p_1^{-1}(L_1)$, and any leaf L' of $\mathcal{F}'$ has the form $(p_1')^{-1}(L_1') = (p_2')^{-1}(L_2')$, and from using the usual property of pullbacks that $(p_0 \circ \mathrm{pr}_1)^{-1}\mathcal{F}_0 = \mathrm{pr}_1^{-1} p_0^{-1}\mathcal{F}_0$, etc. This proves that $(M'', (p_0 \circ \mathrm{pr}_1)^{-1}(\mathcal{F}_0))$, with the surjective submersions $p_0 \circ \mathrm{pr}_1$ and $p_2' \circ \mathrm{pr}_2$, gives a Molino transverse equivalence between $\mathcal{F}_0$ and $\mathcal{F}_2$. $\square$

We now relate Molino transverse equivalence and diffeology. One direction is straightforward: a Molino transverse equivalence induces a diffeological diffeomorphism of the leaf spaces.

PROPOSITION 3.10. *Suppose $p : M \to N$ is a surjective submersion with connected fibers, and let $\mathcal{F}$ be a singular foliation of N. Then the map*

$$\varphi : N/\mathcal{F} \to M/p^{-1}\mathcal{F}, \quad L \mapsto p^{-1}(L)$$

is a diffeological diffeomorphism. Consequently, if $p_i : M \to N_i$ witness a Molino transverse equivalence between $\mathcal{F}_i$, then the map

$$N_0/\mathcal{F}_0 \to N_1/\mathcal{F}_1, \quad L \mapsto p_1(p_0^{-1}(L))$$

is a diffeological diffeomorphism.

PROOF. The map φ is well-defined by definition of $p^{-1}\mathcal{F}$, and its inverse if $p^{-1}(L) \mapsto L$. Both φ and φ^{-1} fit in the following diagram:

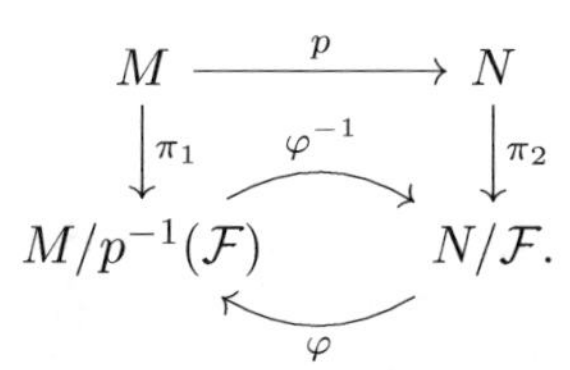

Because π_1, and $\pi_2 \circ p$, are local subductions, so are φ and φ^{-1}. $\square$

EXAMPLE 3.11. Consider the singular foliations induced by the action of $O(n)$ on $\mathbb{R}^n$, from Example 3.3. Because the $\mathbb{R}^n/O(n)$ are not diffeomorphic for different n, the singular foliations of $\mathbb{R}^n$ are not Molino transversely equivalent.

The converse is more subtle. Indeed, there exist two regular foliations, $(N_i, \mathcal{F}_i)$, with diffeomorphic leaf spaces, but which are not Molino transverse equivalent. We will see this by detouring through Morita equivalence of Lie groupoids.

3.2. Molino versus Morita equivalence. We saw in Example 2.8 and Example 2.22 that every Lie groupoid induces a singular foliation on its base manifold. For Lie groupoids, there is a long established notion of transverse, or weak, equivalence, called *Morita equivalence.* In this subsection, we compare the notions of Molino and Morita equivalence. See Appendix A for the relevant language and constructions from the theory of Lie groupoids, which we use freely here.

PROPOSITION 3.12. *Let $\mathcal{G} \rightrightarrows M$ and $\mathcal{H} \rightrightarrows N$ be two source-connected Lie groupoids with Hausdorff arrow spaces. If $\mathcal{G}$ and $\mathcal{H}$ are Morita equivalent, then the singular foliations $\mathcal{F}_{\mathcal{G}}$ and $\mathcal{F}_{\mathcal{H}}$ on M and N are Molino transverse equivalent.*

PROOF. Fix an invertible bibundle $P : \mathcal{G} \to \mathcal{H}$ (see Definition A.3 for details)

$$\begin{array}{ccccc} \mathcal{G} \circlearrowright & & P & & \circlearrowleft \mathcal{H} \\ & \swarrow^{a} & & \searrow^{a'} & \\ M & & & & N. \end{array}$$

We claim that this exhibits a Molino transverse equivalence. First, since $\mathcal{G}$ and $\mathcal{H}$ are Hausdorff, by Lemma A.4, we may choose P to be Hausdorff. Both a and a' are surjective submersions by assumption. Since $\mathcal{G}$ acts freely and transitively on the fibers of a', the fiber of a' over x_0' is diffeomorphic to $s^{-1}(x_0') \cdot p_0$ for any fixed p_0 in the fiber, and this is connected because $\mathcal{G}$ is source-connected. Similarly, the fibers of a are connected.

All that remains is to show the leaves of $a^{-1}\mathcal{F}_{\mathcal{G}}$ coincide with the leaves of $(a')^{-1}(\mathcal{F}_{\mathcal{H}})$. Let $\mathcal{O}$ be the orbit through a fixed x_0 in M. Choose $p_0 \in P$ with $a(p_0) = x_0$, and let $\mathcal{O}'$ be the orbit of $a'(p_0)$ in N. We claim $a^{-1}(\mathcal{O}) = (a')^{-1}(\mathcal{O}')$.

Letting $a(p) \in \mathcal{O}$, we may show $a'(p) \in \mathcal{O}'$; the converse direction is similar. There is an arrow $g : x_0 \to a(p)$. By $\mathcal{G}$-invariance of a', both $g \cdot p_0$ and p are in the same fiber of a'. Since the $\mathcal{H}$ action is transitive on this fiber, there is some arrow h with $g \cdot p_0 \cdot h = p$. This action is only possible if h is an arrow $h : a'(p) \to a'(g \cdot p_0) = a'(p_0)$. Thus $a'(p)$ and $a'(p_0)$ are in the same orbit $\mathcal{O}'$, as required. $\square$

REMARK 3.13. The Hausdorff assumption is material, because if $\mathcal{G}$ and $\mathcal{H}$ are not Hausdorff, it is not clear that we may choose the invertible bibundle P above to be Hausdorff. One possible attempt to treat the case of non-Hausdorff Lie groupoids is to modify the definition of Molino transverse equivalence to allow for non-Hausdorff manifolds. But doing so begs the question of how to define singular foliations on non-Hausdorff manifolds, which we leave for another time. Garmendia and Zambon encounter a similar problem in the case of Androulidakis-Skandalis singular foliations, and suggest it may be remedied by viewing Androulidakis-Skandalis singular foliations as sheaves instead of submodules, as in Remark 2.31; see [**9**, Section 4.1].

Given an arbitrary singular foliation $(N, \mathcal{F})$, it is unknown whether there is a Lie groupoid $\mathcal{G} \rightrightarrows N$ which induces $\mathcal{F}$. Furthermore, it is possible that two non-Morita equivalent Lie groupoids induce the same singular foliation.

EXAMPLE 3.14. Consider the actions of the general linear group $\mathrm{GL}(n)$, and the special linear group $\mathrm{SL}(n)$, on $\mathbb{R}^n$. Both induce the same singular foliation of $\mathbb{R}^n$, whose leaves are the origin and its complement. However, the action groupoids $\mathrm{GL}(n) \ltimes \mathbb{R}^n$ and $\mathrm{SL}(n) \ltimes \mathbb{R}^n$ are not Morita equivalent. This is because their stabilizer groups at the origin are not isomorphic.

For regular foliations $(N, \mathcal{F})$, however, we have a distinguished groupoid inducing the foliation, the holonomy groupoid $\mathrm{Hol}(\mathcal{F})$ (see Example A.5 for some details). It is reasonable to ask whether Molino transverse equivalence of regular foliations implies their holonomy groupoids are Morita equivalent, and indeed this is the case:

PROPOSITION 3.15. *If two regular foliations are Molino transverse equivalent, then their holonomy groupoids are Morita equivalent.*

This coincides with Proposition 3.30 in Garmendia and Zambon's paper [**9**], whose proof in the regular case is outlined after their Theorem 3.21. We recall that Garmendia and Zambon work with Androulidakis-Skandalis singular foliations, and not Stefan singular foliations. However, in the regular case, these notions coincide, so their proof of this statement works in this setting and we refer the reader there for details.

REMARK 3.16. In fact, Garmendia and Zambon prove a stronger statement. As we discussed in Section 2.3, Androulidakis-Skandalis singular foliations are not generally induced by any Lie groupoid. However, Androulidakis and Skandalis in [**1**] constructed an open topological groupoid for every Androulidakis-Skandalis singular foliation $\mathcal{F}$, called the *holonomy groupoid* of $\mathcal{F}$, which coincides with the usual holonomy groupoid in the regular case. Garmendia and Zambon proved that, provided the holonomy groupoids are Hausdorff, they are Morita equivalent (as open topological groupoids) if and only if the associated singular foliations are what they call Hausdorff Morita equivalent.

COROLLARY 3.17. *Two regular foliations with Hausdorff holonomy groupoids are Molino transverse equivalent if and only if their holonomy groupoids are Morita equivalent.*

We are now prepared to illustrate the example alluded to at the end of Section 3.1, of two regular foliations with diffeomorphic leaf spaces, but which are not Molino transverse equivalent. This example also appears in [**17**], in a different context.

EXAMPLE 3.18. Let $h : \mathbb{R} \to \mathbb{R}$ be a smooth non-negative function that is flat[7] at 0 and is positive everywhere else, such that the vector field $X := h\frac{\partial}{\partial x}$ is complete. Let $\psi := \Phi_1^X$ denote the time-1 flow of X, and set

$$\hat{\psi}(x) := \begin{cases} \psi(x) & \text{if } x \geq 0 \\ \psi^{-1}(x) & \text{if } x < 0. \end{cases}$$

[7]this means that h and all its derivatives vanish at the point

Both ψ and $\hat{\psi}$ are smooth. By iterating ψ, we get a $\mathbb{Z}$-action on $\mathbb{R}$, and also one on $\mathbb{R}^2$ given by

$$k \cdot (t, x) := (t + k, \psi^k(x)).$$

This action preserves the foliation of $\mathbb{R}^2$ by horizontal lines (but not the leaves themselves), and the action is free and properly discontinuous. Therefore, passing to the quotient, $M_\psi := \mathbb{R}^2/\psi$ is a manifold and we have the quotient foliation $\mathcal{F}_\psi$. Moreover, $M_\psi/\mathcal{F}_\psi$ is diffeomorphic to $\mathbb{R}/\psi$. Similarly, we can form $M_{\hat{\psi}}/\mathcal{F}_{\hat{\psi}}$, and this is diffeomorphic to $\mathbb{R}/\hat{\psi}$. But $\mathbb{R}/\psi = \mathbb{R}/\hat{\psi}$, so we have two foliations foliation with diffeologically diffeomorphic leaf spaces. However, by [**17**, Proposition 7.1], the étale holonomy groupoids (see Example A.5) associated to these foliations are not Morita equivalent. Therefore their holonomy groupoids are not Morita equivalent, hence by Proposition 3.15, the foliations are not transverse equivalent.

3.3. Diffeology and Molino equivalence for quasifolds. While in general, a diffeological diffeomorphism between the leaf spaces of singular foliations does not induce a Molino transverse equivalence (c.f. Example 3.18), there is a class of regular foliations for which this holds. These are regular foliations whose holonomy groupoids are Morita equivalent to quasifold groupoids, introduced in [**17**].

DEFINITION 3.19. A *n-quasifold groupoid* is a Lie groupoid $\mathcal{G} \rightrightarrows M$, with Hausdorff arrow space, such that: for each $x \in M$, there is an open neighbourhood U of x, a countable group Γ acting affinely on $\mathbb{R}^n$, an open subset V of $\mathbb{R}^n$, and an isomorphism[8] of Lie groupoids $\mathcal{G}|_U \to (\Gamma \ltimes \mathbb{R}^n)|_{\mathsf{V}}$.

PROPOSITION 3.20. *Assume* $(N_i, \mathcal{F}_i)$ *are regular foliations, and their holonomy groupoids* $\mathrm{Hol}(\mathcal{F}_i)$ *are each Morita equivalent to effective quasifold groupoids* $\mathcal{G}_i \rightrightarrows M_i$. *If the leaf spaces* $N_i/\mathcal{F}_i$ *are diffeologically diffeomorphic, then the foliations* $\mathcal{F}_i$ *are Molino transverse equivalent.*

PROOF. Morita equivalent Lie groupoids have diffeomorphic orbit spaces, therefore the $M_i/\mathcal{G}_i$ are diffeomorphic to $N_i/\mathcal{F}_i$, for each $i = 0, 1$. It follows from the assumption that the $M_i/\mathcal{G}_i$ are diffeomorphic to each other. Then, because the $\mathcal{G}_i$ are effective quasifold groupoids, we use [**17**, Proposition 5.4] to conclude that the $\mathcal{G}_i$ are Morita equivalent. Therefore the $\mathrm{Hol}(\mathcal{F}_i)$ are Morita equivalent.

If we show each $\mathrm{Hol}(\mathcal{F}_i)$ is Hausdorff, we may conclude the proof with Corollary 3.17. But to be Hausdorff is invariant under Morita equivalence ([**27**, Proposition 5.3], and the $\mathcal{G}_i$ are Hausdorff by definition of quasifold groupoid, hence the holonomy groupoids are Hausdorff, as required. □

REMARK 3.21. The assumption that the $\mathcal{G}_i$ are effective is unnecessary. Each $\mathrm{Hol}(\mathcal{F}_i)$ is Morita equivalent to its étale holonomy groupoid, as outlined in Example A.5, and these are effective. To be effective is stable under Morita equivalence ([**27**, Example 5.21 (2)], hence the $\mathcal{G}_i$ are necessarily effective.

In Proposition 3.20, we impose a condition on the holonomy groupoid $\mathrm{Hol}(\mathcal{F}_i)$, and not on the foliation itself. There are cases, however, where we can deduce the holonomy groupoid is Morita equivalent to a quasifold groupoid directly from properties of the foliation. We end this brief section with an example.

[8] invertible functor with smooth inverse

EXAMPLE 3.22. A distinguished class of quasifold groupoids are proper étale groupoids, which are sometimes called orbifold groupoids. Their local linear models arise as a consequence of the linearization theorem for proper groupoids, for example see [**4**]; the resulting groups Γ are the stabilizer subgroups $\mathcal{G}_x$, which are finite. If all of the leaves of a regular foliation $\mathcal{F}$ are compact and have finite holonomy groups, by the Reeb stability theorem the holonomy groupoid $\mathrm{Hol}(\mathcal{F})$ is proper ([**27**, Example 5.28 (2)]). Because to be proper is stable under Morita equivalence ([**27**, Proposition 5.26]), we may apply Proposition 3.20 to such foliations $\mathcal{F}$.

4. Basic cohomology of singular foliations

Given a singular foliation $(M,\mathcal{F})$, we have the associated complex of basic differential forms. In this section, we will show that Molino transverse equivalent singular foliations have identical complexes of basic forms, strengthening the notion that the basic complex captures the transverse geometry. We will end with a discussion of the comparison between the complex of basic differential forms, and the complex of diffeological differential forms on the leaf space.

DEFINITION 4.1. A differential form $\alpha \in \Omega^\bullet(M)$ is *$\mathcal{F}$-basic* if, for every vector field X tangent to $\mathcal{F}$, both

$$\iota_X\alpha = 0 \text{ and } \mathcal{L}_X\alpha = 0.$$

The first condition says α is *horizontal.* The second says α is *invariant.* We denote the collection of basic differential forms by $\Omega^\bullet_b(M,\mathcal{F})$. This is a de-Rham subcomplex of $\Omega^\bullet(M)$, with the usual differential. Its cohomology $H^\bullet_b(M,\mathcal{F})$ is the *basic cohomology* associated to $\mathcal{F}$.

EXAMPLE 4.2. If G is a connected Lie group acting smoothly on M, and $\mathcal{F}$ is the associated singular foliation, the $\mathcal{F}$-basic forms are precisely those that are horizontal and G-invariant. More generally, if $\mathcal{G}$ is a source-connected Lie groupoid, and $\mathcal{F}$ is the associated singular foliation, the $\mathcal{F}$-basic forms are those for which $s^*\alpha = t^*\alpha$, see [**26**, Proposition 5.5].

PROPOSITION 4.3. *Suppose $p : M \to N$ is a surjective submersion with connected fibers, and that $(N,\mathcal{F})$ is a singular foliation. Then p^* is an isomorphism from $\Omega^\bullet_b(N,\mathcal{F})$ to $\Omega^\bullet_b(M,p^{-1}\mathcal{F})$.*

PROOF. Because p is a submersion, p^* is injective. It remains to show p^* is into, and onto. First, let α be an $\mathcal{F}$-basic form on N, and take X tangent to $p^{-1}\mathcal{F}$. Then

$$(\iota_X p^*\alpha)_x = \alpha_{p(x)}(p_*X_x,\cdot) = (\iota_{p_*X_x}\alpha)_x = 0,$$

since p_*X_x is tangent to $\mathcal{F}$. As $d\alpha$ is also $\mathcal{F}$-basic, by a similar computation we have $\iota_X(p^*d\alpha) = 0$, so by Cartan's formula,

$$\mathcal{L}_X p^*\alpha = \iota_X(dp^*\alpha) = \iota_X(p^*d\alpha) = 0.$$

Second, let β be a $p^{-1}\mathcal{F}$-basic form on M. Since p is a surjective submersion with connected fibers, its fibers form a regular foliation $\mathcal{F}'$ of M. Any vector field tangent to these fibers is also tangent to $p^{-1}\mathcal{F}$, so the $p^{-1}\mathcal{F}$-basic forms are all $\mathcal{F}'$-basic. In particular, β is $\mathcal{F}'$-basic. Now, we can identify $N \cong M/\mathcal{F}'$, and we can identify the quotient with $p : M \to N$. By [**22**, Exercise 14.9], the pullback p^* is an isomorphism from $\Omega^\bullet(N)$ to $\Omega^\bullet_b(M,\mathcal{F}')$. This provides $\alpha \in \Omega^\bullet(N)$ such that $p^*\alpha = \beta$.

We now check that α is $\mathcal{F}$-basic. Take Y tangent to $\mathcal{F}$, and fix $x \in M$. By Lemma 3.5, lift Y to a p-related vector field X about x, which is tangent to $p^{-1}\mathcal{F}$. Since β is horizontal,

$$0 = (p^*\alpha)_x(X, \cdot) = \alpha_{p(x)}(Y_{p(x)}, p_*\cdot).$$

But p and p_* are onto, so we conclude $\iota_Y \alpha = 0$, and α is horizontal. We also check that since β is invariant,

$$p^*(\mathcal{L}_Y \alpha) = L_X p^* \alpha = 0.$$

But p^* is injective, so $\mathcal{L}_Y \alpha = 0$, and α is invariant. This completes the proof. □

COROLLARY 4.4. *If two singular foliations are Molino transverse equivalent, then their complexes of basic forms, hence their basic cohomologies, are isomorphic.*

Diffeology provides an alternative approach to using differential forms to study the transverse geometry of a singular foliation. Every diffeological space X comes with a de Rham complex of diffeological differential forms, $\Omega^\bullet(X)$. When X is a leaf space of a singular foliation $(N, \mathcal{F})$, we can directly compare $\Omega^\bullet(N/\mathcal{F})$ and $\Omega_b^\bullet(N, \mathcal{F})$.

THEOREM 4.5. *Fix a singular foliation $(N, \mathcal{F})$. The quotient map $\pi : N \to N/\mathcal{F}$ induces a map of chain complexes $\pi^* : \Omega^\bullet(N/\mathcal{F}) \to \Omega^\bullet(N)$. This is injective, and its image is contained in the space of $\mathcal{F}$-basic forms. The pullback π^* is an isomorphism $\Omega^\bullet(N/\mathcal{F}) \to \Omega_b^\bullet(N, \mathcal{F})$ whenever*

- *the set of points in leaves of dimension k is a diffeological submanifold of N, for each k, or*
- *π^* is an isomorphism when we replace $(N, \mathcal{F})$ with $N_{>0}$ (N with the zero-leaves excised) and the induced singular foliation $\mathcal{F}_{>0}$.*

This is the main result of [**26**]. It implies the earlier results from Karshon and Watts [**19**] and Watts [**36**], in the case of connected groups and source-connected groupoids, respectively. For regular foliations, this was proved by Hector, Marcías-Virgós, and Sanmartín-Carbón in [**11**]. The following question remains open:

QUESTION. Is $\pi^* : \Omega^\bullet(N/\mathcal{F}) \to \Omega_b^\bullet(N, \mathcal{F})$ always an isomorphism?

However, Corollary 4.4 lets us show π^* is an isomorphism for an *a priori* larger class of singular foliations than Theorem 4.5.

PROPOSITION 4.6. *Suppose $(N_i, \mathcal{F}_i)$ are Molino transverse equivalent singular foliations, and assume that $\pi_i : N_i \to N_i/\mathcal{F}_i$ induces an isomorphism $\pi_i^* : \Omega^\bullet(N_i/\mathcal{F}_i) \to \Omega_b^\bullet(N_i, \mathcal{F}_i)$ for $i = 0$. Then the same holds for $i = 1$.*

PROOF. Using Proposition 3.10, create the following commutative pentagon.

$$\begin{array}{ccccc} & & M & & \\ & \swarrow^{p_0} & & \searrow^{p_1} & \\ N_0 & & & & N_1 \\ \downarrow{\scriptstyle \pi_0} & & & & \downarrow{\scriptstyle \pi_1} \\ N_0/\mathcal{F}_0 & & \xrightarrow{\varphi} & & N_1/\mathcal{F}_1. \end{array}$$

All of φ^*, π_0^*, p_0^*, and p_1^* are isomorphisms of the relevant complexes. Thus π_1^* is too. □

We could also make the same statement in cohomology.

Appendix A. Review of Lie groupoids

Here we establish some necessary facts and terminology about Lie groupoids. For details, see [**27**] or [**23**]. A *Lie groupoid* is a category $\mathcal{G} \rightrightarrows M$, whose arrows are all invertible, such that: M, called the *base*, is a manifold, $\mathcal{G}$ has a smooth structure (but is not necessarily Hausdorff or second-countable), all the structure maps are smooth, and the source s (hence the target t) is a submersion with Hausdorff fibers. We typically denote an arrow g with source x and target x' by $g : x \to x'$. The relation $x \sim x'$ if there is an arrow $g : X \to x'$ is an equivalence relation, whose classes we call *orbits.*

Example A.1. If G is a Lie group acting smoothly on a manifold M, we can form the *action groupoid* $G \ltimes M$, with arrow space $G \times M$, base space M, source $(g,x) \mapsto x$, target $(g,x) \mapsto g \cdot x$, and multiplication $(g',x')(g,x) = (g'g,x)$. In this way, Lie groupoids subsume Lie group actions.

If the fibers of the source map s are connected, we say $\mathcal{G}$ is *source-connected.* When the arrow space $\mathcal{G}$ is Hausdorff, we call $\mathcal{G}$ itself Hausdorff. If $\mathcal{G}$ is Hausdorff, and the map $(s,t) : \mathcal{G} \to M \times M$ is proper (i.e. pre-images of compact sets are compact), we call $\mathcal{G}$ *proper.*

Given an immersed submanifold $\iota : (N, \sigma_N) \to M$, if $(t,s) : \mathcal{G} \to M \times M$ is transverse to $(\iota,\iota) : N \to M \times M$, then we can form the *pullback* of $\mathcal{G}$ along ι, and obtain $\iota^*\mathcal{G} \rightrightarrows N$, whose arrows $y \to y'$ are those arrows in $\mathcal{G}$ from $\iota(y) \to \iota(y')$. If U is an open subset of M, we denote the pullback of $\mathcal{G}$ to U by $\mathcal{G}|_U \rightrightarrows U$. Its arrows are the arrows of $\mathcal{G}$ with source and target in U.

We call a Lie groupoid *étale* if its arrow space and base space have the same dimension. In this case, the source and target are local diffeomorphisms. If the map

$$g \mapsto \mathrm{germ}_{s(g)}(t \circ s^{-1}),$$

where s^{-1} is the local inverse of s mapping $s(g)$ to g, is injective, we call the étale groupoid $\mathcal{G}$ *effective.*

Definition A.2. A *left action* of a Lie groupoid $\mathcal{G} \rightrightarrows M$ on a manifold P consists of a *multiplication* $\mu : \mathcal{G} \,{}_s\!\times_a P \to P$, which we usually denote by $\cdot$, and an *anchor* $a : P \to M$, such that the following diagram commutes:

$$\begin{array}{ccc} \mathcal{G} \,{}_s\!\times_a P & \xrightarrow{\ \mu\ } & P \\ \big\downarrow{\scriptstyle \mathrm{pr}_1} & & \big\downarrow{\scriptstyle a} \\ \mathcal{G} & \xrightarrow{\ t\ } & M, \end{array}$$

and furthermore

- $g' \cdot (g \cdot p) = g'g \cdot p$ whenever this makes sense, and
- $1_{a(p)} \cdot p = p$ for all $p \in P$.

We can similarly define a right action.

Definition A.3. An *invertible bibundle* from $\mathcal{G}$ to $\mathcal{H}$, denoted $P : \mathcal{G} \to \mathcal{H}$, is a (not-necessarily Hausdorff) manifold P on which $\mathcal{G}$ acts from the left (along anchor a) and $\mathcal{H}$ acts on the right (along anchor a'), such that the anchors are surjective submersions, the actions commute, and

- a is $\mathcal{H}$-invariant, and a' is $\mathcal{G}$-invariant, and

- the maps

$$\mathcal{G}\ {}_s\!\times_a P \to P\ {}_{a'}\!\times_{a'} P, \quad (g,p) \mapsto (g \cdot p, p),$$

and

$$P\ {}_{a'}\!\times_t \mathcal{H} \to P\ {}_a\!\times_a P, \quad (p,h) \mapsto (p, p \cdot h)$$

are diffeomorphisms. Equivalently, $\mathcal{G}$ and $\mathcal{H}$ act freely and transitively on the fibers of a' and a, respectively

We say $\mathcal{G}$ and $\mathcal{H}$ are *Morita equivalent* if there is an invertible bibundle between them.

LEMMA A.4. *If two Hausdorff Lie groupoids are Morita equivalent, there is a Hausdorff invertible bibundle between them.*

For a proof, see [**9**, Corollary A.7].

EXAMPLE A.5. Given a regular foliation $(M, \mathcal{F})$, we can construct the *holonomy groupoid* $\mathrm{Hol}(\mathcal{F}) \rightrightarrows M$. There is an arrow $x \to y$ if and only if x and y are in the same leaf L; in this case, each arrow is a *holonomy class* of a path from x to y within L. Two paths are in the same holonomy class if they induce the same diffeomorphism between small transversal slices (submanifolds which intersect the leaves transversely) at x and y. Paths in the same homotopy class are in the same holonomy class.

The holonomy groupoid is source-connected. Given a complete transversal T to $\mathcal{F}$, i.e. an immersed submanifold (T, σ_T) which intersects each leaf of $\mathcal{F}$ at least once, and always intersects leaves transversely, we can pull back the holonomy groupoid to T, obtaining $\mathrm{Hol}(\mathcal{F})|_T \rightrightarrows T$. This is an étale groupoid. Different choices of complete transversals T yield Morita equivalent pullback groupoids, hence we call $\mathrm{Hol}(\mathcal{F})|_T \rightrightarrows T$ "the" *étale holonomy groupoid* of $\mathcal{F}$. This groupoid is always effective.

A *Lie Algebroid* is a vector bundle $A \to M$, together with a Lie bracket on the space of sections of A, and a bundle map $a : A \to TM$, the *anchor*, satisfying a Leibniz rule with respect to the algebroid bracket. We denote a Lie algebroid by $A \implies M$. To each Lie groupoid $\mathcal{G} \rightrightarrows M$, we can associate a Lie algebroid $\mathrm{Lie}(\mathcal{G}) \to M$. The maximal integral submanifolds of the distribution $a(\mathrm{Lie}(\mathcal{G}))$ are the connected components of the orbits of $\mathcal{G}$.

Acknowledgments

I am grateful to my supervisor, Yael Karshon, for introducing me to singular foliations, and for describing the idea of Example 3.18. I also thank Jordan Watts and Jean-Pierre Magnot for their invitation to speak at the AMS-EMS-SMF Joint International Meeting in Grenoble, and Jean-Pierre Magnot for the invitation to submit this article to the present conference proceedings. Finally, I thank the anonymous reviewer and Alfonso Garmendia for several helpful comments on earlier drafts.

References

[1] Iakovos Androulidakis and Georges Skandalis, *The holonomy groupoid of a singular foliation*, J. Reine Angew. Math. **626** (2009), 1–37, DOI 10.1515/CRELLE.2009.001. MR2492988

[2] Iakovos Androulidakis and Marco Zambon, *Smoothness of holonomy covers for singular foliations and essential isotropy*, Math. Z. **275** (2013), no. 3-4, 921–951, DOI 10.1007/s00209-013-1166-5. MR3127043

[3] A. Clebsch, *Ueber die simultane Integration linearer partieller Differentialgleichungen* (German), J. Reine Angew. Math. **65** (1866), 257–268, DOI 10.1515/crll.1866.65.257. MR1579322

[4] Marius Crainic and Ivan Struchiner, *On the linearization theorem for proper Lie groupoids* (English, with English and French summaries), Ann. Sci. Éc. Norm. Supér. (4) **46** (2013), no. 5, 723–746, DOI 10.24033/asens.2200. MR3185351

[5] F. Deahna, *Ueber die Bedingungen der Integrabilität lineärer Differentialgleichungen erster Ordnung zwischen einer beliebigen Anzahl veränderlicher Größen* (German), J. Reine Angew. Math. **20** (1840), 340–349, DOI 10.1515/crll.1840.20.340. MR1578247

[6] Paul Donato and Patrick Iglésias, *Exemples de groupes difféologiques: flots irrationnels sur le tore* (French, with English summary), C. R. Acad. Sci. Paris Sér. I Math. **301** (1985), no. 4, 127–130. MR799609

[7] Lance D. Drager, Jeffrey M. Lee, Efton Park, and Ken Richardson, *Smooth distributions are finitely generated*, Ann. Global Anal. Geom. **41** (2012), no. 3, 357–369, DOI 10.1007/s10455-011-9287-8. MR2886203

[8] G. Frobenius, *Ueber das Pfaffsche Problem* (German), J. Reine Angew. Math. **82** (1877), 230–315, DOI 10.1515/crll.1877.82.230. MR1579710

[9] Alfonso Garmendia and Marco Zambon, *Hausdorff Morita equivalence of singular foliations*, Ann. Global Anal. Geom. **55** (2019), no. 1, 99–132, DOI 10.1007/s10455-018-9620-6. MR3916125

[10] Thomas Hawkins, *Frobenius, Cartan, and the problem of Pfaff*, Arch. Hist. Exact Sci. **59** (2005), no. 4, 381–436, DOI 10.1007/s00407-004-0095-4. MR2188939

[11] G. Hector, E. Macías-Virgós, and E. Sanmartín-Carbón, *De Rham cohomology of diffeological spaces and foliations*, Indag. Math. (N.S.) **21** (2011), no. 3-4, 212–220, DOI 10.1016/j.indag.2011.04.004. MR2835929

[12] Robert Hermann, *The differential geometry of foliations. II*, J. Math. Mech. **11** (1962), 303–315. MR142131

[13] Patrick Iglesias-Zemmour, *Diffeology*, Mathematical Surveys and Monographs, vol. 185, American Mathematical Society, Providence, RI, 2013, DOI 10.1090/surv/185. MR3025051

[14] Patrick Iglesias-Zemmour and Jean-Pierre Laffineur, *Noncommutative geometry and diffeology: the case of orbifolds*, J. Noncommut. Geom. **12** (2018), no. 4, 1551–1572, DOI 10.4171/JNCG/319. MR3896235

[15] C. G. J. Jacobi, *Ueber die Integration der partiellen Differentialgleichungen erster Ordnung* (German), J. Reine Angew. Math. **2** (1827), 317–329, DOI 10.1515/crll.1827.2.317. MR1577659

[16] Henri Joris, *Une C^∞-application non-immersive qui possède la propriété universelle des immersions* (French), Arch. Math. (Basel) **39** (1982), no. 3, 269–277, DOI 10.1007/BF01899535. MR682456

[17] Yael Karhson and David Miyamoto, *Quasifold groupoids and diffeological quasifolds*, preprint, 2022. https://arxiv.org/abs/2206.14776, to appear in Transformation Groups.

[18] Yael Karshon, David Miyamoto, and Jordan Watts, *Diffeological submanifolds and their friends*, preprint, 2022. https://arxiv.org/abs/2204.10381.

[19] Yael Karshon and Jordan Watts, *Basic forms and orbit spaces: a diffeological approach*, SIGMA Symmetry Integrability Geom. Methods Appl. **12** (2016), Paper No. 026, 19, DOI 10.3842/SIGMA.2016.026. MR3470746

[20] Jan Kubarski, *About Stefan's definition of a foliation with singularities: a reduction of the axioms* (English, with French summary), Bull. Soc. Math. France **118** (1990), no. 4, 391–394. MR1090406

[21] Sylvain Lavau, *A short guide through integration theorems of generalized distributions*, Differential Geom. Appl. **61** (2018), 42–58, DOI 10.1016/j.difgeo.2018.07.005. MR3856749

[22] John M. Lee, *Introduction to smooth manifolds*, 2nd ed., Graduate Texts in Mathematics, vol. 218, Springer, New York, 2013. MR2954043

[23] Eugene Lerman, *Orbifolds as stacks?*, Enseign. Math. (2) **56** (2010), no. 3-4, 315–363, DOI 10.4171/LEM/56-3-4. MR2778793

[24] Claude Lobry, *Contrôlabilité des systèmes non linéaires* (French), SIAM J. Control **8** (1970), 573–605. MR271979

[25] Michihiko Matsuda, *An integration theorem for completely integrable systems with singularities*, Osaka Math. J. **5** (1968), 279–283. MR243555

[26] David Miyamoto, *The basic de Rham complex of a singular foliation*, Int. Math. Res. Not. IMRN **8** (2023), 6364–6401, DOI 10.1093/imrn/rnac044. MR4574377

[27] I. Moerdijk and J. Mrčun, *Introduction to foliations and Lie groupoids*, Cambridge Studies in Advanced Mathematics, vol. 91, Cambridge University Press, Cambridge, 2003, DOI 10.1017/CBO9780511615450. MR2012261

[28] Pierre Molino, *Riemannian foliations*, Progress in Mathematics, vol. 73, Birkhäuser Boston, Inc., Boston, MA, 1988. Translated from the French by Grant Cairns; With appendices by Cairns, Y. Carrière, É. Ghys, E. Salem and V. Sergiescu, DOI 10.1007/978-1-4684-8670-4. MR932463

[29] Tadashi Nagano, *Linear differential systems with singularities and an application to transitive Lie algebras*, J. Math. Soc. Japan **18** (1966), 398–404, DOI 10.2969/jmsj/01840398. MR199865

[30] Bruce L. Reinhart, *Foliated manifolds with bundle-like metrics*, Ann. of Math. (2) **69** (1959), 119–132, DOI 10.2307/1970097. MR107279

[31] Hans Samelson, *Differential forms, the early days; or the stories of Deahna's theorem and of Volterra's theorem*, Amer. Math. Monthly **108** (2001), no. 6, 522–530, DOI 10.2307/2695706. MR1840658

[32] P. Stefan, *Accessible sets, orbits, and foliations with singularities*, Proc. London Math. Soc. (3) **29** (1974), 699–713, DOI 10.1112/plms/s3-29.4.699. MR362395

[33] Héctor J. Sussmann, *Orbits of families of vector fields and integrability of distributions*, Trans. Amer. Math. Soc. **180** (1973), 171–188, DOI 10.2307/1996660. MR321133

[34] Hector J. Sussmann, *Smooth distributions are globally finitely spanned*, Analysis and design of nonlinear control systems, Springer, Berlin, 2008, pp. 3–8, DOI 10.1007/978-3-540-74358-3_1. MR2758930

[35] Roy Wang, *On Integrable Systems & Rigidity for PDEs with Symmetry*, posted on 2017, `https://arxiv.org/abs/1712.00808`, DOI 10.48550/arXiv.1712.00808. PhD Thesis.

[36] Jordan Watts, *The orbit space and basic forms of a proper LIE groupoid*, Current trends in analysis, its applications and computation, Trends Math. Res. Perspect., Birkhäuser/Springer, Cham, 2022, pp. 513–523, DOI 10.1007/978-3-030-87502-2_52. MR4559547

University of Toronto
Email address: `david.miyamoto@mail.utoronto.ca`

Contemporary Mathematics
Volume **794**, 2024
https://doi.org/10.1090/conm/794/15929

Diffeological coarse moduli spaces of stacks over manifolds

Jordan Watts and Seth Wolbert

Abstract. In this paper, we consider diffeological spaces as stacks over the site of smooth manifolds, as well as the "underlying" diffeological space of any stack. More precisely, we consider diffeological spaces as so-called concrete sheaves and show that the Grothendieck construction sending these sheaves to stacks has a left adjoint: the functor sending any stack to its diffeological coarse moduli space. As an application, we restrict our attention to differentiable stacks and examine the geometry behind the coarse moduli space construction in terms of Lie groupoids and their principal bundles. Within this context, we define a "gerbe", and show when a Lie groupoid is such a gerbe (or when a stack is represented by one). Additionally, we define basic differential forms for stacks and confirm in the differentiable case that these agree (under certain conditions) with basic differential forms on a representative Lie groupoid. These basic differentiable forms in turn match the diffeological forms on the orbit space.

1. Introduction

The category of smooth manifolds lacks many structures that naturally arise in geometry. Important examples include (critical) level sets of smooth real-valued functions, orbit spaces of Lie group actions, and function spaces, among many others. Many categories extending the notion of smoothness have been introduced to account for this deficiency; for example, that of diffeological spaces, introduced by Souriau [**23**] (see Definition 2.2). Baez and Hoffnung in [**3**] show that the category of diffeological spaces forms a complete and co-complete quasi-topos; in particular, all subsets, quotients, and function spaces exist.

On the other hand, the language of stacks arose out of the need to better understand orbit and moduli spaces [**6**]. Indeed, the quotient stack associated to a compact Lie group action contains more information than the quotient topology, the ring of smooth functions on the orbit space in the sense of Schwarz [**5, 22**], or even the associated diffeology. For example, consider $\mathrm{SO}(n)$ acting on $\mathbb{R}^n$ via rotations. The orbit space is homeomorphic to $[0,\infty)$ for all n, and the rings of invariant smooth functions for each n are all isomorphic. The diffeology detects which n we started with (this is the "diffeological dimension" at the singularity) [**10**]. However, the orbit space obtained from the action of $\mathrm{SO}(2n)$ on $\mathbb{R}^{2n}$ via rotations is diffeomorphic as a diffeological space to that obtained from the action

2020 *Mathematics Subject Classification.* Primary 58A99; Secondary 18F20, 14D23, 22A22.

The second author was supported by National Science Foundation grant DMS 08-38434 "EMSW21-MCTP: Research Experience for Graduate Students".

of $\mathrm{U}(n)$ on the same space. The corresponding stacks are not isomorphic in these examples, as the stabilizers at the origin are different, a property detected by the stacks.

The deficiencies of manifolds in modeling geometric structures has an effect on stacks over manifolds as well. In particular, in the topological setting, a stack over the site of (compactly-generated) topological spaces admits a natural coarse moduli space [**18**]. This correspondence serves as a left adjoint to the (2-categorical) Yoneda embedding of topological spaces into stacks over topological spaces. Unfortunately, in the case of stacks over manifolds, no such left adjoint exists: there are examples of stacks over manifolds for which no *manifold* coarse moduli space exists [**8**, Section 2]. The purpose of this paper is to show that diffeology remedies this deficiency.

The first part this paper is mostly expository, casting results from category and topos theory known to experts into the languages of stacks over manifolds and diffeology, languages perhaps more familiar to differential geometers having a few references such as [**11, 12**]. The second part explores the coarse moduli space of a stack over manifolds from a geometric perspective using principal bundles and diffeology, and is new.

As a bridge between the language of stacks and that of diffeology, we will utilize so-called concrete sheaves. A concrete sheaf is a sheaf of sets that not only encodes an underlying set, but a family of maps that induces a diffeology on the set; see Definition 2.6. In fact, Baez and Hoffnung in [**3**] show that the category of diffeological spaces is equivalent to the category of concrete sheaves over open subsets of Euclidean spaces with smooth maps between them. Moreover, Baez and Hoffnung show that the inclusion of concrete sheaves into all sheaves of sets has a left adjoint functor called concretization $\hat{\kappa}\colon \mathbf{Shf}(\mathbf{Man}) \to \mathbf{CShf}(\mathbf{Man})$, which sends every sheaf of sets to its "underlying" concrete sheaf (or, equivalently, diffeology); see Definition 3.1.

In this paper we go further, extending this adjunction to stacks over manifolds; note that we do not require the stacks to be differentiable. Indeed, a straightforward computation shows that there is a functor $\mathbf{Coarse}\colon \mathbf{St}(\mathbf{Man}) \to \mathbf{CShf}(\mathbf{Man})$ from the category of stacks over manifolds to the category of concrete sheaves, sending each stack to its diffeological coarse moduli space. This is a left adjoint to $\int\colon \mathbf{CShf}(\mathbf{Man}) \to \mathbf{St}(\mathbf{Man})$, the Grothendieck construction taking concrete sheaves to stacks. **Coarse** is a 2-functor with respect to the 2-categorical structure of stacks and the degenerate 2-categorical structure of concrete sheaves (*i.e.* that for which every 2-arrow the identity). This 2-functor factors through a functor $\hat{\pi}_0\colon \mathbf{St}(\mathbf{Man}) \to \mathbf{Shf}(\mathbf{Man})$ called discretization, which sends each stack to an "underlying" sheaf of sets, before being concretized in the sense described above. Moreover, **Coarse** has the useful property of being "co-continuous" in the sense that it sends (small) 2-colimits to (1-)colimits; see Theorem 3.14.

Since manifolds form a full subcategory of diffeological spaces, they in turn form a full subcategory of concrete sheaves. The Grothendieck construction embedding concrete sheaves into stacks over manifolds restricted to this full subcategory corresponds exactly to the (2-categorical) Yoneda embedding of manifolds into stacks. In other words, for $\iota\colon \mathbf{Man} \to \mathbf{CShf}(\mathbf{Man})$ the aforementioned embedding of manifolds into concrete sheaves, $\int\colon \mathbf{CShf}(\mathbf{Man}) \to \mathbf{St}(\mathbf{Man})$ the Grothendieck construction, and $y\colon \mathbf{Man} \to \mathbf{St}(\mathbf{Man})$ the Yoneda embedding, the diagram

$$\begin{array}{ccc} & \mathbf{CShf}(\mathbf{Man}) & \\ {}^{\iota}\nearrow & & \searrow^{\int} \\ \mathbf{Man} & \xrightarrow[y]{\qquad\qquad} & \mathbf{St}(\mathbf{Man}) \end{array}$$

commutes. For this reason, we may properly think of $\int\colon \mathbf{CShf}(\mathbf{Man}) \to \mathbf{St}(\mathbf{Man})$ as an extension of the Yoneda embedding. As **Coarse** and $\int$ form an adjunction, we recover a result similar to the topological coarse moduli space correspondence mentioned above (hence the name "diffeological coarse moduli space" for the object $\mathbf{Coarse}(\mathcal{X})$ for any stack $\mathcal{X}$). One may notice that ι here is the 1-categorical Yoneda embedding of manifolds into sheaves over manifolds; thus $\int$ can be viewed as extending this embedding to its 2-categorical counterpart.

Diffeology *a priori* is a theory sitting in the realm of differential topology/geometry, not category theory. We therefore apply the categorical theory developed to differentiable stacks, obtaining geometric properties of their diffeological coarse moduli spaces. If $G = (G_1 \rightrightarrows G_0)$ is a Lie groupoid representing a differentiable stack $\mathcal{X}$, we show that $\mathbf{Coarse}(\mathcal{X})$ is (isomorphic to) the concrete sheaf associated to the quotient diffeology on G_0/G_1, the orbit space of the groupoid; see Theorem 4.2. In the case where $\mathcal{X} = BG$, the stack of right principal G-bundles, we show that discretization ($\hat{\pi}_{\mathbf{0}}$) identifies principal G-bundles over each manifold that are locally G-equivariantly bundle-diffeomorphic, and concretization ($\hat{\kappa}$) further identifies such bundles that are fiberwise G-equivariantly bundle-diffeomorphic; see Propositions 4.6 and 4.10.

This paper is organized as follows. Section 2 reviews diffeology and concrete sheaves, and extends their categorical equivalence to concrete sheaves over manifolds (Lemma 2.9). Section 3 reviews that the inclusion of concrete sheaves into sheaves of sets over manifolds has a left adjoint functor (concretization), and that the inclusion of sheaves of sets into stacks over manifolds (also called the Grothendieck construction) has a left adjoint functor (discretization). Together, these left adjoints compose to form **Coarse** (see Definition 3.10 and Theorem 3.12). We prove that it is co-continuous in Theorem 3.14, in the sense that it takes 2-colimits (more prevalently used in stacks) to colimits. As a first application, we restrict our attention to differentiable stacks in Section 4, obtaining the results mentioned above. Here we define a "gerbe" (Definition 4.20) and show that the discretization and concretization of a differentiable stack are equal if and only if it is represented by such a gerbe (Proposition 4.23). In Section 5, we consider basic differential forms as a second application, extending the results of [**14**, **25**] to certain differentiable stacks.

For more details on diffeological spaces, see [**10**]. For more details on stacks, see [**7**] and [**24**], as well as [**15**] for a description of the relationship between Lie groupoids and differentiable stacks. Coarse moduli spaces have been used in a variety of applications. For example, for topological stacks, the *topological coarse moduli space* is used in [**2**] to develop string topology for differentiable stacks over manifolds; in particular, the coarse moduli space is used to find "hidden loops" in the loop space of a stack. In [**17**], the authors use the topological coarse moduli space to extend the Kashiwara index formula from manifolds to orbifolds. In [**8**] and [**9**], the *underlying topological space of a stack* (similar to but fundamentally different from a coarse moduli space) is used to develop Morse theory and flows of vector fields for stacks over manifolds. For more details on the topological coarse moduli space, see [**18**]. For this paper, we do not focus on topology; however, we do make note that the topology of the underlying topological space of a stack as defined by Hepworth in [**8**] matches the so-called D-topology of the diffeological coarse moduli space (see Remark 3.16).

Prerequisites. This paper uses a lot of language from category theory. Besides the basics, the reader is assumed to be familiar with the following. We give an intuitive idea of what each of these objects is and pinpoint an actual reference for a definition or statement.

- ◇ Given a category C, a *coverage* on C is a generalization of open covers from topology. Essentially, it is a specification of the open covers of each object of C satisfying some coherence conditions; see [**12**, A2.1.9]. A *site* is a category equipped with a coverage. (Note that we do not require the underlying category of a site to be small, whereas some references do.) We will be dealing with the site of open subsets of Euclidean spaces with smooth maps as arrows (which is small), as well as the site of smooth manifolds (which is essentially small). In both of these cases, the coverage is given by usual open covers by open sets.
- ◇ Given a category C, a *presheaf of sets* on C is a contravariant functor $C^{op} \to \mathbf{Set}$. If C is a site, then a *sheaf of sets* is a presheaf of sets on C that obeys a "sheaf condition" with respect to the coverage on C; see [**12**, A2.1.9]. For example, given the site of manifolds, the sheaf $\Omega^k(\cdot)$ assigns to each manifold M its set of k-forms $\Omega^k(M)$. Presheaves of sets over C form a category (and sheaves a subcategory) with arrows given by natural transformations.
- ◇ A presheaf of sets P on a site C has a unique *sheafification*, a sheaf of sets S over C such that there is a natural transformation $\alpha\colon P \to S$ satisfying the following universal property: if T is any sheaf of sets over C and $\beta\colon P \to T$ a natural transformation, then β factors through α; see [**24**, Definition 2.63 and Theorem 2.64]. Note that sheafification is the left adjoint to the inclusion of sheaves of sets into presheaves of sets.
- ◇ **Comparison Lemma:** Let C be a locally small site, and D a small dense subsite. Then the restriction functor sending sheaves of sets over C to sheaves of sets over D is an equivalence of categories; see [**12**, C2.2.3].
- ◇ Given a category C, a *category fibered in groupoids (CFG)* over C is a functor $\pi\colon D \to C$ whose fibers are groupoids, satisfying some conditions; see [**24**, Definition 3.5]. CFGs over C form a 2-category, where an arrow $F\colon (\pi\colon D \to C) \to (\pi'\colon D' \to C)$ is a functor from D to D' such that $\pi' \circ F = \pi$, and 2-arrows are given by natural transformations.
- ◇ A *stack* over a site C is a CFG that satisfies a "descent condition", which is a 2-categorical version of the sheaf condition; see [**24**, Definition 4.6]. Stacks over C form a sub-2-category of CFGs over C.
- ◇ Let $P\colon C^{op} \to \mathbf{Set}$ be a presheaf of sets. Then the *category of elements* of P, denoted $\pi\colon \int P \to C$, is a CFG defined as follows. The objects of $\int P$ are pairs (x, c) where c is an object in C and $x \in P(c)$, and the arrows $\tilde{f}\colon (x, c) \to (x', c')$ are arrows $f\colon c \to c'$ in C such that $P(f)(c') = c$. Define π to be the obvious projection. Note that in this case, the resulting

CFG is *discrete*; that is, the arrows in each fiber are trivial, and so we think of these groupoids as just sets. See [**1**, page 203] for more details. It is not hard to show that $\int$ is a fully faithful functor from sheaves to CFGs, known as the *Grothendieck construction*, sending a map of presheaves to the obvious map of CFGs. Note that if C is a site and P a sheaf, then $\pi\colon \int P \to C$ is in fact a stack.

2. From diffeology to concrete sheaves

In this section we review the definition of a diffeology and a concrete sheaf, and establish an equivalence of categories between diffeological spaces and concrete sheaves over manifolds (see Corollary 2.10 and Remark 2.11). Our main resource for diffeology is the book by Iglesias-Zemmour [**10**]. For more details on concrete sheaves, see [**3**].

DEFINITION 2.1. Let **Open** be the category of open subsets of $\mathbb{R}^n$ for any n and their smooth maps. Let **Man** be the category of manifolds. Equipped with coverages generated by standard open covers, both **Open** and **Man** are sites. Note that the singleton $\mathbb{R}^0 := \{0\}$ is an object of **Open** and **Man**.

DEFINITION 2.2 (Diffeological Space). Let X be a set. A *parametrization* $p\colon U \to X$ is a function from an object of **Open** to X. A *diffeology* $\mathcal{D}$ on X is a family of parametrizations satisfying the following three axioms:

- ◇ **(Covering Axiom)** All constant maps $p\colon U \to X$ are contained in $\mathcal{D}$.
- ◇ **(Gluing Axiom)** If $p\colon U \to X$ is a parametrization such that there exist an open cover $\{U_\alpha\}_{\alpha\in A}$ of U and a family $\{p_\alpha\colon U_\alpha \to X\}_{\alpha\in A} \subseteq \mathcal{D}$ of parametrizations that satisfy $p_\alpha = p|_{U_\alpha}$, then p is contained in $\mathcal{D}$.
- ◇ **(Smooth Compatibility Axiom)** If $p\colon U \to X$ is in $\mathcal{D}$ and $f\colon V \to U$ is a smooth map between open sets of Euclidean spaces, then $f^*p = p\circ f$ is also contained in $\mathcal{D}$.

The pair $(X, \mathcal{D})$ is called a *diffeological space*, and the elements of $\mathcal{D}$ are called *plots*.

DEFINITION 2.3 (Smooth Map). Let $(X, \mathcal{D}_X)$ and $(Y, \mathcal{D}_Y)$ be diffeological spaces, and let $F\colon X \to Y$ be a map. Then F is *(diffeologically) smooth* if for every plot $p \in \mathcal{D}_X$, the composition $F \circ p$ is a plot in $\mathcal{D}_Y$.

REMARK 2.4. Diffeological spaces with smooth maps between them form a category, denoted by **Diffeol**. As mentioned in the introduction, this category is complete and cocomplete [**3**], and contains the category of smooth manifolds with smooth maps **Man** as a full subcategory.

NOTATION 2.5. For C a site, we will use the symbols $\mathbf{PShf}(C)$, $\mathbf{Shf}(C)$, $\mathbf{CFG}(C)$, and $\mathbf{St}(C)$ to denote the categories of presheaves of sets, sheaves of sets, categories fibered in groupoids, and stacks over C, respectively.

DEFINITION 2.6 (Concrete Sheaves, [**3**, Definition 4.10]). Let C be the site **Open** or **Man**. A presheaf $P\colon C^{op} \to \mathbf{Set}$ is *concrete at* $U \in C_0$ if the map "$\underline{\ }$" sending each element $p \in P(U)$ to the function

$$\underline{p}\colon \mathrm{Hom}(*, U) \to P(*) : u \mapsto u^*p \tag{2.1}$$

is injective. We say that P is a *concrete presheaf* if it is concrete at each object U. If in addition P satisfies the sheaf axiom, then we call P a *concrete sheaf*. We will refer to the category of concrete sheaves over **Open** and **Man** (with natural transformations as arrows) by **CShf**(**Open**) and **CShf**(**Man**), respectively.

LEMMA 2.7 (Baez-Hoffnung [**3**], Prop. 4.15). *There is an equivalence of categories from* **Diffeol** *to* **CShf**(**Open**). *This equivalence sends a diffeological space* X *to the sheaf over* **Open** *assigning to* $U \in \mathbf{Open}_0$ *the set of all plots of* X *with domain* U.

The proof of the following lemma is a straightforward calculation and hence is omitted.

LEMMA 2.8. *Let* C *be the site* **Open** *or* **Man** *and* $S\colon C^{op} \to \mathbf{Set}$ *a sheaf. Let* $U \in C_0$ *and* $\{U_i \to U\}$ *a cover of* U. *If* S *is concrete at each* U_i, *then it is concrete at* U.

LEMMA 2.9 (Concrete Comparison Lemma). *The categories* **CShf**(**Open**) *and* **CShf**(**Man**) *are equivalent.*

PROOF. Recall that the Comparison Lemma establishes an equivalence of categories

$$\mathbf{Shf}(\mathbf{Open}) \underset{\text{res}}{\overset{\text{ext}}{\rightleftarrows}} \mathbf{Shf}(\mathbf{Man}) \tag{2.2}$$

where res(S) is the restriction of $S\colon \mathbf{Man}^{op} \to \mathbf{Set}$ to elements of **Open** in **Man** and ext takes $S'\colon \mathbf{Open}^{op} \to \mathbf{Set}$ to the Kan extension against the inclusion of **Open** into **Man**.

The functor res takes **CShf**(**Man**) to **CShf**(**Open**). Since any n-dimensional manifold M has an open cover by open subsets of $\mathbb{R}^n$, it follows from Lemma 2.8 that the sheaf $S\colon \mathbf{Man}^{op} \to \mathbf{Set}$ must be concrete if res(S) is concrete. So res takes *only* concrete sheaves of **Shf**(**Man**) to concrete sheaves of **Shf**(**Open**). Therefore, since res $\circ$ ext must take **CShf**(**Open**) to **CShf**(**Open**) (concreteness is preserved under isomorphism), it follows that ext takes **CShf**(**Open**) to **CShf**(**Man**). Thus the equivalence of categories (Equation (2.2)) restricts to our desired equivalence between **CShf**(**Open**) and **CShf**(**Man**). □

COROLLARY 2.10. *The categories* **Diffeol** *and* **CShf**(**Man**) *are equivalent.*

REMARK 2.11. The construction of a functor I from **Diffeol** to **CShf**(**Man**) that is part of the equivalence in Corollary 2.10 is given as follows. Given a diffeological space $(X, \mathcal{D})$ and a manifold M of dimension m, define a sheaf $I(X, \mathcal{D})$ over **Man** by setting $I(X, \mathcal{D})(M)$ to be the set of all maps $p\colon M \to X$ such that there exist an open cover $\{U_\alpha\}$ of M in which $U_\alpha \subseteq \mathbb{R}^m$ for each α and plots $q_\alpha\colon U_\alpha \to X$ in $\mathcal{D}$ satisfying $p|_{U_\alpha} = q_\alpha$. It follows that $I(X, \mathcal{D})(M)$ is the set of all diffeologically smooth maps $M \to X$; we will refer to these as *plots* of X. This is the concrete sheaf over **Man** associated to $\mathcal{D}$. A diffeologically smooth map F is sent via I to a map of sheaves sending a plot p to $F \circ p$.

REMARK 2.12. The fully faithful embedding of manifolds into diffeologies takes any manifold M to the diffeological space built on the underlying set of M with plots all smooth maps from open subsets of Euclidean space to M. So the concrete sheaf $I(M)$ (for I the functor of the previous remark) is just the sheaf of smooth functions to M: $I(M)(N) = C^\infty(N, M)$. Write ι for the composition of the inclusion of the embedding of manifolds into diffeologies with I. By the definition of the Yoneda embedding y: **Man** $\to$ **St**(**Man**) (see, for instance, [**11**, Section 8.2]), the following diagram commutes

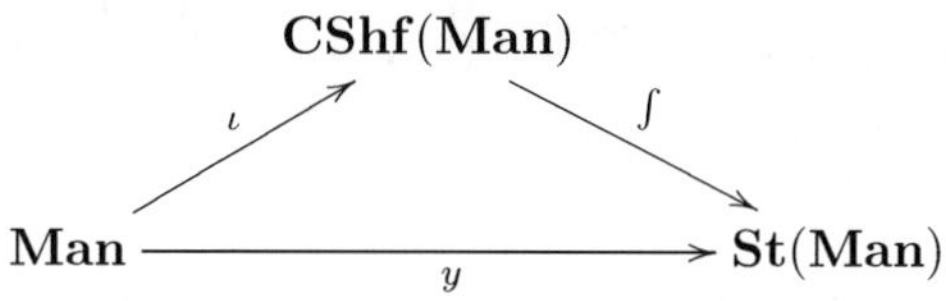

where $\int$: **CShf**(**Man**) $\to$ **St**(**Man**) is the Grothendieck construction. Thus, we may think of $\int$ as an extension of the Yoneda embedding.

3. From concrete sheaves to sheaves to stacks

In this section we begin by reviewing concretization. This is a functor introduced by Baez and Hoffnung in [**3**], where they prove that it is left adjoint to the inclusion **CShf**(**Man**) $\to$ **Shf**(**Man**). We then define discretization following [**18**], and show that it is left adjoint to the Grothendieck construction. These two functors compose to form the functor **Coarse**. It follows that **Coarse** is left adjoint to the restricted Grothendieck construction (Theorem 3.12) – since **CShf**(**Man**) is a 1-category, we need only consider this left adjunction in a strict sense and not a 2-categorical sense. We similarly give the same treatment to $\hat{\pi}_0$ in Lemma 3.7. Moreover, **Coarse** is co-continuous, taking 2-colimits to colimits (Theorem 3.14).

Definition 3.1 (Concretization). Let P be a presheaf over **Man**. Define $\kappa(P)$: $\mathbf{Man}^{op} \to \mathbf{Set}$ to be the presheaf assigning to each manifold M the quotient set $P(M)/\sim$, where $p_1 \sim p_2$ if $\underline{p_1} = \underline{p_2}$ (see Equation (2.1) for a definition of $\underline{p_i}$). Define the *concretization* of P, denoted by $\hat{\kappa}(P)$: $\mathbf{Man}^{op} \to \mathbf{Set}$, to be the sheafification of $\kappa(P)$.

Remark 3.2. It follows from the definition of sheafification that $\hat{\kappa}(P)$ is a concrete sheaf.

Lemma 3.3 (Baez-Hoffnung, [**3**, Lemma 5.20]). $\hat{\kappa}$: **Shf**(**Man**) $\to$ **CShf**(**Man**) *is a left adjoint functor to the inclusion functor* **CShf**(**Man**) $\hookrightarrow$ **Shf**(**Man**).

We'll see below that concretization can have interesting effects, especially on sheaves coming from differentiable stacks. However, some sheaves of sets are practically destroyed by the process:

Example 3.4 ($\hat{\kappa}$ Applied to Ω^k). Let Ω^k: $\mathbf{Man}^{op} \to \mathbf{Set}$ be the sheaf of differential k-forms. If $k = 0$, then $\hat{\kappa}(\Omega^k)$ is precisely equal to $C^\infty(-, \mathbb{R})$, the Yoneda embedding of $\mathbb{R}$ in **Shf**(**Man**), or equivalently, $I(\mathbb{R})$ where I is as in Remark 2.11. If $k > 0$, then for every manifold M, we have $\hat{\kappa}(\Omega^k)(M) = \{0\}$. //

Definition 3.5 (Discretization). Let $\mathcal{X}$ be a stack over **Man**. Define $\pi_0(\mathcal{X})$ to be the presheaf over **Man** assigning to each manifold M the set of isomorphism classes in $\mathcal{X}(M)$. Define $\hat{\pi}_0(\mathcal{X})$ to be the sheafification of $\pi_0(\mathcal{X})$. For any manifold M and $\xi \in \mathcal{X}(M)$, denote by $[\xi]$ the image of the isomorphism class of ξ in $\hat{\pi}_0(\mathcal{X})(M)$ via the natural map $\pi_0(\mathcal{X}) \to \hat{\pi}_0(\mathcal{X})$.

REMARK 3.6.

(1) In general, sheafification is necessary in the definition of discretization; see Example 4.7.
(2) Note that for any stack $\mathcal{X}$, $\pi_0(\mathcal{X})(*) = \hat{\pi}_0(\mathcal{X})(*)$.

LEMMA 3.7. $\hat{\pi}_0\colon \mathbf{St}(\mathbf{Man}) \to \mathbf{Shf}(\mathbf{Man})$ *is left adjoint to the Grothendieck construction* $\int\colon \mathbf{Shf}(\mathbf{Man}) \to \mathbf{St}(\mathbf{Man})$.

REMARK 3.8. It is already known that $\int$ has a left adjoint functor, but this is typically described in much more generality (see [**1**]).

PROOF. We will show $\int\colon \mathbf{Shf}(\mathbf{Man}) \to \mathbf{St}(\mathbf{Man})$ has the universal property of a right adjoint. Fix a stack $\mathcal{X} \in \mathbf{St}(\mathbf{Man})$. Then there exists a natural map of CFGs $\phi_{\mathcal{X}}\colon \mathcal{X} \to \int \pi_0\mathcal{X}$ with $\phi_{\mathcal{X}}(\xi) = (M, [\xi])$ for each $\xi \in \mathcal{X}(M)$.

Suppose $S\colon \mathbf{Man}^{op} \to \mathbf{Set}$ is a sheaf of sets and $\psi\colon \mathcal{X} \to \int S$ is any map of stacks. Since the fibers of $\int S$ are sets, ψ must be constant on isomorphism classes of the fibers of $\mathcal{X}$. It readily follows that ψ factors uniquely through $\phi_{\mathcal{X}}$.

Since $\int$ is fully faithful, the unique map of CFGs $\bar{\psi}\colon \int \pi_0\mathcal{X} \to \int S$ satisfying $\psi = \bar{\psi} \circ \phi_{\mathcal{X}}$ is equal to $\int$ applied to a unique map of presheaves $f\colon \pi_0\mathcal{X} \to S$. Furthermore, the map f must factor uniquely through the sheafification $\eta_{\mathcal{X}}\colon \pi_0\mathcal{X} \to \hat{\pi}_0\mathcal{X}$. In other words, there exists a unique map of sheaves $g\colon \hat{\pi}_0\mathcal{X} \to S$ with $f = g \circ \eta$. Thus, $\int(g)\colon \int \hat{\pi}_0\mathcal{X} \to \int S$ is the unique map of stacks such that $\int(g) \circ (\int(\eta_{\mathcal{X}}) \circ \phi_{\mathcal{X}}) = \psi$.

It follows $\int\colon \mathbf{Shf}(\mathbf{Man}) \to \mathbf{St}(\mathbf{Man})$ satisfies the universal property of a right adjoint with unit of adjunction $\mathcal{X} \mapsto \int(\eta_{\mathcal{X}}) \circ \phi_{\mathcal{X}}$ and left adjoint $\hat{\pi}_0$. □

REMARK 3.9. One can check that $\pi_0\colon \mathbf{St}(\mathbf{Man}) \to \mathbf{PShf}(\mathbf{Man})$ sends 2-commuting diagrams in $\mathbf{St}(\mathbf{Man})$ to strictly commuting diagrams in $\mathbf{PShf}(\mathbf{Man})$. It follows that $\hat{\pi}_0$ is a 2-functor, where $\mathbf{PShf}(\mathbf{Man})$ inherits a 2-category structure with trivial 2-arrows.

DEFINITION 3.10 (The Functor **Coarse**). Define the *coarse (2-)functor* **Coarse** : $\mathbf{St}(\mathbf{Man}) \to \mathbf{CShf}(\mathbf{Man})$ as the composition $\hat{\kappa} \circ \hat{\pi}_0$. For any stack $\mathcal{X}$, we will call the concrete sheaf $\mathbf{Coarse}(\mathcal{X})$ *the diffeological coarse moduli space of* $\mathcal{X}$ (see Remark 3.13).

REMARK 3.11. A straightforward computation shows that one only needs to sheafify once when applying **Coarse**; indeed, we can define **Coarse** to be the composition $\hat{\kappa} \circ \pi_0$.

THEOREM 3.12 (**Coarse** is a Left Adjunction). **Coarse** *is left adjoint to the Grothendieck construction restricted to* $\mathbf{CShf}(\mathbf{Man})$.

PROOF. This follows directly from the composition of the left adjoint functors of Lemma 3.3 and Lemma 3.7. □

REMARK 3.13 (Diffeological Coarse Moduli Space). As noted in Remark 2.12, the functor $\int\colon \mathbf{CShf}(\mathbf{Man}) \to \mathbf{St}(\mathbf{Man})$ is an extension of the Yoneda embedding of manifolds into stacks and, as shown in Theorem 3.12, the functor **Coarse** is a left adjoint to this extension. This justifies our reference to $\mathbf{Coarse}(\mathcal{X})$ as the diffeological coarse moduli space of the stack $\mathcal{X}$, following the nomenclature of Hepworth (see [**8**, Section 2]). We append the adjective "diffeological" to "coarse moduli space" as a standard coarse moduli space should be an object in the underlying site of a stack, but $\mathbf{Coarse}(\mathcal{X})$ need not be a manifold in general.

THEOREM 3.14 (Coarse is Co-continuous). *If a stack $\mathcal{X}$ is equivalent to a 2-colimit* $\operatorname{colim} F$ *in the 2-category of stacks, then* $\mathbf{Coarse}(\mathcal{X})$ *is isomorphic to the 1-colimit* $\operatorname{colim}(\mathbf{Coarse} \circ F)$.

PROOF. It follows from Remark 3.9 that **Coarse** takes 2-commutative diagrams in $\mathbf{St}(\mathbf{Man})$ to strictly commuting diagrams in $\mathbf{CShf}(\mathbf{Man})$. Therefore, the 2-cocones used to define colimits in $\mathbf{St}(\mathbf{Man})$ are sent to cocones in $\mathbf{CShf}(\mathbf{Man})$. The proof now follows from a standard argument for co-continuity of left adjoints (for instance, see the dual argument in [**16**, Theorem 1, page 118]). □

REMARK 3.15. Baez and Hoffnung generalize "concreteness" by defining what it means to be a concrete site (see [**3**, Definition 20]). In this way, concretization (Definition 3.1) generalizes to any sheaf of sets over a site satisfying these concreteness conditions for which Lemma 3.7 is still valid. Similarly, discretization (Definition 3.5) generalizes to stacks over any site, as does Lemma 3.7. Therefore, Theorem 3.12 generalizes as well, providing an adjunction to the Yoneda embedding of any concrete site C into the category of stacks over C. However, for this paper we are interested in diffeology, and so for the sake of brevity, we restrict our attention to the site **Man**.

REMARK 3.16 (Topological Considerations). Given a diffeological space $(X, \mathcal{D})$ the *D-topology* of X is defined to be the strongest topology on X for which all plots in $\mathcal{D}$ are continuous. That is, $A \subseteq X$ is open if and only if for every $p \in \mathcal{D}$, the set $p^{-1}(A)$ is open in the domain of p.

Let the diffeological coarse moduli space of the stack $\mathcal{X}$ be identified with the diffeological space $(X, \mathcal{D})$. In particular, $\pi_0(\mathcal{X}(*))$ is identified with the set X. Then $A \subseteq X$ is open if and only if for every manifold M and object $\xi \in \mathcal{X}(M)$, we have $\underline{[\xi]}^{-1}(A) \subseteq M$ is open. Applying the Yoneda embedding y, we have that A is open if and only if $\underline{[\varphi]}^{-1}(A) \subseteq M$ is open for every manifold M and every map of stacks $\varphi\colon y(M) \to \mathcal{X}$. This matches the topology of *the underlying space of a stack* used by Hepworth in [**8**].

4. Differentiable stacks

This section assumes that the reader is familiar with Lie groupoids, their actions and principal bundles, the localization of the 2-category of Lie groupoids via bibundles and its equivalence to differentiable stacks over manifolds. (One could also consider localizations à la Pronk [**19**] using generalized morphisms, or à la Roberts [**21**] using anafunctors; to keep in the spirit of principal bundles, however, we choose bibundles.) Our main resource for this theory is Lerman's survey on the topic, [**15**]; we refer to this for definitions of bibundles, BG, etc. In this section, we show that the quotient diffeology on the orbit space G_0/G_1 of a Lie groupoid G matches the diffeology associated to $\mathbf{Coarse}(BG)$ (Theorem 4.2). In the remainder of the section, we discuss the relationship between principal G-bundles and plots of the orbit space G_0/G_1, and an application to the "relation space" of a Lie groupoid (Proposition 4.23) that allows us to define a "gerbe".

The following is a special case of a folklore theorem well-known to experts; we refer to the proof given in [**4**].

LEMMA 4.1. *Let* $G = \left(G_1 \overset{s}{\underset{t}{\rightrightarrows}} G_0\right)$ *be a Lie groupoid,* BG *its associated stack, and* $\mathcal{X}$ *a discrete stack (see the end of the Introduction for a definition; informally, this means that it is a sheaf of sets). There is a bijection between maps of stacks* $BG \to \mathcal{X}$ *and objects* $\xi \in \mathcal{X}(G_0)$ *satisfying* $s^*\xi = t^*\xi$.

PROOF. By [**4**, Proposition 6.6], there exists an equivalence of categories between cocycles $(\xi, f\colon s^*\xi \to t^*\xi)$ and maps of stacks $BG \to \mathcal{X}$. Since $\mathcal{X}$ is discrete, f is trivial and $\mathrm{Hom}(BG, \mathcal{X})$ has trivial arrows. Thus, the equivalence reduces to a bijection. □

THEOREM 4.2 (The Orbit Space of a Lie Groupoid). *Let* $\mathcal{X}$ *be a differentiable stack. Let* $G = (G_1 \rightrightarrows G_0)$ *be a Lie groupoid representing* $\mathcal{X}$ *(that is,* $\mathcal{X} \cong BG$*). Then* $\mathbf{Coarse}(\mathcal{X})$ *is isomorphic to the concrete sheaf associated to the quotient diffeology on the orbit space* G_0/G_1 *of* G.

REMARK 4.3. In particular, the above theorem shows that the diffeology associated to the diffeological coarse moduli space $\mathbf{Coarse}(BG)$ is not only independent of the choice of an atlas for BG, but only dependent on the equivalence class of BG in the category of *all* stacks over manifolds.

PROOF. The quotient map $\pi\colon G_0 \to G_0/G_1$ corresponds to the associated plot in $\int I(G_0/G_1)(G_0)$ via the 2-Yoneda Lemma (see [**11**, Section 8.2]), where I is the functor of Remark 2.11. By Lemma 4.1, there is a corresponding map $\theta \in \mathrm{Hom}(BG, \int I(G_0/G_1))$. Applying **Coarse** and Corollary 2.10, and using the universal property of diffeological quotients, we obtain an isomorphism $\mathbf{Coarse}(BG) \cong I(G_0/G_1)$. □

Compare the following corollary with [**13**] and [**25**].

COROLLARY 4.4. *Given differentiable stacks* $\mathcal{X}$ *and* $\mathcal{Y}$ *with a map of stacks* $F\colon \mathcal{X} \to \mathcal{Y}$*, the map* $\mathbf{Coarse}(F)$ *corresponds to a smooth map between the corresponding quotient diffeologies of representative Lie groupoid orbit spaces. Moreover, if* F *is an equivalence (and hence Lie groupoids representing* $\mathcal{X}$ *and* $\mathcal{Y}$ *are Morita equivalent), then* $\mathbf{Coarse}(F)$ *is a diffeomorphism.*

Let G be a Lie groupoid, and let BG be its corresponding differentiable stack. Given principal G-bundles P and Q over a manifold B, we have that P and Q are isomorphic (*i.e.* G-equivariantly bundle-diffeomorphic) if and only if they represent the same element in $\pi_0(BG)(B)$. This equivalence no longer holds upon sheafification of $\pi_0(BG)$ to $\hat{\pi}_0(BG)$ (see Example 4.7).

DEFINITION 4.5 (Locally Isomorphic Bundles). Let G be a Lie groupoid, and let P and Q be principal G-bundles over a manifold B. Then P and Q are *locally isomorphic* if there exists an open cover $\{i_\alpha\colon B_\alpha \to B\}$ of B such that for each α, the principal G-bundles i_α^*P and i_α^*Q are isomorphic.

PROPOSITION 4.6 (Discretization for Differentiable Stacks). *Let* G *be a Lie groupoid, and let* P *and* Q *be principal* G*-bundles over a manifold* B*. Then,* P *and* Q *are locally isomorphic if and only if* $[P] = [Q]$ *in* $\hat{\pi}_0(BG)(B)$.

PROOF. P and Q are locally isomorphic if and only if there exists an open cover $\{i_\alpha\colon B_\alpha \to B\}$ of B such that for each α, there is an isomorphism $i_\alpha^*P \cong i_\alpha^*Q$. This is equivalent to the equality of the isomorphism class of i_α^*P with the isomorphism class of $i_\alpha^* Q$ in $\pi_0(BG)(B_\alpha)$, for each α. By definition of sheafification, this is equivalent to $[P] = [Q]$. □

The following example illustrates the difference between $\pi_0(BG)$ and $\hat{\pi}_{\mathbf{0}}(BG)$.

EXAMPLE 4.7 ($B\mathbb{Z}_2$). Let $\mathcal{X}$ be the differentiable stack $B\mathbb{Z}_2$, where $\mathbb{Z}_2 = \mathbb{Z}/2\mathbb{Z}$. Then $B\mathbb{Z}_2(\mathbb{S}^1)$ contains two isomorphism classes of bundles: one represented by the trivial $\mathbb{Z}_2$-bundle which we denote by P, and the other represented by the boundary of the Möbius band which we denote by Q. While P and Q do not represent the same elements in $\pi_0(BG)$, by Proposition 4.6 we know that $[P] = [Q]$ in $\hat{\pi}_{\mathbf{0}}(BG)$. //

REMARK 4.8. The above example is not unique; given *any* Lie group Γ, local triviality of principal Γ bundles ensures that all such bundles are locally isomorphic and hence $\hat{\pi}_{\mathbf{0}}(B\Gamma)(M)$ is a singleton set for each $M \in \mathbf{Man}_0$.

Given a Lie groupoid G and two principal G-bundles P and Q, we give conditions under which $[P]$ and $[Q]$ yield the same objects after applying the functor κ to $\hat{\pi}_{\mathbf{0}}(BG)$.

DEFINITION 4.9 (Fiberwise Isomorphic Bundles). Let G be a Lie groupoid, and let $\rho_P \colon P \to B$ and $\rho_Q \colon Q \to B$ be principal G-bundles over a manifold B. Then P and Q are *fiberwise isomorphic* if for every $b \in B$, there exists a G-equivariant diffeomorphism from $\rho_P^{-1}(b)$ to $\rho_Q^{-1}(b)$.

PROPOSITION 4.10 (Concretization for Differentiable Stacks). *Let G be a Lie groupoid, and let $\rho_P \colon P \to B$ and $\rho_Q \colon Q \to B$ be principal G-bundles over a manifold B. Then, P and Q are fiberwise isomorphic if and only if P and Q descend to the same elements in* **Coarse**$(BG)(B)$.

PROOF. P and Q descend to the same elements of $\kappa(\hat{\pi}_{\mathbf{0}}(BG))(B)$ if and only if $\underline{[P]}$ and $\underline{[Q]}$ are the same. But this is equivalent to $[b^*P] = [b^*Q]$ for each $b \in B$. That is, for each $b \in B$, there is a G-equivariant diffeomorphism between b^*P and b^*Q. But these two objects are exactly the fibers $\rho_P^{-1}(b)$ and $\rho_Q^{-1}(b)$, respectively. The result now follows from the fact that any concrete presheaf P is *separated* [**3**, Lemma 5.21]; *i.e.* its sheafification induces an injection $\kappa(P)(M) \to \hat{\kappa}(P)(M)$ for each manifold M. □

Not every pair of fiberwise isomorphic principal bundles are locally isomorphic, as illustrated by the following example.

EXAMPLE 4.11 (Reflection in $\mathbb{R}$). Let G be the action groupoid $(\mathbb{Z}_2 \times \mathbb{R}) \rightrightarrows \mathbb{R}$ obtained from the action of $\mathbb{Z}_2$ on $\mathbb{R}$ sending $(\pm 1, x)$ to $\pm x$. Consider the following two plots $p_1, p_2 \colon \mathbb{R} \to \mathbb{R}$ defined as:

$$p_1(\tau) = \begin{cases} -e^{-1/\tau^2} & \text{if } \tau < 0, \\ 0 & \text{if } \tau = 0, \\ e^{-1/\tau^2} & \text{if } \tau > 0, \end{cases}$$

$$p_2(\tau) = \begin{cases} -e^{-1/\tau^2} & \text{if } \tau \neq 0, \\ 0 & \text{if } \tau = 0. \end{cases}$$

Computing the pullback bundles $P_i := p_i^* t$ $(i = 1, 2)$, where $t\colon G_1 \to G_0$ is the unit bundle (in fact, this is a trivial $\mathbb{Z}_2$-bundle), we have

$$\begin{aligned} P_1 &= \{(\tau, a, x) \in \mathbb{R} \times \mathbb{Z}_2 \times \mathbb{R} \mid p_1(\tau) = a \cdot x\} \\ &= \{(\tau, 1, -e^{-1/\tau^2}) \mid \tau < 0\} \cup \{(0, \pm 1, 0)\} \cup \{(\tau, 1, e^{-1/\tau^2}) \mid \tau > 0\} \\ &\quad \cup \{(\tau, -1, e^{-1/\tau^2}) \mid \tau < 0\} \cup \{(\tau, -1, -e^{-1/\tau^2}) \mid \tau > 0\}, \end{aligned}$$

and similarly,

$$\begin{aligned} P_2 = &\{(\tau, 1, -e^{-1/\tau^2}) \mid \tau < 0\} \cup \{(0, \pm 1, 0)\} \cup \{(\tau, 1, -e^{-1/\tau^2}) \mid \tau > 0\} \\ &\cup \{(\tau, -1, e^{-1/\tau^2}) \mid \tau < 0\} \cup \{(\tau, -1, e^{-1/\tau^2}) \mid \tau > 0\}. \end{aligned}$$

If $\phi\colon P_1 \to P_2$ is a bundle-diffeomorphism, then in particular it must send τ to τ; that is, ϕ must fix the first argument of each triple in P_1. The anchor map $\alpha_i\colon P_i \to \mathbb{R}$ is the projection to the third argument of each triple, and so if ϕ is equivariant, then it must fix the third argument of each triple in P_1. Thus, ϕ must be the identity map when restricted to fibers for which $\tau < 0$, and it switches connected components of the fibers for $\tau > 0$. Hence, ϕ cannot be continuous at $\tau = 0$, and so we conclude that there is no equivariant diffeomorphism of bundles; in fact, not even a local one exists. However, it is clear that the fibers of P_1 and P_2 are equivariantly diffeomorphic, and so by Proposition 4.10 we have $\underline{[P_1]} = \underline{[P_2]}$. This corresponds to the fact that $\pi_G \circ p_1 = \pi_G \circ p_2$ where $\pi_G\colon G_0 \to G_0/G_1$ is the quotient map (see Proposition 4.12). //

In light of Remark 2.11, we show in the following proposition how elements of **Coarse**(BG) and plots of the corresponding diffeology are related.

PROPOSITION 4.12 (From Principal Bundles to Plots). *Let G be a Lie groupoid, and let $\rho\colon P \to B$ be a principal G-bundle, where B is an open subset of some Euclidean space. Then, P induces a plot $p\colon B \to G_0/G_1$ defined as*

$$p(b) = \pi_G \circ \alpha \circ \sigma(b) \tag{4.1}$$

where $\pi_G\colon G_0 \to G_0/G_1$ is the quotient map, $\alpha\colon P \to G_0$ is the anchor map, and σ is a local section of ρ about b. Moreover, every plot in the quotient diffeology locally arises in this way. In particular, $\pi_0(BG(*))$ is in bijection with the underlying set of G_0/G_1. Finally, two principal G-bundles are fiberwise isomorphic if and only if the corresponding plots are equal.

REMARK 4.13. The statement that $\pi_0(BG(*))$ is in bijection with G_0/G_1 is not new, but we emphasize it due to it being crucial to understanding the diffeological coarse moduli space of a differentiable stack from the standpoint of diffeology. Also, if B is a manifold, but not an open subset of some Euclidean space, then the proposition still holds where p is a plot in the sense that it is contained in $I(\mathcal{D})(B)$ where $\mathcal{D}$ is the quotient diffeology on G_0/G_1 and I is as given in Remark 2.11.

PROOF. Suppose that σ and σ' are two local sections of ρ about b. There exists $g \in G_1$ such that $\sigma(b) \cdot g = \sigma'(b)$. Consequently, $\alpha(\sigma(b))$ and $\alpha(\sigma'(b))$ lie in the same G-orbit, and so descend to the same point via π_G. The map p is thus independent of the local section chosen, and so is well-defined. Moreover, p is diffeologically smooth, and so is a plot. To show that p and $\underline{[P]}$ are the same, consider the bundle b^*P, which is the fiber of P over b. The anchor map restricted to b^*P has an image contained in exactly one G-orbit of G_0, which by definition of G_0/G_1 is exactly $p(b)$. But $[b^*P] = \underline{[P]}(b)$, and this proves the claim.

On the other hand, given any plot $p\colon U \to G_0/G_1$ in the quotient diffeology, there exist an open cover $\{U_\alpha\}$ of U and a family $\{q_\alpha\colon U_\alpha \to G_0\}$ of smooth maps such that $p|_{U_\alpha} = \pi_G \circ q_\alpha$. Taking the pullback bundles $q_\alpha^*(t\colon G_1 \to G_0)$ proves the claim.

The last statement follows immediately from Equation (4.1). □

Given the above discussion, a natural question arises: for BG a differentiable stack, when is $\tilde{\pi}_{\mathbf{0}}(BG)$ a concrete sheaf; *i.e.* when does $\mathbf{Coarse}(BG) = \tilde{\pi}_{\mathbf{0}}(BG)$? It turns out that this condition is equivalent to asking that BG be a differentiable gerbe over its quotient space. To define this, we first need to define subductions, the natural generalization of a surjective submersion to the category of diffeological spaces.

DEFINITION 4.14. A map $F\colon X \to Y$ of diffeological spaces is a *subduction* if it is a smooth surjection such that for every plot $p\colon U \to Y$ and every $u \in U$ there is an open neighborhood V of u and a plot $q\colon V \to X$ such that $p|_V = F \circ q$.

EXAMPLE 4.15. Given a diffeological space X and an equivalent relation $\sim$, the quotient map $X \to X/\sim$ is a subduction. In particular, given a Lie groupoid $G_1 \rightrightarrows G_0$, the quotient map to the orbit space $G_0 \to G_0/G_1$ is a subduction. //

REMARK 4.16. Suppose that $G = (G_1 \to G_0)$ is a Lie groupoid and X is a diffeological space. Then by identifying X with the trivial groupoid $X \rightrightarrows X$, we may treat X as a *diffeological groupoid*; that is, a groupoid in which the object and arrow spaces are diffeological spaces, and all structure maps are smooth. With this in mind, suppose that $\pi : G \to X$ is smooth functor (i.e., a functor which is smooth on objects and morphisms). As a slight abuse of notation, we will also use π to refer to the associated map $\pi\colon G_0 \to X$. Via the basic properties of functors and pullbacks, such a functor $\pi\colon G \to X$ induces a unique smooth map

$$(s,t)\colon G_1 \to G_0 \times_X G_0, \quad (s,t)(g) := (s(g), t(g))$$

As a particular example of Remark 4.16, note that the quotient map $G_0 \to G_0/G_1$ discussed in Example 4.15 induces a smooth functor $G \to G_0/G_1$ which, due to its importance for what follows, we define separately with its own special notation.

DEFINITION 4.17. Let $G = (G_1 \rightrightarrows G_0)$ be a Lie groupoid. Then the *relation space of G* is the diffeological pullback

$$\begin{array}{ccc} R & \longrightarrow & G_0 \\ \big\downarrow & & \big\downarrow{\scriptstyle \pi} \\ G_0 & \xrightarrow[\pi]{} & G_0/G_1 \end{array}$$

for $\pi\colon G_0 \to G_0/G_1$ the diffeological quotient map.

We can now define what it means to be a gerbe over a diffeological space. Note this is an extension of [**20**, Definition 6.1], where we have replaced surjective submerions with subductions.

DEFINITION 4.18. Let $G = (G_1 \to G_0)$ be a Lie groupoid and X be a diffeological space. A smooth functor $\pi\colon G \to X$ is a *gerbe over* X if the maps $\pi\colon G_0 \to X$ and $(s,t)\colon G_1 \to G_0 \times_X G_0$ are both subductions.

While Definition 4.18 is a "correct" definition from the perspective of thinking about gerbes as objects over a base space, it turns out that a smooth functor $\pi\colon G \to X$ can only be (up to diffeomorphism) a gerbe over the quotient space G_0/G_1 of G. Indeed, we have the following:

PROPOSITION 4.19. *Let $G = (G_1 \to G_0)$ be a Lie groupoid, $\pi\colon G \to X$ be a smooth functor, and $q\colon G_0 \to G_0/G_1$ be the diffeological quotient map. If $\pi\colon G_0 \to X$ is a subduction and $(s,t)\colon G_1 \to G_0 \times_X G_0$ is surjective, then π is also a diffeological quotient map with respect to the relation on G_0 induced by G_1; that is, there exists a unique diffeomorphism $f\colon G_0/G_1 \to X$ satisfying $f \circ q = \pi$.*

In particular, if $\pi\colon G \to X$ is a gerbe over X, then $\pi\colon G_0 \to X$ is a diffeological quotient map with respect to the relation on G_0 induced by G_1.

PROOF. First, note that π being a functor means that π is invariant under the relation on G_0 induced by G_1. Thus, via the universal property of quotient spaces, there exists a unique smooth map $f\colon G_0/G_1 \to X$ for which $f \circ q = \pi$.

Next, note that f is bijective: surjectivity follows from the surjectivity of π. For injectivity, suppose that $q(x), q(y) \in G_0/G_1$ satisfy $f(q(x)) = f(q(y))$. Then we have $\pi(x) = \pi(y)$ and thus $(x,y) \in G_0 \times_X G_0$. Therefore, by the surjectivity of $(s,t)\colon G_1 \to G_0 \times_X G_0$, there exists $g \in G_1$ with $s(g) = x$ and $t(g) = y$ and thus $q(x) = q(y)$.

Finally, note that f is a subduction: as π is a subduction, any plot $p\colon U \to X$ locally lifts to a plot of G_0 which, after composing with q, descends to a local lift of p to a plot of G_0/G_1. Therefore, f is a diffeomorphism (as is any injective subduction, see [**10**, Lemma 5.20]). □

Thus, by the above proposition, a gerbe $\pi\colon G \to X \cong G_0/G_1$ over a diffeological space X can be thought of as a property intrinsic to the groupoid itself and hence one may simply call the groupoid G a gerbe and suppress the reference to the base space X.

DEFINITION 4.20. A Lie groupoid $G = (G_1 \rightrightarrows G_0)$ is a *gerbe* if the map $(s,t)\colon G_1 \to R$ is a subduction, where R is the relation space of G. Call a stack $\mathcal{X}$ a *differentiable gerbe* if it is presented by a gerbe (i.e., if it is isomorphic to BG as a stack).

EXAMPLE 4.21 (Irrational Torus). Fix $\alpha \in \mathbb{R} \smallsetminus \mathbb{Q}$. Consider the action groupoid of the Lie group action of $\mathbb{Z}^2$ on $\mathbb{R}$ given by $(m,n) \cdot x := x + m + n\alpha$. The relation space R is a dense subset of $\mathbb{R}^2$ (a dense collection of translates of the line $y = x$), and $(s,t)\colon \mathbb{Z}^2 \times \mathbb{R} \to R$ is a subduction. Note that $T_\alpha := \mathbb{R}/\mathbb{Z}^2$ is *not* a manifold, and so our concept of gerbe indeed generalizes [**20**, Definition 6.1]. //

Finally, we will show that we can tie the property of being a differentiable gerbe to our functor **Coarse**. To do so, we will need the following lemma on the local nature of principal G-bundles and their morphisms.

LEMMA 4.22. *Let $G = (G_1 \rightrightarrows G_0)$ be a Lie groupoid and let $f_i\colon M \to G_0$ $(i = 1, 2)$ be two smooth functions. Then the pullback bundles $P_i := f_i^* G_1$ are* *isomorphic if and only if there exists a smooth function $\varphi\colon M \to G_1$ satisfying $s \circ \varphi = f_1$ and $t \circ \varphi = f_2$.*

PROOF. Given a function $\varphi\colon M \to G_1$ satisfying $s \circ \varphi = f_1$ and $t \circ \varphi = f_2$, we have the isomorphism

$$\tilde{\varphi}\colon P_1 \to P_2,\ (p,g) \mapsto (p, \varphi(p) \cdot g),\ \text{for } (p,g) \in P_1 = M \times_{f_1,G_0,t} G_1.$$

Conversely, recall that for any principal G-bundle P over M, there exists a smooth division map

$$d\colon P \times_M P \to G_1,\ (p,q) \mapsto g,\ \text{for } g \text{ satisfying } p \cdot g = q$$

(this is a consequence of the requirement that the map $P \times_{G_0} G_1 \to P \times_M P$ sending (p,g) to $(p, p \cdot g)$ be a diffeomorphism).

Let $\psi\colon P_1 \to P_2$ be an isomorphism of principal G-bundles, let d designate the division map for P_2, and denote by $u\colon G_0 \to G_1$ the unit map. Let $\tilde{f}_i\colon M \to P_i$ be the maps $x \mapsto (x, u(f_i(x)))$. Then

$$\varphi\colon M \to G_1 : x \mapsto d\left(\tilde{f}_2(x), \psi(\tilde{f}_1(x))\right)$$

yields the required map with $s \circ \varphi = f_1$ and $t \circ \varphi = f_2$. □

PROPOSITION 4.23. *Let G be a Lie groupoid. Then $\hat{\pi}_0(BG)$ is a concrete sheaf of sets if and only if G is a gerbe.*

PROOF. By Proposition 4.6, $\hat{\pi}_0(BG)$ is the sheaf of local isomorphism classes of principal G-bundles. By Proposition 4.10, we have that any two principal G-bundles descend to the same plot of $\mathbf{Coarse}(BG) = \hat{\kappa}(\hat{\pi}_0(BG))$ exactly when they are fiberwise isomorphic. Since $\hat{\pi}_0(BG)$ is concrete exactly when $\hat{\kappa}(\hat{\pi}_0(BG)) = \hat{\pi}_0(BG)$, it is enough to prove that the map (s,t) is a subduction exactly when any two principal G-bundles which are fiberwise isomorphic are locally isomorphic.

By definition, (s,t) is a subduction if any plot p of R locally lifts to plots of G_1. By definition of R, this is equivalent to the following: given a pair of plots $p_i\colon M \to G_0$ $(i = 1,2)$ satisfying $\pi \circ p_1 = \pi \circ p_2$, the plot (p_1,p_2) locally lifts to plots of G_1. Let $\pi\colon G_0 \to G_0/G_1$ be the quotient map.

Suppose (s,t) is a subduction. Let P_i $(i = 1,2)$ be fiberwise isomorphic principal G-bundles. There exist an open cover $\{U_\alpha\}$ of M and for each α and $i = 1,2$, a plot $p_i^\alpha\colon U_\alpha \to G_0$ such that $P_i|_{U_\alpha} \cong (p_i^\alpha)^* G_1$. Fix α. Then $(p_1^\alpha)^* G_1$ is fiberwise isomorphic to $(p_2^\alpha)^* G_1$. By Proposition 4.12, $\pi \circ p_1^\alpha = \pi \circ p_2^\alpha$. Thus, (p_1^α, p_2^α) locally lifts to G_1. By Lemma 4.22, these local lifts to G_1 yield local isomorphisms from $P_1|_{U_\alpha}$ to $P_2|_{U_\alpha}$. Therefore P_1 and P_2 are locally isomorphic.

Conversely, suppose any fiberwise isomorphic principal G-bundles are locally isomorphic. Let $p_i\colon M \to G_0$ $(i = 1,2)$ be plots satisfying $\pi \circ p_1 = \pi \circ p_2$. By Proposition 4.12, $p_1^* G_1$ is fiberwise isomorphic to $p_2^* G_2$. By hypothesis, these are locally isomorphic; that is, by Lemma 4.22, there is an open cover $\{U_\alpha\}$ of M and plots $g_\alpha\colon U_\alpha \to G_1$ such that $s \circ g_\alpha = p_1|_{U_\alpha}$ and $t \circ g_\alpha = p_2|_{U_\alpha}$, for each α. So $(s,t) \circ g_\alpha = (p_1,p_2)|_{U_\alpha}$ for each α, and we conclude that (s,t) is a subduction. □

REMARK 4.24. The equivalence noted in Proposition 4.23 is proven in the context of stacks over topological spaces in [**18**, Proposition 5.3], albeit with respect to a more general definition of a gerbe.

5. Basic differential forms

Recall that given a Lie groupoid G with source and target map s and t, resp., a "basic differential form" α on G_0 is one such that $s^*\alpha = t^*\alpha$ (see [**25**, Definition 3.1]). In this section we give a notion of a basic differential form on a stack, and show that it extends the definition for a Lie groupoid (see Proposition 5.5). In the case of a stack equal to the Grothendieck construction applied to a concrete sheaf, the definition matches that for a corresponding diffeological space (see Remark 5.4). It turns out that this definition only depends on the discretization of the stack, and in the case that the stack is differentiable and represented by a specific type of Lie groupoid, it only depends on the underlying diffeological coarse moduli space (see Theorem 5.6). It remains an open question whether this dependence on the diffeological coarse moduli space holds for all stacks.

DEFINITION 5.1 (Differential Forms). Let $\Omega^k\colon \mathbf{Man}^{op} \to \mathbf{Set}$ be the sheaf of differential k-forms over manifolds. Then, for a sheaf $S\colon \mathbf{Man}^{op} \to \mathbf{Set}$, we will define a *basic differential k-form* to be a map of sheaves $\mu\colon S \to \Omega^k$. Define $\Omega^k(S) := \mathrm{Hom}_{\mathbf{Shf(Man)}}(S, \Omega^k)$. Similarly, for a stack $\mathcal{X}$, define a *basic k-form* to be a map of stacks $\beta\colon \mathcal{X} \to \int \Omega^k$ and let $\Omega^k(\mathcal{X}) := \mathrm{Hom}_{\mathbf{St(Man)}}(\mathcal{X}, \int \Omega^k)$.

REMARK 5.2. Note that $\Omega^k(\mathcal{X})$ is discrete; *i.e.* for each manifold M, $\Omega^k(\mathcal{X})(M)$ is a set. In fact $\Omega^k(\mathcal{X})(M)$ has a vector space structure, but we do not need this for our purposes.

DEFINITION 5.3 (Differential Forms in Diffeology, see [**10**, Article 6.28]). A *differential k-form* μ on a diffeological space X is an assignment to each plot $p\colon U \to X$ a differential form $\mu(p) \in \Omega^k(U)$ satisfying the following compatibility condition: if $f\colon V \to U$ is a smooth map between open subsets of Euclidean spaces, then $\mu(p \circ f) = f^*\mu(p)$.

REMARK 5.4. When considering a diffeological space $(X, \mathcal{D})$ as a concrete sheaf $I(\mathcal{D})\colon \mathbf{Man}^{op} \to \mathbf{Set}$, this definition says precisely that a differential k-form for a diffeological space is a map of sheaves $\mu\colon I(\mathcal{D}) \to \Omega^k$, matching Definition 5.1.

PROPOSITION 5.5 (Basic Forms of a Lie Groupoid). *Let $G = (G_1 \rightrightarrows G_0)$ be a Lie groupoid, and let y be the Yoneda embedding of* **Man** *into* **St**(**Man**). *An atlas $\alpha\colon y(G_0) \to BG$ induces a bijection from $\Omega^k(BG)$ to basic k-forms on G_0; that is, differential k-forms μ such that $s^*\mu = t^*\mu$ where s and t are the source and target maps of G, respectively.*

PROOF. Let $\mu \in \Omega^k(BG)$, and let $\alpha\colon y(G_0) \to BG$ be an atlas. Then

$$\mu \circ \alpha \circ y(s) = \mu \circ \alpha \circ y(t).$$

Thus, $\mu \circ \alpha$ corresponds to a basic differential form on G_0.

In the other direction, any basic form η on G_0 induces a map of stacks $\tilde{\eta}\colon y(G_0) \to \int \Omega^k$ such that

$$\tilde{\eta} \circ y(s) = \tilde{\eta} \circ y(t).$$

By Lemma 4.1 and the 2-Yoneda Lemma, there is a map $\theta\colon BG \to \int \Omega^k$ such that $\tilde{\eta} = \theta \circ \alpha$. Since $\int \Omega^k$ is discrete, θ is unique. $\square$

THEOREM 5.6 (Differential Forms and **Coarse**). *Let $\mathcal{X}$ be a stack. Then, pre-composition with $\hat{\pi}_0$ induces a bijection from $\Omega^k(\int \hat{\pi}_0(\mathcal{X}))$ to $\Omega^k(\mathcal{X})$, and pre-composition with* **Coarse** *induces an injection from $\Omega^k(\int \mathbf{Coarse}(\mathcal{X}))$ to $\Omega^k(\mathcal{X})$. Moreover, if G is*

1. *a finite-dimensional, Hausdorff, paracompact, proper Lie groupoid; or*
2. *the action groupoid of a Lie group on a manifold whose identity component acts properly; or*
3. *a gerbe over G_0/G_1,*

then the injection from $\Omega^k(\int \mathbf{Coarse}(BG))$ to $\Omega^k(BG)$ is in fact a bijection.

PROOF. The first claim is immediate from the fact that the Grothendieck construction is a fully faithful functor, along with Lemma 3.7. The second claim follows from the fact that any differential form $\mu\colon \mathbf{Coarse}(\mathcal{X}) \to \Omega^k$ is completely determined by it values on elements of the form $\underline{[\xi]}$ where ξ are objects in $\mathcal{X}$. Finally, the last claim follows from Lemma 2.7; Remark 5.4; and [**25**] in the case of Item 1, or [**14**] in the case of Item 2, or Proposition 4.23 in the case of Item 3. □

Acknowledgments

The authors thank Eugene Lerman for many discussions regarding stacks, and introducing us to various categorical notions required to put the entire puzzle that is this paper together. The authors also thank David M. Roberts for many useful suggestions, which resulted in a better paper. The first author would also like to thank Patrick Iglesias-Zemmour and Yael Karshon for many illuminating diffeological examples, and Pierre Albin for discussions on the presentation.

References

[1] S. Awodey, *Category theory*, Oxford Logic Guides, vol. 49, The Clarendon Press, Oxford University Press, New York, 2006, DOI 10.1093/acprof:oso/9780198568612.001.0001. MR2229319

[2] K. Behrend, G. Ginot, B. Noohi, and P. Xu, *String topology for stacks* (English, with English and French summaries), Astérisque **343** (2012), xiv+169. MR2977576

[3] J. C. Baez and A. E. Hoffnung, *Convenient categories of smooth spaces*, Trans. Amer. Math. Soc. **363** (2011), no. 11, 5789–5825, DOI 10.1090/S0002-9947-2011-05107-X. MR2817410

[4] B. Collier, E. Lerman, and S. Wolbert, *Parallel transport on principal bundles over stacks*, J. Geom. Phys. **107** (2016), 187–213, DOI 10.1016/j.geomphys.2016.05.010. MR3521476

[5] R. Cushman and J. Śniatycki, *Differential structure of orbit spaces*, Canad. J. Math. **53** (2001), no. 4, 715–755, DOI 10.4153/CJM-2001-029-1. MR1848504

[6] A. Grothendieck, "Revêtements étales et groupe fondamental", In: *Séminaire de Géométrie Algébrique du Bois Marie 1960-1961 (SGA 1), Lecture Notes in Math.* **224**, Springer-Verlag, Berlin, 1971.

[7] J. Heinloth, *Notes on differentiable stacks*, Mathematisches Institut, Georg-August-Universität Göttingen: Seminars Winter Term 2004/2005, Universitätsdrucke Göttingen, Göttingen, 2005, pp. 1–32. MR2206877

[8] R. Hepworth, *Morse inequalities for orbifold cohomology*, Algebr. Geom. Topol. **9** (2009), no. 2, 1105–1175, DOI 10.2140/agt.2009.9.1105. MR2511141

[9] R. Hepworth, *Vector fields and flows on differentiable stacks*, Theory Appl. Categ. **22** (2009), 542–587. MR2591949

[10] P. Iglesias-Zemmour, *Diffeology*, Mathematical Surveys and Monographs, vol. 185, American Mathematical Society, Providence, RI, 2013, DOI 10.1090/surv/185. MR3025051

[11] N. Johnson and D. Yau, *2-dimensional categories*, Oxford University Press, Oxford, 2021, DOI 10.1093/oso/9780198871378.001.0001. MR4261588

[12] P. T. Johnstone, *Sketches of an elephant: a topos theory compendium. Vol. 1*, Oxford Logic Guides, vol. 43, The Clarendon Press, Oxford University Press, New York, 2002. MR1953060

[13] Y. Karshon and D. Miyamoto, *Quasifold groupoids and diffeological quasifolds*, Preprint, arXiv:2206.14776.

[14] Y. Karshon and J. Watts, *Basic forms and orbit spaces: a diffeological approach*, SIGMA Symmetry Integrability Geom. Methods Appl. **12** (2016), Paper No. 026, 19 pp., DOI 10.3842/SIGMA.2016.026. MR3470746

[15] E. Lerman, *Orbifolds as stacks?*, Enseign. Math. (2) **56** (2010), no. 3-4, 315–363, DOI 10.4171/LEM/56-3-4. MR2778793

[16] S. Mac Lane, *Categories for the working mathematician*, 2nd ed., Graduate Texts in Mathematics, vol. 5, Springer-Verlag, New York, 1998. MR1712872

[17] D. Maulik and D. Treumann, *Constructible functions and Lagrangian cycles on orbifolds*, 2012, Preprint, arXiv:1110.3866v2

[18] B. Noohi, *Foundations for Topological Stacks I*, 2005, Preprint, arXiv:math/0503247

[19] D. A. Pronk, *Etendues and stacks as bicategories of fractions*, Compositio Math. **102** (1996), no. 3, 243–303. MR1401424

[20] D. M. Roberts and R. F. Vozzo, *Smooth loop stacks of differentiable stacks and gerbes* (English, with English and French summaries), Cah. Topol. Géom. Différ. Catég. **59** (2018), no. 2, 95–141. MR3727316

[21] D. M. Roberts, *The elementary construction of formal anafunctors*, Categ. Gen. Algebr. Struct. Appl. **15** (2021), no. 1, 183–229. MR4357762

[22] G. W. Schwarz, *Smooth functions invariant under the action of a compact Lie group*, Topology **14** (1975), 63–68, DOI 10.1016/0040-9383(75)90036-1. MR370643

[23] J.-M. Souriau, *Groupes différentiels* (French), Differential geometrical methods in mathematical physics (Proc. Conf., Aix-en-Provence/Salamanca, 1979), Lecture Notes in Math., vol. 836, Springer, Berlin, 1980, pp. 91–128. MR607688

[24] A. Vistoli, *Grothendieck topologies, fibered categories and descent theory*, Fundamental algebraic geometry, Math. Surveys Monogr., vol. 123, Amer. Math. Soc., Providence, RI, 2005, pp. 1–104. MR2223406

[25] J. Watts, *The orbit space and basic forms of a proper LIE groupoid*, Current trends in analysis, its applications and computation, Trends Math. Res. Perspect., Birkhäuser/Springer, Cham, 2022, pp. 513–523, DOI 10.1007/978-3-030-87502-2_52. MR4559547

Department of Mathematics, Central Michigan University, Mount Pleasant, Michigan

Email address: jordan.watts@cmich.edu

Department of Mathematics and Statistics, Minnesota State University, Mankato, Minnesota

Email address: seth.wolbert@mnsu.edu

Contemporary Mathematics
Volume **794**, 2024
https://doi.org/10.1090/conm/794/15927

Generalized Laurent monomials in nonrational toric geometry

Fiammetta Battaglia and Elisa Prato

ABSTRACT. We generalize Laurent monomials to toric quasifolds, a special class of highly singular spaces that extend simplicial toric varieties to the nonrational setting.

Introduction

Toric quasifolds are highly singular spaces that extend simplicial toric varieties to nonrational fans. They were first introduced in the symplectic category [**17**, **18**], by means of a generalization of the classical Delzant construction of symplectic toric manifolds [**10**]. Complex toric quasifolds, on the other hand, first appeared in [**2**] and follow the construction of simplicial toric varieties as complex quotients given by Audin in [**1**]. Toric quasifolds provide an interesting class of examples of diffeological spaces: the notion of diffeological quasifold was introduced, jointly with Iglesias–Zemmour, in [**14**] and related to that of quasifold groupoid by Karshon–Miyamoto in [**15**].

Quasifolds are locally the quotient of a manifold modulo the action of a countable group. If the countable groups are all finite, quasifolds are orbifolds; if they are all trivial, quasifolds are manifolds. Similarly to what happens for manifolds, in order to define a quasifold structure we need an atlas, namely a cover by mutually compatible local models. We have shown in [**2**] that every n–dimensional complex toric quasifold has a finite atlas whose local models are given by $\mathbb{C}^n$ modulo the action of a countable subgroup of $(S^1)^n$. Now, recall from Danilov [**9**, Section 0.2] that a smooth toric variety is characterized by the fact that it is covered by a finite number of copies of $\mathbb{C}^n$, with changes of charts expressed in terms of Laurent monomials. In this article, we introduce generalized Laurent monomials and we prove that the changes of charts of toric quasifolds can be expressed, similarly to what happens in the smooth case, in terms of generalized Laurent monomials.

This allows for the possibility of defining toric quasifolds directly, by starting with a collection of complex quasifold models and by gluing them together by

2020 *Mathematics Subject Classification.* Primary 14M25; Secondary 52B20, 53D20.

Key words and phrases. Toric variety, toric quasifold, nonrational fan, nonrational polytope, Laurent monomial.

The authors were partially supported by the PRIN Project "Real and Complex Manifolds: Topology, Geometry and Holomorphic Dynamics" (MIUR, Italy) and by GNSAGA (INdAM, Italy).

means of generalized Laurent monomials, thus extending what was originally done for smooth toric varieties [**11**]. We plan to address this matter and its implications in future work.

The article is structured as follows. In Section 1, we define generalized Laurent monomials for local quasifold models. In Section 2, we recall the construction of a complex toric quasifold, we describe the canonical affine atlas and write the chart transitions in terms of generalized Laurent monomials. Finally, in Section 3, we compute the monomials in a number of interesting examples in dimensions 1, 2, and 3.

1. Generalized Laurent monomials

The goal of this section is to show that, in the quasifold setting, we can make sense of the notion of Laurent monomials with real exponents.

Let us consider, first of all, an example of countable group that will be of particular relevance for us, that of a *quasilattice*. We recall from [**17, 18**] that a quasilattice Q in $\mathbb{R}^n$ is an additive subgroup of $\mathbb{R}^n$ given by the $\mathbb{Z}$–span of a finite generating subset of $\mathbb{R}^n$. Notice that a quasilattice is a lattice if, and only if, it is generated by a basis of $\mathbb{R}^n$.

Let us now describe a quasifold of one chart which provides the local model for toric quasifolds. First of all, write the exponential mapping from the Lie algebra $\mathbb{C}^m$ to the torus $(\mathbb{C}^*)^m$ as follows:

$$\begin{array}{rccc} \exp\colon & \mathbb{C}^m & \longrightarrow & (\mathbb{C}^*)^m \\ & (z_1,\dots,z_m) & \longmapsto & (e^{2\pi i z_1},\dots,e^{2\pi i z_m}). \end{array}$$

We denote exp also its restriction to the Lie algebra $\mathbb{R}^m$ of $\mathbb{R}^m/\mathbb{Z}^m = (S^1)^m$.

If $Q \subset \mathbb{R}^n$ is a quasilattice, then $\Gamma = \exp(Q)$ is a countable subgroup of $\mathbb{R}^n/\mathbb{Z}^n = (S^1)^n$. We consider the quasifold of one chart

$$\mathbb{C}^n/\Gamma.$$

We are now ready to introduce the generalized Laurent monomials. Consider $\underline{a} = (a_1,\dots,a_n) \in Q$. If $l \in \{1,\dots,n\}$ and $(z_1,\dots,z_n) \in (\mathbb{C}^*)^n$, then the class $[z_l^{a_1} : \cdots : z_l^{a_n}]$ is well defined in $(\mathbb{C}^*)^n/\Gamma$. In fact, if we write

$$z_l = e^{2\pi i(x_l+h_l+iy_l)},$$

with $h_l \in \mathbb{Z}$, then

$$z_l^{a_j} = e^{2\pi i a_j(x_l+h_l+iy_l)}.$$

Notice now that $\exp(h_l\underline{a}) \in \Gamma$.

Similarly, consider a subset of indices $I \subseteq \{1,\dots,n\}$ of cardinality k and take $\underline{a}^l = (a_{1l},\dots,a_{nl}) \in Q$, $l \in I$. Then the class

$$\left[\prod_{l\in I} z_l^{a_{1l}} : \cdots : \prod_{l\in I} z_l^{a_{nl}}\right]$$

is well defined in $(\mathbb{C}^*)^n/\Gamma$. We call the single terms

$$\prod_{l\in I} z_l^{a_{jl}}, \quad j=1,\dots,n$$

generalized Laurent monomials. We can write them briefly as $\underline{z}^{\underline{a}_j}$, where $\underline{a}_j = (a_{jl_1},\dots,a_{jl_k})$ and $\underline{z} = (z_{l_1},\dots,z_{l_k})$.

2. The canonical atlas and the generalized Laurent monomials

2.1. Fans. Let us recall some basic facts on fans. For further details, we refer the reader to Cox et al. [**8**] and Ziegler [**21**].

A fan Σ in $\mathbb{R}^n$ is a collection of strongly convex polyhedral cones in $\mathbb{R}^n$, such that each nonempty face of a cone in Σ is itself a cone in Σ, and such that the intersection of any two cones in Σ is a face of each.

The cones of dimension n are called *maximal cones*, while the cones of dimension 1 are called *generating rays*.

An important example of fan is the *normal fan* Σ_Δ of a convex polytope Δ. It is the fan whose generating rays are inward pointing and orthogonal to the polytope facets. The rays are grouped to generate all the cones of the fan following the combinatorics of the polytope faces and we have an inclusion–reversing bijection between cones in Σ_Δ and faces of Δ.

Each cone of a fan is spanned by a subset of a set of generating vectors $X_1, \ldots, X_d$. The fan is called *simplicial* if, for each cone, these vectors are linearly independent. The normal fan of a convex polytope Σ_Δ is simplicial if, and only if, the polytope Δ is simple. Consider a simplicial fan. Then, for each maximal cone σ, there is a subset $I_\sigma \subset \{1, \ldots, d\}$ of n indices, such that this maximal cone is generated by $X_i, i \in I_\sigma$. The faces of σ are in bijective correspondence with the subsets of the set of indices I_σ, the 0–dimensional cone of Σ corresponding to the empty set.

The support $|\Sigma|$ of a fan Σ is the union in $\mathbb{R}^n$ of its cones. We will be interested in fans that have convex support of full dimension; this implies that $|\Sigma|$ is the union of the maximal cones of Σ. A notable example is the case of a *complete* fan, namely a fan whose support is the whole space $\mathbb{R}^n$ like, for instance, the normal fan of a convex polytope.

A fan Σ in $\mathbb{R}^n$ is said to be *rational* if there exists a lattice $L \subset \mathbb{R}^n$ such that the intersection of L with each generating ray of Σ is nonempty. A convex polytope Δ is said to be rational if its normal fan Σ_Δ is.

We are interested in fans and polytopes that are not rational. In this context, it is convenient to introduce a different notion. A fan Σ in $\mathbb{R}^n$ is said to be *quasirational* with respect to a quasilattice $Q \subset \mathbb{R}^n$ if the intersection of Q with each generating ray of Σ is nonempty. A convex polytope Δ is said to be quasirational if its normal fan Σ_Δ is.

2.2. The construction of complex toric quasifolds. We recall from [**2**, Theorem 2.2] the construction of toric quasifolds in the complex category. In the original paper, we had considered simple convex polytopes, but the same argument goes through in the more general setting of *simplicial fans with convex support of full dimension*.

Let Q be a quasilattice in $\mathbb{R}^n$, and assume that Σ is quasirational with respect to Q. Let $X_1, \ldots, X_d \in \mathbb{R}^n$ be a choice of generators for the d rays of Σ and assume that they lie in Q. In the case of the normal fan of a convex polytope Δ, these vectors are each orthogonal to a different codimension one face (*facet*) of Δ and point inwards; we will be referring to them as *normals* for Δ.

The *fundamental triple*

$$(\Sigma, Q, \{X_1, \ldots, X_d\})$$

effectively replaces the standard triple

$$\text{(fan, lattice, primitive generators)}$$

that is needed for the construction of toric varieties in the rational case. We will in fact construct the n–dimensional complex toric quasifold, X_Σ, relatively to the fundamental triple above.

Before we proceed, we remark that, in this framework, tori are generalized by *quasitori* [**17**, **18**, **2**]. They are given by the quasifolds of one chart

$$\mathbb{R}^n/Q$$

and

$$\mathbb{C}^n/Q.$$

The first is real, while the second is complex. They are both abelian groups and, when Q is a true lattice, $\mathbb{R}^n/Q$ and $\mathbb{C}^n/Q$ are true tori. Their Lie algebras are given by $\mathbb{R}^n$ and $\mathbb{C}^n$, respectively.

Let now σ be a maximal cone, and let I_σ denote the corresponding index set. Consider the open subset of $\mathbb{C}^d$ given by

$$U_\sigma = \{\, \underline{z} \in \mathbb{C}^d \mid z_j \neq 0, \;\; j \notin I_\sigma \,\}.$$

Notice that we have

$$U_\sigma = \bigsqcup_{J \subset I_\sigma} \{\, \underline{z} \in \mathbb{C}^d \mid z_j = 0 \;\text{ iff }\; j \in J \,\}.$$

Take now

$$U_\Sigma = \bigcup_{\sigma \text{ maximal cone}} U_\sigma.$$

This open subset of $\mathbb{C}^d$ is acted upon by a subgroup $N_\mathbb{C} \subset (\mathbb{C}^*)^d$ that we construct in the following way.

Let $\{e_1, \ldots, e_d\}$ denote the standard basis of $\mathbb{R}^d$ and $\mathbb{C}^d$; consider the surjective linear mapping

$$\begin{array}{rccc} \pi : & \mathbb{R}^d & \longrightarrow & \mathbb{R}^n \\ & e_j & \longmapsto & X_j \end{array}$$

and its complexification

$$\begin{array}{rccc} \pi_\mathbb{C} : & \mathbb{C}^d & \longrightarrow & \mathbb{C}^n \\ & e_j & \longmapsto & X_j. \end{array}$$

Consider the quasitorus $\mathbb{R}^n/Q$ and its complexification $\mathbb{C}^n/Q$. The mappings π and $\pi_\mathbb{C}$ induce group homomorphisms

$$\Pi : \, (S^1)^d = \mathbb{R}^d/\mathbb{Z}^d \longrightarrow \mathbb{R}^n/Q$$

and

$$\Pi_\mathbb{C} : \, (\mathbb{C}^*)^d = \mathbb{C}^d/\mathbb{Z}^d \longrightarrow \mathbb{C}^d/Q.$$

We define N to be the kernel of the mapping Π and $N_\mathbb{C}$ to be the kernel of the mapping $\Pi_\mathbb{C}$. Notice that neither N nor $N_\mathbb{C}$ are honest tori unless Q is a honest lattice.

Then the complex toric quasifold corresponding the fundamental triple $(\Sigma, Q, \{X_1, \ldots, X_d\})$ is given by the quotient

$$X_\Sigma = \frac{U_\Sigma}{N_\mathbb{C}}.$$

It is an n–dimensional complex quasifold. If Σ is the normal fan of a convex polytope Δ, we will call this quasifold X_Δ.

Notice that the mapping $\Pi_{\mathbb{C}}$ induces an isomorphism

$$(\mathbb{C}^*)^d/N_{\mathbb{C}} \longrightarrow \mathbb{C}^n/Q.$$

Thus the complex quasitorus $\mathbb{C}^n/Q$ acts on X_Σ. This action is holomorphic and, as in the rational case, X_Σ is the disjoint union of the k–dimensional orbits, each corresponding to an $(n-k)$–dimensional cone of Σ. In particular, there is a dense open orbit, corresponding to the 0–dimensional cone. Also, there are a finite number of fixed points, one for each maximal cone σ. They are given by the points $[z_1 : \cdots : z_d]$, with $z_j = 0$, for $j \in I_\sigma$ and $z_j = 1$ otherwise.

REMARK 2.1. In recent years, nonrational toric geometry has been investigated from several different, though interrelated, points of view. See [**6**] for a detailed account.

2.3. The symplectic picture. It was shown in [**17, 18**] that, to each fundamental triple

$$(\Sigma_\Delta, Q, \{X_1, \dots, X_d\}),$$

with Δ a simple convex polytope, there corresponds a $2n$–dimensional symplectic toric quasifold M_Δ, endowed with the effective Hamiltonian action of the real quasitorus $\mathbb{R}^n/Q$. The image of the corresponding moment mapping Φ is the initial polytope Δ. The quasifold M_Δ is obtained explicitly via symplectic reduction with respect to the subgroup $N \subset (S^1)^d$ that we have introduced above, thereby extending the classical construction of symplectic toric manifolds by Delzant [**10**]. It is shown in [**2**, Theorem 3.2] that M_Δ and X_Δ are equivariantly diffeomorphic, and that the complex and symplectic structures are compatible, thus defining a Kahler structure on X_Δ, exactly as in the rational case (for which we refer to Audin [**1**] and Guillemin [**13**]). Moreover, the fixed points of the action of $\mathbb{C}^n/Q$ on X_Δ correspond to the fixed points of the action of $\mathbb{R}^n/Q$ on M_Δ, and the latter are mapped bijectively via Φ to the vertices of the polytope.

2.4. The canonical affine atlas and the generalized Laurent monomials. We adapt from [**2**] the construction of the canonical atlas for X_Σ. The novelty here, besides the weaker assumptions on Σ, is that we write the changes of charts explicitly in terms of generalized Laurent monomials.

Before we do so, notice that we can write the groups N and $N_{\mathbb{C}}$ as follows:

$$N = \{\exp(X) \mid X \in \mathbb{R}^d,\ \ \pi(X) \in Q\}$$

$$N_{\mathbb{C}} = \{\exp(Z) \mid Z \in \mathbb{C}^d,\ \ \pi(Z) \in Q\}.$$

The respective Lie algebras are $\mathfrak{n} = \ker(\pi)$ and $\mathfrak{n}_{\mathbb{C}} = \ker(\pi_{\mathbb{C}})$, which are linear subspaces of dimension $d-n$ of $\mathbb{R}^d$ and $\mathbb{C}^d$, respectively. The mapping exp restricts to both.

Fix a maximal cone, σ, with its subset of n indices, I_σ. Consider the restrictions π_σ and $(\pi_\sigma)_{\mathbb{C}}$ of the mappings π and $\pi_{\mathbb{C}}$ to the n–dimensional subspaces $\prod_{i\in I_\sigma} \mathbb{R}e_i \subset \mathbb{R}^d$ and $\prod_{i\in I_\sigma} \mathbb{C}e_i \subset \mathbb{C}^d$ respectively. Both maps are isomorphisms. We have the following lemma, that, in the complex setting, is proven in [**2**, Lemma 2.3]. We give a new proof that is tailored to our purposes:

LEMMA 2.2. *For each maximal cone σ, we have that $N = \Gamma_\sigma \exp(\mathfrak{n})$ and $N_{\mathbb{C}} = \Gamma_\sigma \exp(\mathfrak{n}_{\mathbb{C}})$, where*

$$\Gamma_\sigma = \left\{ \exp(X) \mid X \in \prod_{j \in I_\sigma} \mathbb{R}e_j,\ \ \pi(X) \in Q \right\}.$$

PROOF. Let $X \in \mathbb{R}^d$ such that $\pi(X) \in Q$ and let $Y = \pi_\sigma^{-1}(\pi(X))$. Then $W = X - Y \in \mathfrak{n}$ and therefore $\exp(X) = \exp(Y)\exp(W)$, with $\exp(Y) \in \Gamma_\sigma$, $\exp(W) \in \exp(\mathfrak{n})$. The same argument applies to the complexified group and Lie algebra. □

REMARK 2.3. For each $j \notin I_\sigma$ denote $\underline{a}^j = \pi_\sigma^{-1}(X_j) \in \prod_{i \in I_\sigma} \mathbb{R}e_i \subset \mathbb{R}^d$. The $d - n$ linearly independent vectors $e_j - \underline{a}^j$ form a basis of $\mathfrak{n}$ and $\mathfrak{n}_{\mathbb{C}}$. Notice that $\exp(\underline{a}^j) \in \Gamma_\sigma$.

We are finally ready to prove the following

THEOREM 2.4. *The complex toric quasifold X_Σ corresponding to the fundamental triple $(\Sigma, Q, \{X_1, \ldots, X_d\})$ is endowed with a generalized canonical atlas, having a chart around each of the fixed points of the quasitorus action. The transition mappings are written in terms of generalized Laurent monomials.*

PROOF. We define a chart for each maximal cone in the fan Σ. Given such a cone, σ, we can assume, up to renumbering, that the corresponding index set, I_σ, is given by $\{1, \ldots, n\}$. Then the mapping

$$\begin{array}{ccc} \mathbb{C}^n/\Gamma_\sigma & \xrightarrow{\eta_\sigma} & V_\sigma = U_\sigma/N_{\mathbb{C}} \\ [z_1 : \cdots : z_n] & \longmapsto & [z_1 : \cdots : z_n : 1 : \cdots : 1] \end{array}$$

is a homeomorhism and defines a chart around the fixed point

$$[0 : \cdots : 0 : 1 : \cdots : 1].$$

We will be referring to this chart as the quadruple $(\mathbb{C}^n, \Gamma_\sigma, V_\sigma, \eta_\sigma)$.

Consider now a second chart $(\mathbb{C}^n, \Gamma_\tau, V_\tau, \eta_\tau)$, corresponding to another maximal cone τ. Notice that the intersection $V_\sigma \cap V_\tau$ is always nonempty. Let us restrict our attention to the case $I_\sigma \cap I_\tau \neq \emptyset$; the case $I_\sigma \cap I_\tau = \emptyset$ is simpler and can be treated similarly. Assume, again up to renumbering, that we have the following partition

$$\{1, \ldots, h, h+1, \ldots, n, n+1, \ldots, n+h, n+h+1, \ldots, d\},\ \ 1 \leq h \leq n-1,$$

with

$$\begin{aligned} I_\sigma &= \{1, \ldots, h, h+1, \ldots, n\}, \\ I_\tau &= \{h+1, \ldots, n, n+1, \ldots, n+h\}, \\ I_\sigma \cap I_\tau &= \{h+1, \ldots, n\}. \end{aligned}$$

Let us compute the transition mapping between these two charts. Take

$$[z_{h+1} : \cdots : z_{n+h}] \xrightarrow{\eta_\tau} [1 : \cdots : 1 : z_{h+1} : \cdots : z_{n+h} : 1 : \cdots : 1].$$

Let us now move the representative on the right–hand side within the $N_{\mathbb{C}}$–orbit by acting via

$$\left(\prod_{j=n+1}^{n+h} z_j^{a_{1j}}, \ldots, \prod_{j=n+1}^{n+h} z_j^{a_{nj}}, z_{n+1}^{-1}, \ldots, z_{n+h}^{-1}, 1, \ldots, 1 \right) \in \exp(\mathfrak{n}_{\mathbb{C}}),$$

where $\underline{a}^j = \pi_\sigma^{-1}(X_j) = (a_{1j}, \ldots, a_{nj})$. We obtain

$$\left[\prod_{j=n+1}^{n+h} z_j^{a_{1j}} : \cdots : \prod_{j=n+1}^{n+h} z_j^{a_{hj}} : \prod_{j=n+1}^{n+h} z_j^{a_{h+1,j}} z_{h+1} : \cdots : \prod_{j=n+1}^{n+h} z_j^{a_{nj}} z_n : 1 : \cdots : 1\right].$$

Write now $\prod_{j=n+1}^{n+h} z_j^{a_{ij}} = \underline{z}^{\underline{a}_i}$, where $\underline{z} = (z_{n+1}, \ldots, z_{n+h}) \in \mathbb{C}^h$ and

$$\underline{a}_i = (a_{i,n+1}, \ldots, a_{i,n+h}) \in \mathbb{R}^h, \quad i = 1, \ldots, n.$$

Then the transition mapping from $(\mathbb{C}^n, \Gamma_\tau, V_\tau, \eta_\tau)$ to $(\mathbb{C}^n, \Gamma_\sigma, V_\sigma, \eta_\sigma)$ can be finally written in terms of generalized Laurent monomials as

$$\begin{array}{ccc} \eta_\tau^{-1}(V_\sigma \cap V_\tau) & \xrightarrow{\eta_\sigma^{-1}\circ\eta_\tau} & \eta_\sigma^{-1}(V_\sigma \cap V_\tau) \\ [z_{h+1} : \cdots : z_{n+h}] & \longmapsto & [\underline{z}^{\underline{a}_1} : \cdots : \underline{z}^{\underline{a}_h} : \underline{z}^{\underline{a}_{h+1}} z_{h+1} : \cdots : \underline{z}^{\underline{a}_n} z_n]. \end{array}$$

□

3. Examples

We compute the generalized Laurent monomials in a number of interesting examples. It will be convenient to label the charts relatively to the index sets I_σ instead of the maximal cones σ themselves.

3.1. The quasisphere. The quasisphere is a toric quasifold that was first introduced in [**17, 18**] and can be constructed as follows. Let a be a positive irrational number and consider the unit interval $[0,1]$. Then the quasisphere is the toric quasifold obtained from the fundamental triple

$$(\Sigma_{[0,1]}, \mathbb{Z} + a\mathbb{Z}, \{X_1 = a, X_2 = -1\}).$$

It is given by

$$X_a = \frac{\mathbb{C}^2 \setminus \{0\}}{\{\, (e^{2\pi i r}, e^{2\pi i a r}) \in (\mathbb{C}^*)^2 \mid r \in \mathbb{C} \,\}}$$

and it is endowed with the holomorphic action of the complex dimension one quasitorus (quasicircle)

$$\mathbb{C}/(\mathbb{Z} + a\mathbb{Z}).$$

Notice that this quasicircle is the complexification of the *irrational torus* of Donato–Iglesias [**12**]. The canonical atlas of the complex quasisphere is made of two charts, one around the point $[0:1]$, the other around the point $[1:0]$. The first is given by $(\mathbb{C}, \Gamma_1, V_1, \eta_1)$, where

$$\Gamma_1 = \left\{ e^{2\pi i \frac{h}{a}} \in S^1 \mid h \in \mathbb{Z} \right\},$$

$$V_1 = \{[z_1 : z_2] \in X_a \mid z_2 \neq 0\},$$

$$\begin{array}{ccc} \mathbb{C}/\Gamma_1 & \xrightarrow{\eta_1} & V_1 \\ [z] & \longmapsto & [z:1]. \end{array}$$

While the second is given by $(\mathbb{C}, \Gamma_2, V_2, \eta_2)$, where

$$\Gamma_2 = \left\{ e^{2\pi i a h} \in S^1 \mid h \in \mathbb{Z} \right\},$$

$$V_2 = \{[w_1 : w_2] \in X_a \mid w_1 \neq 0\},$$

$$\begin{array}{ccc} \mathbb{C}/\Gamma_2 & \xrightarrow{\eta_2} & V_2 \\ [w] & \longmapsto & [1:w]. \end{array}$$

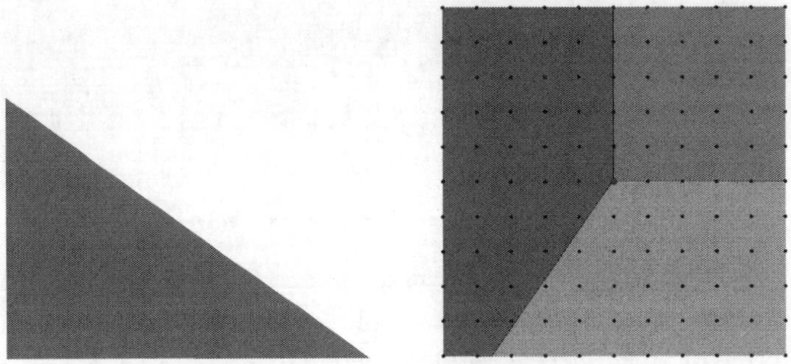

FIGURE 1. The right triangle T_a and its normal fan.

The transition map between these two charts is given by

$$\begin{array}{ccc} \eta_1^{-1}(V_1 \cap V_2) & \stackrel{\eta_2^{-1}\circ\eta_1}{\longrightarrow} & \eta_2^{-1}(V_1 \cap V_2) \\ [z] & \longmapsto & [z^{-a}] \,. \end{array}$$

Here we get the simplest possible Laurent monomial.

A thorough description of the quasisphere from the symplectic viewpoint can be found in [**20**].

3.2. Generalized weighted projective space. Consider, for any positive real number a, the right triangle T_a of vertices $(0,0)$, $(a,0)$, and $(0,1)$ (see Figure 1).

Though this triangle is rational with respect to the lattice spanned by $(1,0)$ and $(0,a)$, it is interesting to view it in a nonrational setting, as follows. Let us take normals given by $X_1 = (1,0)$, $X_2 = -(1,a)$ and $X_3 = (0,1)$. They span the quasilattice

$$\mathbb{Z} \times (\mathbb{Z} + a\mathbb{Z}) \supseteq \mathbb{Z}^2.$$

The toric quasifold corresponding to the fundamental triple

$$(\Sigma_{T_a}, \mathbb{Z} \times (\mathbb{Z} + a\mathbb{Z}), \{X_1, X_2, X_3\})$$

is given by

$$\mathbb{CP}^2_{(1,1,a)} = \frac{\mathbb{C}^3 \setminus \{0\}}{\{\, (e^{2\pi i u}, e^{2\pi i u}, e^{2\pi i a u}) \in (\mathbb{C}^*)^3 \mid u \in \mathbb{C} \,\}}.$$

It is endowed with the holomorphic action of the complex quasitorus

$$\mathbb{C}^2/[\mathbb{Z} \times (\mathbb{Z} + a\mathbb{Z})] \simeq \mathbb{C}^* \times (\mathbb{C}/(\mathbb{Z} + a\mathbb{Z})).$$

Notice that for $a = 1$ we get ordinary complex projective space $\mathbb{CP}^2$, while for $a = n$, a positive integer, we get the weighted projective space $\mathbb{CP}^2_{(1,1,n)}$. This complex toric quasifold admits an atlas with three charts. Consider the charts around the fixed points $[1:0:0]$ and $[0:1:0]$. The first is given by $(\mathbb{C}^2, \Gamma_{23}, V_{23}, \eta_{23})$, where

$$\Gamma_{23} = \left\{\, (1, e^{2\pi i a h}) \in (S^1)^2 \mid h \in \mathbb{Z} \right\},$$

$$V_{23} = \{[z_1 : z_2 : z_3] \in \mathbb{CP}^2_{(1,1,a)} \mid z_1 \neq 0\},$$

$$\begin{array}{ccc} \mathbb{C}^2/\Gamma_{23} & \stackrel{\eta_{23}}{\longrightarrow} & V_{23} \\ [z_2 : z_3] & \longmapsto & [1 : z_2 : z_3] \,. \end{array}$$

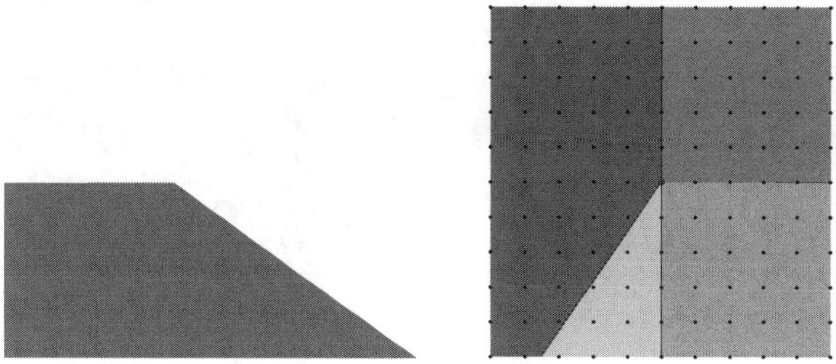

FIGURE 2. The trapezoid $\mathcal{T}_a$ and its normal fan.

The second is given by is given by $(\mathbb{C}^2, \Gamma_{13}, V_{13}, \eta_{13})$, where

$$\Gamma_{13} = \left\{ (1, e^{2\pi i a h}) \in (S^1)^2 \mid h \in \mathbb{Z} \right\},$$

$$V_{13} = \{[w_1 : w_2 : w_3] \in \mathbb{CP}^2_{(1,1,a)} \mid w_2 \neq 0\},$$

$$\begin{array}{ccc} \mathbb{C}^2/\Gamma_{13} & \xrightarrow{\eta_{13}} & V_{13} \\ [w_1 : w_3] & \longmapsto & [w_1 : 1 : w_3]. \end{array}$$

The transition map and Laurent monomials between these two charts are given by

$$\begin{array}{ccc} \eta_{23}^{-1}(V_{23} \cap V_{13}) & \xrightarrow{\eta_{13}^{-1} \circ \eta_{23}} & \eta_{13}^{-1}(V_{23} \cap V_{13}) \\ [z_2 : z_3] & \longmapsto & \left[z_2^{-1} : z_2^{-a} z_3\right]. \end{array}$$

3.3. Generalized Hirzebruch surfaces. We recall the construction of a one–parameter family of toric quasifolds generalizing Hirzebruch surfaces which was first introduced jointly with Zaffran in [**7**].

Consider, for any positive real number a, the trapezoid $\mathcal{T}_a$ of vertices $(0,0)$, $(1,0)$, $(1,1)$, and $(a+1,0)$ (see Figure 2). We recall that $\mathcal{T}_n$, for n a positive integer, is the polytope that corresponds to the Hirzebruch surface $\mathbb{H}_n$. Similarly to what happens with the right triangle T_a, the trapezoid $\mathcal{T}_a$ is always rational with respect to the lattice generated by $(1,0)$, $(0,a)$. However, again as above, we view $\mathcal{T}_a$ in a nonrational setting. To do so, we choose normals given by $X_1 = (1,0)$, $X_2 = -(1,a)$, $X_3 = (0,1)$, and $X_4 = (0,-1)$; they too span the quasilattice $\mathbb{Z} \times (\mathbb{Z} + a\mathbb{Z})$. The toric quasifold corresponding to the fundamental triple

$$(\Sigma_{\mathcal{T}_a}, \mathbb{Z} \times (\mathbb{Z} + a\mathbb{Z}), \{X_1, X_2, X_3, X_4\}).$$

is given, if $\underline{z} = (z_1, z_2, z_3, z_4) \in \mathbb{C}^4$, by

$$\mathbb{H}_a = \frac{\{\,\underline{z} \mid z_1 z_3 \neq 0\,\} \cup \{\,\underline{z} \mid z_1 z_4 \neq 0\,\} \cup \{\,\underline{z} \mid z_2 z_3 \neq 0\,\} \cup \{\,\underline{z} \mid z_2 z_4 \neq 0\,\}}{\{\,(e^{2\pi i u}, e^{2\pi i u}, e^{2\pi i(au+v)}, e^{2\pi i v}) \in (\mathbb{C}^*)^4 \mid u, v \in \mathbb{C}\,\}} =$$

$$= \frac{(\mathbb{C}^2 \setminus \{0\}) \times (\mathbb{C}^2 \setminus \{0\})}{\{\,(e^{2\pi i u}, e^{2\pi i u}, e^{2\pi i(au+v)}, e^{2\pi i v}) \in (\mathbb{C}^*)^4 \mid u, v \in \mathbb{C}\,\}}.$$

It is also endowed with the holomorphic action of the complex quasitorus

$$\mathbb{C}^* \times (\mathbb{C}/(\mathbb{Z} + a\mathbb{Z})).$$

Notice that for $a = n$ we obtain the Hirzebruch surface $\mathbb{H}_n$. This quasifold is covered by 4 complex charts; let us describe the ones around the points $[1:0:0:1]$ and

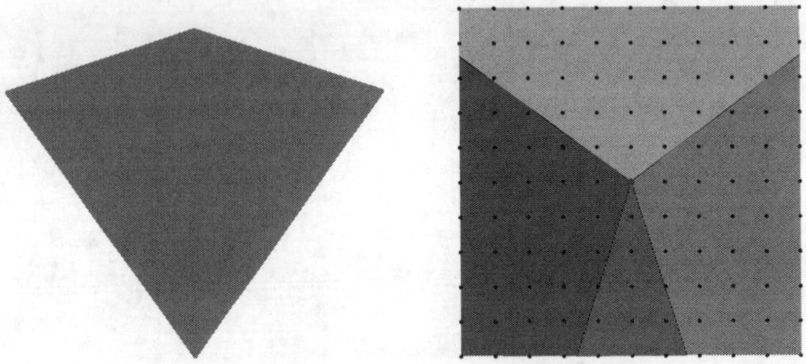

FIGURE 3. The Penrose kite and its normal fan.

$[0:1:0:1]$. The chart around the point $[1:0:0:1]$ is given by $(\mathbb{C}^2, \Gamma_{23}, V_{23}, \eta_{23})$, where

$$\Gamma_{23} = \left\{ (1, e^{2\pi i a h}) \in (S^1)^2 \mid h \in \mathbb{Z} \right\},$$

$$V_{23} = \{[z_1 : z_2 : z_3 : z_4] \in \mathbb{H}_a \mid z_1 z_4 \neq 0\},$$

$$\begin{array}{ccc} \mathbb{C}^2/\Gamma_{23} & \xrightarrow{\eta_{23}} & V_{23} \\ [z_2 : z_3] & \longmapsto & [1 : z_2 : z_3 : 1]\,. \end{array}$$

On the other hand, the chart around the point $[0:1:0:1]$ is given by $(\mathbb{C}^2, \Gamma_{13}, V_{13}, \eta_{13})$, where

$$\Gamma_{13} = \left\{ (1, e^{2\pi i a h}) \in (S^1)^2 \mid h \in \mathbb{Z} \right\},$$

$$V_{13} = \{[w_1 : w_2 : w_3 : w_4] \in \mathbb{H}_a \mid w_2 w_4 \neq 0\},$$

$$\begin{array}{ccc} \mathbb{C}^2/\Gamma_{13} & \xrightarrow{\eta_{13}} & V_{13} \\ [w_1 : w_3] & \longmapsto & [w_1 : 1 : w_3 : 1]\,. \end{array}$$

The transition map between these two charts and the corresponding Laurent monomials are given by

$$\begin{array}{ccc} \eta_{23}^{-1}(V_{23} \cap V_{13}) & \xrightarrow{\eta_{13}^{-1} \circ \eta_{23}} & \eta_{13}^{-1}(V_{23} \cap V_{13}) \\ [z_2 : z_3] & \longmapsto & [z_2^{-1} : z_2^{-a} z_3]\,. \end{array}$$

Notice that this change of charts is the same as the one for the weighted projective space above. This is not a coincidence, but rather a manifestation of the fact that, analogously to what happens in the smooth case, the generalized Hirzebruch surface $\mathbb{H}_a$ is the blow–up of the generalized weighted projective space $\mathbb{CP}^2_{(1,1,a)}$ at its third fixed point $[0:0:1]$ (see [**5, 7**]).

3.4. The Penrose kite. The kite and dart tiling is an aperiodic tiling of the plane introduced by Penrose [**16**]. The kite is a quadrilateral having three of its angles equal to $\frac{2\pi}{5}$, the fourth equal to $\frac{4\pi}{5}$. Its long edges are ϕ times the short edges, where $\phi = \frac{1+\sqrt{5}}{2}$ is the *golden ratio* (see Figure 3). The kite is not rational, however it is quasirational with respect to the pentagonal quasilattice Q_5 generated by the fifth–roots of unity:

$$\begin{array}{l} Y_0 = (1,0) \\ Y_1 = (\cos\frac{2\pi}{5}, \sin\frac{2\pi}{5}) = \frac{1}{2}(\frac{1}{\phi}, \sqrt{2+\phi}) \\ Y_2 = (\cos\frac{4\pi}{5}, \sin\frac{4\pi}{5}) = \frac{1}{2}(-\phi, \frac{1}{\phi}\sqrt{2+\phi}) \\ Y_3 = (\cos\frac{6\pi}{5}, \sin\frac{6\pi}{5}) = \frac{1}{2}(-\phi, -\frac{1}{\phi}\sqrt{2+\phi}) \\ Y_4 = (\cos\frac{8\pi}{5}, \sin\frac{8\pi}{5}) = \frac{1}{2}(\frac{1}{\phi}, -\sqrt{2+\phi}). \end{array}$$

Choose normals $X_1 = -Y_1$, $X_2 = -Y_3$, $X_3 = Y_2$ and $X_4 = Y_4$. Notice that

$$\phi = 1 + \frac{1}{\phi}$$

and that

$$\begin{cases} Y_2 = - \quad Y_1 \quad - \quad \phi Y_4 \\ Y_3 = - \quad \phi Y_1 \quad - \quad Y_4. \end{cases}$$

The toric quasifold corresponding to the fundamental triple

$$(\Sigma_\Delta, Q_5, \{X_1, X_2, X_3, X_4\})$$

was first introduced and studied in [**3**]. It is given by

$$X_\Delta = \frac{(\mathbb{C}^2 \setminus \{0\}) \times (\mathbb{C}^2 \setminus \{0\})}{\left\{ \left(e^{2\pi i(\phi u - v)}, e^{2\pi i u}, e^{2\pi i v}, e^{2\pi i(\phi v - u)}\right) \in (\mathbb{C}^*)^4 \mid u, v \in \mathbb{C} \right\}}.$$

It is shown in [**3**, Theorem 6.1] that *it is not* a global quasifold, namely not the quotient of a manifold modulo the action of a countable group.

Again, here we have an atlas made of 4 charts; let us describe the ones around the fixed points $[0:1:1:0]$ and $[1:0:1:0]$. The first is given by $(\mathbb{C}^2, \Gamma_{14}, V_{14}, \tau_{14})$, where

$$\Gamma_{14} = \left\{ (e^{2\pi i\phi h}, e^{2\pi i\phi k}) \in (S^1)^2 \mid h, k \in \mathbb{Z} \right\},$$

$$V_{14} = \{[z_1 : z_2 : z_3 : z_4] \in X_\Delta \mid z_2 z_3 \neq 0\},$$

$$\begin{array}{ccc} \mathbb{C}^2/\Gamma_{14} & \xrightarrow{\tau_{14}} & V_{14} \\ [z_1 : z_4] & \longmapsto & [z_1 : 1 : 1 : z_4]. \end{array}$$

The second is given by $(\mathbb{C}^2, \Gamma_{24}, V_{24}, \tau_{24})$, where

$$\Gamma_{24} = \left\{ (e^{2\pi i\phi h} e^{2\pi i\phi k}, e^{-2\pi i\phi h}) \in (S^1)^2 \mid h, k \in \mathbb{Z} \right\},$$

$$V_{24} = \{[z_1 : z_2 : z_3 : z_4] \in X_\Delta \mid z_1 z_3 \neq 0\},$$

$$\begin{array}{ccc} \mathbb{C}^2/\Gamma_{24} & \xrightarrow{\tau_{24}} & V_{24} \\ [w_2 : w_4] & \longmapsto & [1 : w_2 : 1 : w_4]. \end{array}$$

Their transition and Laurent monomials are given by

$$\begin{array}{ccc} \eta_{14}^{-1}(V_{14} \cap V_{24}) & \xrightarrow{\eta_{24}^{-1} \circ \eta_{14}} & \eta_{24}^{-1}(V_{14} \cap V_{24}) \\ [z_1 : z_4] & \longmapsto & \left[z_1^{-\frac{1}{\phi}} : z_1^{\frac{1}{\phi}} z_4 \right]. \end{array}$$

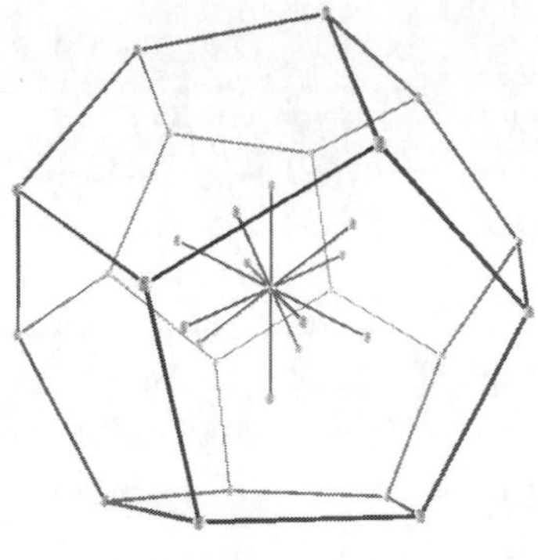

FIGURE 4. The regular dodecahedron and its normals.

3.5. The regular dodecahedron. Let us consider the regular dodecahedron Δ having vertices

$$\begin{array}{l}(\pm 1, \pm 1, \pm 1)\\ (0, \pm\phi, \pm\frac{1}{\phi})\\ (\pm\frac{1}{\phi}, 0, \pm\phi)\\ (\pm\phi, \pm\frac{1}{\phi}, 0).\end{array}$$

Σ_Δ is not rational, but it is quasirational with respect to the quasilattice $P \subset \mathbb{R}^3$ generated by the vectors

$$\begin{array}{l}Y_1 = (\frac{1}{\phi}, 1, 0)\\ Y_2 = (0, \frac{1}{\phi}, 1)\\ Y_3 = (1, 0, \frac{1}{\phi})\\ Y_4 = (-\frac{1}{\phi}, 1, 0)\\ Y_5 = (0, -\frac{1}{\phi}, 1)\\ Y_6 = (1, 0, -\frac{1}{\phi}),\end{array}$$

which is known in physics as the *simple icosahedral lattice.* In fact, the vectors $X_j = Y_j$, $X_{6+j} = -Y_j$, $j = 1, \ldots, 6$, are normals for Δ (see Figure 4). Notice that we have the following relations

$$\left\{\begin{array}{lllllll} Y_4 & = & \frac{1}{\phi}Y_1 & + & \frac{1}{\phi}Y_2 & - & Y_3\\ Y_5 & = & -Y_1 & + & \frac{1}{\phi}Y_2 & + & \frac{1}{\phi}Y_3\\ Y_6 & = & \frac{1}{\phi}Y_1 & - & Y_2 & + & \frac{1}{\phi}Y_3. \end{array}\right. \tag{3.1}$$

Let us apply the generalized toric construction to the fundamental triple

$$(\Sigma_\Delta, P, \{X_1, X_2, X_3, X_4, X_5, X_6, X_7, X_8, X_9, X_{10}, X_{11}, X_{12}\}).$$

The corresponding toric quasifold was first introduced in [**19**] and further studied, together with the toric spaces corresponding to all the other regular polyhedra, in [**4**]. It is given by

$$X_\Delta = \frac{\bigcup_{j=1}^{20}\{\, \underline{z} \in \mathbb{C}^{12} \mid z_j \neq 0,\ \ j \notin I_{\sigma_j}\,\}}{N_{\mathbb{C}}},$$

where the σ_j's are the maximal cones of Σ_Δ; we identify each of them in Table 1, together with the corresponding vertex, fixed point, and index set I_{σ_j}. On the other

TABLE 1. The maximal cones for the dodecahedron

σ	vertex	fixed point	I_σ
σ_1	$(-1,-1-1)$	$[0:0:0:1:1:1:1:1:1:1:1:1]$	$1,2,3$
σ_2	$(0,-\phi,-\frac{1}{\phi})$	$[0:0:1:0:1:1:1:1:1:1:1:1]$	$1,2,4$
σ_3	$(-\phi,-\frac{1}{\phi},0)$	$[0:1:0:1:1:0:1:1:1:1:1:1]$	$1,3,6$
σ_4	$(0,-\phi,\frac{1}{\phi})$	$[0:1:1:0:1:1:1:1:1:1:0:1]$	$1,4,11$
σ_5	$(-1,-1,1)$	$[0:1:1:1:1:0:1:1:1:1:0:1]$	$1,6,11$
σ_6	$(-\frac{1}{\phi},0,-\phi)$	$[1:0:0:1:0:1:1:1:1:1:1:1]$	$2,3,5$
σ_7	$(1,-1,-1)$	$[1:0:1:0:1:1:1:1:1:1:1:0]$	$2,4,12$
σ_8	$(\frac{1}{\phi},0,-\phi)$	$[1:0:1:1:0:1:1:1:1:1:1:0]$	$2,5,12$
σ_9	$(-1,1,-1)$	$[1:1:0:1:0:1:1:1:1:0:1:1]$	$3,5,10$
σ_{10}	$(-\phi,\frac{1}{\phi},0)$	$[1:1:0:1:1:0:1:1:1:0:1:1]$	$3,6,10$
σ_{11}	$(1,-1,1)$	$[1:1:1:0:1:1:1:1:0:1:0:1]$	$4,9,11$
σ_{12}	$(\phi,-\frac{1}{\phi},0)$	$[1:1:1:0:1:1:1:1:0:1:1:0]$	$4,9,12$
σ_{13}	$(0,\phi,-\frac{1}{\phi})$	$[1:1:1:1:0:1:0:1:1:0:1:1]$	$5,7,10$
σ_{14}	$(1,1,-1)$	$[1:1:1:1:0:1:0:1:1:1:1:0]$	$5,7,12$
σ_{15}	$(-1,1,1)$	$[1:1:1:1:1:0:1:0:1:0:1:1]$	$6,8,10$
σ_{16}	$(-\frac{1}{\phi},0,\phi)$	$[1:1:1:1:1:0:1:0:1:1:0:1]$	$6,8,11$
σ_{17}	$(1,1,1)$	$[1:1:1:1:1:1:0:0:0:1:1:1]$	$7,8,9$
σ_{18}	$(0,\phi,\frac{1}{\phi})$	$[1:1:1:1:1:1:0:0:1:0:1:1]$	$7,8,10$
σ_{19}	$(\phi,\frac{1}{\phi},0)$	$[1:1:1:1:1:1:0:1:0:1:1:0]$	$7,9,12$
σ_{20}	$(\frac{1}{\phi},0,\phi)$	$[1:1:1:1:1:1:1:0:0:1:0:1]$	$8,9,11$

hand, $N_{\mathbb{C}} = \exp(\mathfrak{n}_{\mathbb{C}}) \subset T^{12}_{\mathbb{C}}$, where $\mathfrak{n}_{\mathbb{C}}$ is the 9–dimensional vector subspace of $\mathbb{C}^{12}$ having equations

$$\begin{cases} u_1 = u_7 + u_5 - u_{11} + \frac{1}{\phi}(u_{10} - u_4 + u_{12} - u_6) \\ u_2 = u_8 + u_6 - u_{12} + \frac{1}{\phi}(u_{10} - u_4 + u_{11} - u_5) \\ u_3 = u_9 + u_4 - u_{10} + \frac{1}{\phi}(u_{11} - u_5 + u_{12} - u_6). \end{cases}$$

We have 20 charts, one around each different fixed point.

Let us first consider the first two appearing in Table 1. Notice that the two index sets, $\{1,2,3\}$ and $\{1,2,4\}$, have two indices in common. This entails that the corresponding vertices belong to the same edge of Δ.

The first chart is given by $(\mathbb{C}^3, \Gamma_{123}, V_{123}, \eta_{123})$, where

$$\Gamma_{123} = \left\{ (e^{2\pi i\phi(h+k)}, e^{2\pi i\phi(h+l)}, e^{2\pi i\phi(l+k)}) \in (S^1)^3 \mid h,k,l \in \mathbb{Z} \right\},$$

$$V_{123} = \{[\underline{z}] \in X_\Delta \mid z_i \neq 0, i \neq 1,2,3\},$$

$$\begin{array}{ccc} \mathbb{C}^3/\Gamma_{123} & \stackrel{\eta_{123}}{\longrightarrow} & V_{123} \\ [z_1 : z_2 : z_3] & \longmapsto & [z_1 : z_2 : z_3 : 1 : 1 : 1 : 1 : 1 : 1 : 1 : 1 : 1]. \end{array}$$

It is now useful to notice that the relations (3.1) can be rewritten as follows

$$\begin{cases} Y_3 = \frac{1}{\phi}Y_1 + \frac{1}{\phi}Y_2 - Y_4 \\ Y_5 = -\frac{1}{\phi}Y_1 + Y_2 - \frac{1}{\phi}Y_4 \\ Y_6 = Y_1 - \frac{1}{\phi}Y_2 - \frac{1}{\phi}Y_4. \end{cases}$$

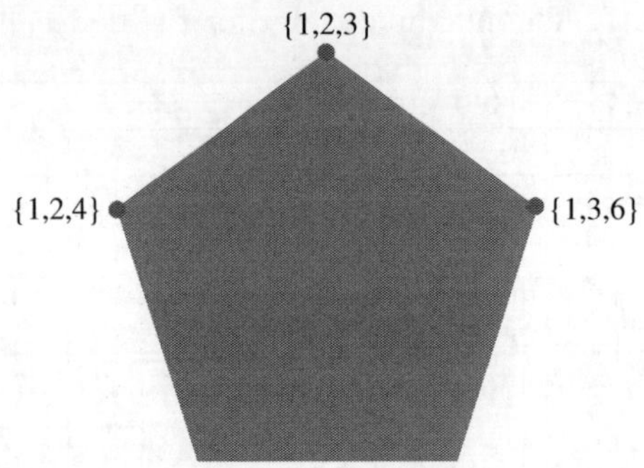

FIGURE 5. The three vertices on their common facet.

The second chart is given by $(\mathbb{C}^3, \Gamma_{124}, V_{124}, \eta_{124})$, where

$$\Gamma_{124} = \left\{ (e^{2\pi i\phi(h+k)}, e^{2\pi i\phi(h+l)}, e^{2\pi i\phi(l+k)}) \in (S^1)^3 \mid h,k,l \in \mathbb{Z} \right\},$$

$$V_{124} = \{[\underline{z}] \in X_\Delta \mid z_i \neq 0, i \neq 1,2,4\},$$

$$\begin{array}{ccc} \mathbb{C}^3/\Gamma_{124} & \stackrel{\eta_{124}}{\longrightarrow} & V_{124} \\ [w_1 : w_2 : w_4] & \longmapsto & [w_1 : w_2 : 1 : w_4 : 1 : 1 : 1 : 1 : 1 : 1 : 1 : 1]. \end{array}$$

The chart transition and generalized Laurent monomials are given by

$$\begin{array}{ccc} \eta_{123}^{-1}(V_{123} \cap V_{124}) & \stackrel{\eta_{124}^{-1} \circ \eta_{123}}{\longrightarrow} & \eta_{124}^{-1}(V_{123} \cap V_{124}) \\ [z_1 : z_2 : z_3] & \longmapsto & \left[z_1 z_3^{\frac{1}{\phi}} : z_2 z_3^{\frac{1}{\phi}} : z_3^{-1} \right]. \end{array}$$

Let us now compare the second chart with the one around the third fixed point in Table 1. The corresponding index sets, $\{1,2,4\}$ and $\{1,3,6\}$, intersect in only one index, which entails that the vertices belong to the same facet, but not the same edge of the dodecahedron (see Figure 5 for the mutual placement of all three vertices on their common facet). Let us show how this reflects on further complexity of the Laurent monomials. Notice first that we have

$$\left\{ \begin{array}{rcrcrcr} Y_2 & = & \frac{1}{\phi}Y_1 & + & \frac{1}{\phi}Y_3 & - & Y_6 \\ Y_4 & = & Y_1 & - & \frac{1}{\phi}Y_3 & - & \frac{1}{\phi}Y_6 \\ Y_5 & = & -\frac{1}{\phi}Y_1 & + & Y_3 & - & \frac{1}{\phi}Y_6. \end{array} \right.$$

The third chart is given by $(\mathbb{C}^3, \Gamma_{136}, V_{136}, \eta_{136})$, where

$$\Gamma_{136} = \left\{ (e^{2\pi i\phi(h+k)}, e^{2\pi i\phi(h+l)}, e^{2\pi i\phi(l+k)}) \in (S^1)^3 \mid h,k,l \in \mathbb{Z} \right\},$$

$$V_{136} = \{[\underline{z}] \in X_\Delta \mid z_i \neq 0, i \neq 1,3,6\},$$

$$\begin{array}{ccc} \mathbb{C}^3/\Gamma_{136} & \stackrel{\eta_{136}}{\longrightarrow} & V_{136} \\ [w_1 : w_3 : w_6] & \longmapsto & [w_1 : 1 : w_3 : 1 : 1 : w_6 : 1 : 1 : 1 : 1 : 1 : 1]. \end{array}$$

The change of charts and generalized Laurent monomials are now given by

$$\begin{array}{ccc} \eta_{124}^{-1}(V_{124} \cap V_{136}) & \stackrel{\eta_{136}^{-1} \circ \eta_{124}}{\longrightarrow} & \eta_{136}^{-1}(V_{124} \cap V_{136}) \\ [z_1 : z_2 : z_4] & \longmapsto & \left[z_1 z_2^{\frac{1}{\phi}} z_4 : z_2^{\frac{1}{\phi}} z_4^{-\frac{1}{\phi}} : z_2^{-1} z_4^{-\frac{1}{\phi}} \right]. \end{array}$$

References

[1] M. Audin, *The topology of torus actions on symplectic manifolds*, Progress in Mathematics, vol. 93, Birkhäuser Verlag, Basel, 1991. Translated from the French by the author, DOI 10.1007/978-3-0348-7221-8. MR1106194

[2] F. Battaglia and E. Prato, *Generalized toric varieties for simple nonrational convex polytopes*, Internat. Math. Res. Notices **24** (2001), 1315–1337, DOI 10.1155/S1073792801000629. MR1866747

[3] F. Battaglia and E. Prato, *The symplectic Penrose kite*, Comm. Math. Phys. **299** (2010), no. 3, 577–601, DOI 10.1007/s00220-010-1103-y. MR2718924

[4] F. Battaglia and E. Prato, *Toric geometry of the regular convex polyhedra*, J. Math., posted on 2017, Art. ID 2542796, 15 pp., DOI 10.1155/2017/2542796. MR3630682

[5] F. Battaglia and E. Prato, *Nonrational symplectic toric cuts*, Internat. J. Math. **29** (2018), no. 10, 1850063, 19 pp., DOI 10.1142/S0129167X18500635. MR3861905

[6] F. Battaglia, E. Prato, *Nonrational polytopes and fans in toric geometry*, Riv. Math. Univ. Parma **14** (2023), 67–86.

[7] F. Battaglia, E. Prato, and D. Zaffran, *Hirzebruch surfaces in a one-parameter family*, Boll. Unione Mat. Ital. **12** (2019), no. 1-2, 293–305, DOI 10.1007/s40574-018-0181-1. MR3936308

[8] D. A. Cox, J. B. Little, and H. K. Schenck, *Toric varieties*, Graduate Studies in Mathematics, vol. 124, American Mathematical Society, Providence, RI, 2011, DOI 10.1090/gsm/124. MR2810322

[9] V. I. Danilov, *The geometry of toric varieties*, Russian Math. Surveys, vol. 33, Russ. Acad. Sci., Moscow (1978), 97–154.

[10] T. Delzant, *Hamiltoniens périodiques et images convexes de l'application moment* (French, with English summary), Bull. Soc. Math. France **116** (1988), no. 3, 315–339. MR984900

[11] M. Demazure, *Sous-groupes algébriques de rang maximum du groupe de Cremona* (French), Ann. Sci. École Norm. Sup. (4) **3** (1970), 507–588. MR284446

[12] P. Donato and P. Iglésias, *Exemples de groupes difféologiques: flots irrationnels sur le tore* (French, with English summary), C. R. Acad. Sci. Paris Sér. I Math. **301** (1985), no. 4, 127–130. MR799609

[13] V. Guillemin, *Moment maps and combinatorial invariants of Hamiltonian T^n-spaces*, Progress in Mathematics, vol. 122, Birkhäuser Boston, Inc., Boston, MA, 1994, DOI 10.1007/978-1-4612-0269-1. MR1301331

[14] P. Iglesias-Zemmour and E. Prato, *Quasifolds, diffeology and noncommutative geometry*, J. Noncommut. Geom. **15** (2021), no. 2, 735–759, DOI 10.4171/jncg/419. MR4325720

[15] Y. Karshon, D. Miyamoto, *Quasifold groupoids and diffeological quasifolds*, Preprint, `arXiv:2206.14776`, 2022.

[16] R. Penrose, *Pentaplexity: a class of nonperiodic tilings of the plane*, Math. Intelligencer **2** (1979/80), no. 1, 32–37, DOI 10.1007/BF03024384. MR558670

[17] E. Prato, *Sur une généralisation de la notion de V-variété* (French, with English and French summaries), C. R. Acad. Sci. Paris Sér. I Math. **328** (1999), no. 10, 887–890, DOI 10.1016/S0764-4442(99)80291-2. MR1689861

[18] E. Prato, *Simple non-rational convex polytopes via symplectic geometry*, Topology **40** (2001), no. 5, 961–975, DOI 10.1016/S0040-9383(00)00006-9. MR1860537

[19] E. Prato, *Symplectic toric geometry and the regular dodecahedron*, J. Math., posted on 2015, Art. ID 967417, 5 pp., DOI 10.1155/2015/967417. MR3426654

[20] E. Prato, *Toric Quasifolds*, Math. Intelligencer **45** (2023), no. 2, 133–138, DOI 10.1007/s00283-022-10212-y. MR4600262

[21] G. M. Ziegler, *Lectures on polytopes*, Graduate Texts in Mathematics, vol. 152, Springer-Verlag, New York, 1995, DOI 10.1007/978-1-4613-8431-1. MR1311028

Dipartimento di Matematica e Informatica "U. Dini", Università degli Studi di Firenze, Viale Morgagni 67/A, 50134 Firenze, Italy

Email address: `fiammetta.battaglia@unifi.it`

Dipartimento di Matematica e Informatica "U. Dini", Università degli Studi di Firenze, Viale Morgagni 67/A, 50134 Firenze, Italy

Email address: `elisa.prato@unifi.it`

Contemporary Mathematics
Volume **794**, 2024
https://doi.org/10.1090/conm/794/15924

On a remark by Alan Weinstein

Iakovos Androulidakis

ABSTRACT. Alan Weinstein remarked that, working in the framework of diffeology, a construction from Noncommutative Differential Geometry might provide the non-trivial representations required for the geometric quantisation of a symplectic structure which is not integral. In this note we show that the construction we gave with P. Antonini does indeed provide non-trivial representations.

Introduction

Let (M, ω) be a simply connected symplectic manifold. The first step towards geometric quantization (cf. [**Sou80**]) is prequantization: The group Γ_ω of periods of $[\omega] \in H^2_{dR}(M, \mathbb{R})$ is a subgroup of the abelian group $(\mathbb{R}, +)$. When Γ_ω is cyclic (hence discrete), the quotient $G(M, \omega) = \mathbb{R}/\Gamma_\omega$ can be identified with the circle $U(1)$ and there exists a principal $U(1)$-bundle Q over M with a connection whose curvature is the original symplectic form ω. Then, using a non-trivial representation of $G(M, \omega)$, one constructs a Hilbert space which is the crucial ingredient of the quantisation process.

Alan Weinstein pointed out in [**Wei89**, Rem. 5.1] that, when Γ_ω is a non-closed dense subgroup of $\mathbb{R}$, the construction of the Hilbert space is no longer possible. That is because the topology of the group $G(M, \omega)$ is trivial, which allows continuity only for the trivial representation. In the same remark, Weinstein proposed that some construction from Noncommutative Differential Geometry might allow to "quantize the unquantizable". In other words, to obtain non-trivial representations. Note that when Γ_ω is not cyclic we can still make sense of the prequantizing bundle, using the notion of *diffeology* which was introduced by Souriau [**Sou70**]. Actually, the results of [**Wei89**] hold in this setting as well.

The purpose of the present note is to confirm that, looking at the group $G(M, \omega)$ as a diffeological space, Noncommutative Differential Geometry does indeed allow us to find non-trivial representations. The geometric construction suggested in [**Wei89**, Rem. 5.1] is given in [**AA18**]: There we showed that when the group $G(M, \omega)$ is not a circle, adding a few extra dimensions allows us to realise it as the leaf space of an irrational rotation foliation on a torus. As we will discuss here, this torus generates the atlas of bisubmersions of this foliation, in the sense of [**AS09**]. These bisubmersions give rise to the diffeological structure of $G(M, \omega)$.

2020 *Mathematics Subject Classification.* Primary 22A10; Secondary 53C12.
Key words and phrases. Diffeological group, C^*-algebra.

What remains is to explain why this construction provides non-trivial representations. Recall that the representations ring of a Lie group can be recovered from the K-theory of its reduced C^*-algebra. Here we first explain how to make sense of $C^*(G(M,\omega))$ by applying the C^*-algebra construction we gave in [**AS09**] to the diffeological structure generated by the torus. In order to show that the K-theory of this C^*-algebra is not trivial, we prove in Theorem 1.2 that $C^*(G(M,\omega))$ corresponds to the irrational rotation algebra. This isomorphism should be thought of as an analogue of Thom isomorphism.

1. The Almeida-Molino groupoid as a diffeological groupoid

We will use the convenient setting of Lie algebroids and Lie groupoids throughout this note. In this section, first we recall this setting and then justify our construction of the convolution algebra $G(M,\omega)$ using diffeology.

1.1. Integrality, Integrability and lifting. Let us recall briefly how the integrality problem of the symplectic manifold (M,ω) can be viewed as an integrability problem for transitive Lie algebroids. This was first explained by Mackenzie in [**Mac05**], see also [**Cra04**].

Consider the vector bundle $TM \oplus (M \times \mathbb{R})$ over M. Since the 2-form ω is closed, the following Lie bracket on the $C^\infty(M)$-module of smooth sections of this bundle, satisfies the Jacobi identity:

$$[X \oplus f, Y \oplus g] = [X,Y] \oplus \{X(g) - Y(f) - \omega(X,Y)\}$$

Here $X, Y \in \mathcal{X}(M)$ and $f, g \in C^\infty(M)$. This bracket also satisfies the Leibniz identity with respect to the anchor map $X \oplus f \mapsto X$, therefore it defines a transitive Lie algebroid which we denote $TM \oplus_\omega (M \times \mathbb{R})$. It integrates to a transitive Lie groupoid with isotropy group $U(1)$ if and only if the group Γ_ω of periods of ω is discrete. In this case, the group $G(M,\omega)$ is a Lie group; it is the isotropy group of the connected and simply connected Lie groupoid which integrates $TM \oplus_\omega (M \times \mathbb{R})$, at an arbitrary point $x \in M$. The associated principal $U(1)$-bundle, together with the connection arising from the splitting $X \mapsto X \oplus 0$ of the anchor map, constitute the prequantization data of (M,ω).

The group Γ_ω of periods may not be closed though. For example, put $M = S^2 \times S^2$ and $\omega = p_1^*(V) \oplus \lambda p_2^*(V)$, where V is a volume form of S^2, $p_i : M \to S_2$ the projection maps, $i = 1, 2$ and λ an irrational real number. Then $\Gamma_\omega = \mathbb{Z} + \lambda\mathbb{Z}$ is a dense and non-closed subgroup of $\mathbb{R}$. Hence, the isotropy group $G(M,\omega) = \mathbb{R}/\Gamma_\omega$ has a quite pathological quotient topology.

In fact, the Lie algebroid $TM \oplus_\omega (M \times \mathbb{R})$ associated to this symplectic structure is the first example of a non-integrable (transitive) Lie algebroid; it was given by Almeida and Molino in [**AM85**]. As shown in [**AA18**], this example encompasses all the "unquantizable" (equivalently, non-integrable) symplectic structures.

Let us recall from [**AA18**, Ex. 2.7] a construction which, as we will explain in §1.2, allows us to view the Almeida-Molino example, within foliation theory and realise the topologically ill-behaved group $G(M,\omega)$ as a leaf space. This starts from observing that the reason for the pathology of $G(M,\omega)$ is really that the dimension of the vector space $\mathbb{R}$ is too small to accommodate an appropriate lattice. So, naively, we add one more dimension. Namely, we consider the $\mathbb{R}^2$-valued 2-form $\overline{\omega} = (p_1^*(V), \lambda p_2^*(V)) : TM \times TM \to M \times \mathbb{R}^2$. That is to say, the form defined by

$$\overline{\omega}(X_{(x,y)}, Y_{(x,y)}) = ((x,y), (p_1^*(V)(X_{(x,y)}, Y_{(x,y)}), \lambda p_2^*(V)(X_{(x,y)}, Y_{(x,y)}))$$

for all $(x, y) \in S^2 \times S^2$ and $X, Y \in \mathcal{X}(S^2 \times S^2)$. Its group of periods $\Gamma_{\overline{\omega}}$ is the lattice $\mathbb{Z} \times \lambda\mathbb{Z}$ of $\mathbb{R}^2$, whence $TM \oplus_{\overline{\omega}} (M \times \mathbb{R}^2)$ is an integrable transitive Lie algebroid. The associated isotropy group $G(M, \overline{\omega}) = \mathbb{R}^2/\Gamma_{\overline{\omega}}$ is diffeomorphic to the torus $\mathbb{T}^2$. Moreover, we have the "addition" map; that is to say, the surjective morphism of Lie algebroids

$$TM \oplus_{\overline{\omega}} (M \times \mathbb{R}^2) \to TM \oplus_\omega (M \times \mathbb{R}) \qquad (X, (a, b)) \mapsto (X, a + b)$$

for every $X \in TM$ and $a, b \in \mathbb{R}$. Integrating this map and restricting to the isotropy groups at an arbitrary point $x \in M$, we obtain the short exact sequence of continuous group morphisms

$$0 \to \mathbb{R} \xrightarrow{\iota_\lambda} \mathbb{T}^2 \xrightarrow{\pi} G(M, \omega) \to 1 \tag{1.1}$$

The inclusion map ι_λ is the irrational rotation with slope $\lambda \in \mathbb{R} \setminus \mathbb{Q}$ on the torus. Of course, in the exact sequence (1.1), $(\mathbb{R}, +)$ and $\mathbb{T}^2$ are Lie groups and ι_λ is a (smooth) morphism of Lie groups.

In [**AA18**, Thm. 3.6] it was shown that the above lifting construction can be done systematically for an arbitrary simply connected[1] symplectic manifold (M, ω) with finitely generated second fundamental group $\pi_2(M)$. Let us recall briefly from [**AA18**, §3.1, 3.2] the general construction of the integrable lift:

(1) The familiar de Rham isomorphism identifies $Hom(H^2_{dR}(M), \mathbb{R})$ with the bidual $H_2(M, \mathbb{R})^{**}$. Hence the natural map

$$\mathcal{J}_M : H_2(M, \mathbb{R}) \to Hom(H^2_{dR}(M); \mathbb{R}), \mathcal{J}_M([\gamma])([\omega]) = \int_\gamma \omega$$

can be seen as an element $\mathcal{J}_M$ in $Hom(H_2(M, \mathbb{R}); H_2(M, \mathbb{R})^{**})$ and the latter can be identified with $Hom(H_2(M, \mathbb{R}); H_2(M, \mathbb{R}))$. The de Rham theorem implies that this element defines a class in de Rham cohomology $H^2_{dR}(M, H_2(M, \mathbb{R}))$ with coefficients in the homology vector space $H_2(M, \mathbb{R})$. Let Θ in $\Omega^2(M, H_2(M, \mathbb{R}))$ be a closed 2-form which represents this class.

(2) The transitive Lie algebroid $TM \oplus_\Theta (M \times H_2(M, \mathbb{R}))$ is integrable, thanks to the Hurewicz theorem and the assumption that M is simply connected.

(3) Now consider the closed 2-form $\Theta \oplus \omega$ in $\Omega^2(M, H_2(M, \mathbb{R}) \times \mathbb{R})$. The previous integrability result implies the integrability of the transitive Lie algebroid

$$TM \oplus_{\Theta \oplus \omega} (M \times (H_2(M, \mathbb{R}) \times \mathbb{R}))$$

(4) The projection $p : H_2(M, \mathbb{R}) \times \mathbb{R} \to \mathbb{R}$ induces the surjective morphism of transitive Lie algebroids

$$q_\omega = id_{TM} \oplus p : TM \oplus_{\Theta \oplus \omega} (M \times (H_2(M, \mathbb{R}) \times \mathbb{R})) \longrightarrow TM \oplus_\omega (M \times \mathbb{R})$$

with kernel the trivial bundle $M \times H_2(M, \mathbb{R})$.

Integrating the map q_ω and restricting to the isotropy group at an arbitrary point $x \in M$ we obtain a short exact sequence of topological groups

$$0 \to H_2(M, \mathbb{R}) \xrightarrow{\iota} \mathbb{T}^{dim(H_2(M, \mathbb{R}))+1} \xrightarrow{\pi} G(M, \omega) \to 1 \tag{1.2}$$

[1] In [**AA18**, §4] it is shown that the simple connectivity assumption can be lifted, under a mild assumption.

1.2. Beyond the naive $*$-algebra of $G(M,\omega)$. In sake of simplicity, throughout the sequel we focus on the exact sequence (1.1). The injective homomorphism of Lie groups

$$\iota_\lambda : (\mathbb{R},+) \to (\mathbb{T}^2,\cdot) \quad t \mapsto (e^{2\pi i t}, e^{2\pi i \lambda t})$$

induces a surjective linear map $(\iota_\lambda)^* : C^\infty(\mathbb{T}^2) \to C^\infty(\mathbb{R})$. Since the map $\pi : \mathbb{T}^2 \to G(M,\omega)$ is a surjective homomorphism of groups, it makes sense to define the naive function space $C^\infty(G(M,\omega))$ to be $\ker(\iota_\lambda)^*$. We might then consider restricting the $*$-algebra structure of $C^\infty(\mathbb{T}^2)$ induced by the group structure of $\mathbb{T}^2$ to $C^\infty(G(M,\omega))$. However, as it happens, the $*$-algebra arising this way does not carry any information[2].

PROPOSITION 1.1. *The vector subspace* $\ker(\iota_\lambda)^*$ *is trivial.*

PROOF. By the diffeomorphism $\mathbb{T}^2 = S^1 \times S^1$ we have $\iota_\lambda(\mathbb{R}) = \{(e^{2\pi i t}, e^{2\pi i \lambda t}) : t \in \mathbb{R}\}$. Since $\lambda \in \mathbb{R}/\mathbb{Q}$, this set is dense in $\mathbb{T}^2$. Now $\ker(\iota_\lambda)^*$ is the ideal of smooth functions on $\mathbb{T}^2$ which vanish on $\iota_\lambda(\mathbb{R})$. Continuity implies that such a function vanishes everywhere in $\mathbb{T}^2$. □

Proposition 1.1 shows that the group structure of the exact sequence (1.1) is not the correct way to attach a C^*-algebra to the group $G(M,\omega)$. We really need to change our point view.

A second reading of the group homomorphism $\pi : \mathbb{T}^2 \to G(M,\omega)$ in (1.1), is that it provides $G(M,\omega)$ with the structure of a diffeological space in the sense of [**Sou70**] [**Sou80**]. Using this structure, in §2 we will attach a space of compactly supported functions $C_c^\infty(G(M,\omega))$ to $G(M,\omega)$.

However, there is yet another way to read (1.1): It really describes the foliation on $\mathbb{T}^2$ by irrational rotation, namely the foliation by the orbits of the following action:

$$\mathbb{T}^2 \times \mathbb{R} \to \mathbb{T}^2 \quad ((e^{2\pi i x}, e^{2\pi i \lambda y}), t) \mapsto (e^{2\pi i (x+t)}, e^{2\pi i (y+\lambda t)}) = (e^{2\pi i x}, e^{2\pi i \lambda y}) \cdot \iota_\lambda(t)$$

To see this, fix a point $p = (e^{2\pi i x}, e^{2\pi i \lambda y})$ in $\mathbb{T}^2$. Then the submanifold $\iota_\lambda(\mathbb{R})$ can be identified with the orbit of p by the irrational rotation action. This implies that the quotient $G(M,\omega) = \mathbb{T}^2/\iota_\lambda(\mathbb{R})$ arising from (1.1) can be identified with the orbit space of this action. In other words, the leaf space[3] of the irrational rotation foliation.

On the other hand, in foliation theory, it is well known that the leaf space can always be described by symmetries, both internal and external: Recall that, given a foliation (M,F), a model for the leaf space M/F is the holonomy groupoid $H(M,F)$. In the case of the irrational rotation on the torus, the holonomy groupoid is the action groupoid $\mathbb{T}^2 \rtimes_\lambda \mathbb{R} \rightrightarrows \mathbb{T}^2$ associated with the previous action. Let us recall the structure of this groupoid. For every $g = (e^{2\pi i \theta}, e^{2\pi i \eta}) \in \mathbb{T}^2$ and $t \in \mathbb{R}$ we have:

$$s(g,t) = g, \quad r(g,t) = gt, \quad (g_1,t)(g_2,s) = (g_2, s+t), \quad (g,t)^{-1} = (gt,-t)$$

---

[2]Proposition 1.1 shows that $C^\infty(\mathbb{T}^2)$ and $C^\infty(\mathbb{R})$ are isomorphic only as vector spaces. However, they cannot be isomorphic as topological vector spaces.

[3]Remarkably, extension (1.1) shows that this leaf space carries a group structure as well.

Also recall that the C^*-algebra $\mathbb{T}^2 \rtimes_\lambda \mathbb{R}$ is the crossed product[4] $C(\mathbb{T}^2) \rtimes_\lambda \mathbb{Z}$. It is well-known that this crossed product is the irrational rotation algebra A_λ, namely the universal C^*-algebra generated by two unitaries u, v such that $uv = e^{2\pi i\lambda}vu$.

The purpose of this sequel is to prove the following result:

THEOREM 1.2. *The space $C_c^\infty(G(M,\omega))$ admits a completion to a C^*-algebra, which is isomorphic to the irrational rotation C^*-algebra A_λ.*

2. The function space of $G(M,\omega)$

In this section, we attach a $\mathbb{C}$-linear space of smooth, compactly supported densities on the group $G(M,\omega)$. To this end, we use the diffeological structure of $G(M,\omega)$, which we describe in §2.1. Our construction of this function space in §2.2 adapts to the diffeology of $G(M,\Omega)$ the construction of the convolution algebra of the holonomy groupoid of a foliation, given in [**AS09**].

2.1. The diffeology of $G(M,\omega)$. Consider the following family of plots:

$$P(G(M,\omega)) = \{\pi|_{\mathcal{O}} : \mathcal{O} \to G(M,\omega), \text{ where } \mathcal{O} \text{ is an open subset of } \mathbb{T}^2\}$$

Of course, here "open" is understood with respect to the well known differential structure of the torus. Whence, each $\mathcal{O}$ may also be thought of as an open subset of $\mathbb{R}^2$.

Now put $D(G(M,\omega))$ the diffeology generated by $P(G(M,\omega))$, namely the smallest diffeology containing these plots. It is characterised as follows: For $k = 1, 2, \ldots,$ a k-plot $\chi : \mathcal{O}_\chi \to G(M,\omega)$ belongs to $D(G(M,\omega))$ if and only if for every point $x \in \mathcal{O}_\chi$ there is a neighbourhood $\mathcal{O}_x$ of x such that either $\chi|_{\mathcal{O}_x}$ is constant, or $\chi|_{\mathcal{O}_x} = \pi|_{\mathcal{O}} \circ h$, where $\pi|_{\mathcal{O}} : \mathcal{O} \to G(M,\omega)$ is a plot in $P(G(M,\omega))$ and $h : \mathcal{O}_x \to \mathcal{O}$ is a smooth map. In remarks 2.1 we give a few properties of this diffeology.

REMARK 2.1.

(1) Recall that the diffeology $D(G(M,\omega))$ induces a topology on $G(M,\omega)$ ("D-topology"). A subset U of $G(M,\omega)$ is open in this topology if and only if for each plot χ in $P(G(M,\omega))$ the inverse image $\chi^{-1}(W)$ is open in $\mathcal{O}_\chi$. It is the same as the quotient topology defined from (1.1).

(2) We may assume that the diffeology $D(G(M,\omega))$ is generated by a finite number of open sets. Indeed, using compactness, let us fix a finite open cover $\mathcal{O}_1, \ldots, \mathcal{O}_n$ of $\mathbb{T}^2$. For every $i = 1, \ldots, n$ put $\pi_i = \pi|_{\mathcal{O}_i}$ and define $P(G(M,\omega)) = \{\pi_i : \mathcal{O}_i \to G(M,\omega) : i = 1, \ldots, n\}$. Let $\mathcal{O}$ be an arbitrary open subset of $\mathbb{T}^2$ and x a point in $\mathcal{O}$. Since $\bigcup_{i=1}^n \mathcal{O}_i = \mathbb{T}^2$, x belongs to $\mathcal{O}_i$ for some $1 \leq i \leq n$. Since $\mathcal{O}$ is open, there is a small enough neighbourhood $\mathcal{O}_x$ of x in $\mathcal{O}$ such that the identity $h = \mathrm{id}_{\mathcal{O}_x}$ is a map $h : \mathcal{O}_x \to \mathcal{O}_i$. We obviously have $\pi|_{\mathcal{O}_x} = \pi_i \circ h$. Whence, for every open subset $\mathcal{O}$, the plot $\pi|_{\mathcal{O}} \to G(M,\omega)$ belongs to $D(G(M,\omega))$. Moreover, $D(G(M,\omega))$ does not depend on the choice of finite cover of $\mathbb{T}^2$.

(3) Consider the family of plots $P(G(M,\omega)) = \{(\pi|_{\mathcal{O}_i}, \mathcal{O}_i)\}_{i\in I}$, where $\mathcal{O}_i$ is an open subset of $\mathbb{T}^2$. Put $\mathcal{O} = \coprod_{i\in I} \mathcal{O}_i$ and let $\overline{\pi} : \mathcal{O} \to G(M,\omega)$ be the map whose restriction to each $\mathcal{O}_i$ is $\pi|_{\mathcal{O}_i}$. Obviously, $(\overline{\pi}, \mathcal{O})$ is a plot in $D(G(M,\omega))$.

[4]Recall that the full crossed product C^*-algebra associated with $\mathbb{T}^2 \rtimes_\lambda \mathbb{R}$ is isomorphic with the reduced one.

(4) If $\mathcal{O}_i$ are plots in $P(G(M,\omega))$ for $i = 1, 2$, then the cartesian product $\mathcal{O}_1 \times \mathcal{O}_2$ is a plot in the diffeology $D(G(M,\omega))$. Indeed, define $\chi : \mathcal{O}_1 \times \mathcal{O}_2 \to G(M,\omega)$ by $\chi = m_G \circ (\pi|_{\mathcal{O}_1} \times \pi|_{\mathcal{O}_2})$, where m_G is the product map in $G(M,\omega)$. Now put $h = m_{\mathbb{T}^2}|_{\mathcal{O}_1 \times \mathcal{O}_2} : \mathcal{O}_1 \times \mathcal{O}_2 \to \mathbb{T}^2$ the restriction of the product map of the Lie group $\mathbb{T}^2$. Since $m_{\mathbb{T}^2} : \mathbb{T}^2 \times \mathbb{T}^2 \to \mathbb{T}^2$ is a submersion, the image of h is an open subset $\mathcal{O}$ of $\mathbb{T}^2$. The map π is a morphism of groups, therefore we have $\pi|_{\mathcal{O}} \circ h = \chi$.

(5) Let $\iota_{\mathbb{T}^2} : \mathbb{T}^2 \to \mathbb{T}^2$ be the inversion map. For every plot $\mathcal{O}$ in $P(G(M,\omega))$, the open subset $\iota(\mathcal{O}) \subseteq \mathbb{T}^2$ is also a plot. Put $\chi : \iota(\mathcal{O}) \to G(M,\omega)$ the map $\pi|_{\mathcal{O}} \circ \iota$ and $h = \iota : \iota(\mathcal{O}) \to \mathcal{O}$. We have $\chi = \pi|_{\mathcal{O}} \circ \iota$.

(6) It follows that the diffeology $D(G(M,\omega))$ is closed by finite cartesian products and inverses of plots in $P(G(M,\omega))$.

(7) Let $\chi_i : \mathcal{O}_{\chi_i} \to G(M,\omega)$, $i = 1, 2$ be two plots in $D(G(M,\omega))$. Put $\chi = m_G \circ (\chi_1 \times \chi_2) : \mathcal{O}_{\chi_1} \times \mathcal{O}_{\chi_2} \to G(M,\omega)$. We call this plot the *product* of the plots $(\chi_i, \mathcal{O}_i)_{i=1,2}$. Let us show that this is a plot in $D(G(M,\omega))$. Indeed, let $\mathcal{O}_i$ open subsets of $\mathbb{T}^2$ and $h_i : \mathcal{O}_{\chi_i} \to \mathcal{O}_i$ smooth maps such that $\chi_i = \pi|_{\mathcal{O}_i} \circ h$. Since π is a morphism of groups we have $\pi \circ m_{\mathbb{T}^2} = m_G \circ (\pi \times \ldots \times \pi)$. This implies that $\chi = \pi|_{\mathcal{O}_{12}} \circ (h_1 \times h_2)$, where $\mathcal{O}_{12}$ is the image of $\mathcal{O}_1 \times \mathcal{O}_2$ by the product map $m_{\mathbb{T}^2}$.

(8) In view of the previous item, denote $\mathcal{O}_{\chi_{12}}$ the image of the restriction of product map $m_{\mathbb{T}^2}$ to $\mathcal{O}_{\chi_1} \times \mathcal{O}_{\chi_2}$. Since $m_{\mathbb{T}^2}$ is a submersion, $\mathcal{O}_{\chi_{12}}$ is an open subset of $\mathbb{T}^2$. Choose a local section $\sigma : \mathcal{O}_{\chi_{12}} \to \mathcal{O}_{\chi_1} \times \mathcal{O}_{\chi_2}$ (shrinking $\mathcal{O}_{\chi_{12}}$ if necessary). Define $\chi_{12} : \mathcal{O}_{\chi_{12}} \to G(M,\omega)$ by $\chi_{12} = m_G \circ (\chi_1 \times \chi_2) \circ \sigma = \chi \circ \sigma = \pi \circ ((h_1 \times h_2) \circ \sigma)$. Whence $(\chi_{12}, \mathcal{O}_{\chi_{12}})$ is a plot in $D(G(M,\omega))$.

(9) Let $\chi : \mathcal{O}_\chi \to G(M,\omega)$ be a plot in $D(G(M,\omega))$. Put $\chi_{-1} = \iota_G \circ \chi$, where ι_G is the inversion map of $G(M,\omega)$. We call this plot the *inverse* of the plot $(\chi, \mathcal{O}_\chi)$. Let $\mathcal{O}$ be an open subset of $\mathbb{T}^2$ and $h : \mathcal{O}_\chi \to \mathcal{O}$ a smooth map such that $\chi = \pi|_{\mathcal{O}} \circ h$. The identity $\iota_G \circ \pi = \pi \circ \iota_{\mathbb{T}^2}$ implies that $\chi_{-1} = \pi|_{\iota_{\mathbb{T}^2}(\mathcal{O})} \circ (\iota_{\mathbb{T}^2} \circ h)$.

(10) Alternatively, we can define the inverse $(\chi^{-1}, \mathcal{O}_{\chi^{-1}})$ of the plot $(\chi, \mathcal{O}_\chi)$ by $\mathcal{O}_{\chi^{-1}} = \iota_{\mathbb{T}^2}(\mathcal{O}_\chi)$ and $\chi^{-1} = \chi \circ \iota_{\mathbb{T}^2}^{-1} = \pi|_{\mathcal{O}} \circ (h \circ \iota_{\mathbb{T}^2}^{-1})$. The pairs $(\chi_{-1}, \mathcal{O}_\chi)$ and $(\chi^{-1}, \mathcal{O}_{\chi^{-1}})$ are equivalent, because $\iota_{\mathbb{T}^2} : \mathcal{O}_\chi \to \mathcal{O}_{\chi^{-1}}$ is a diffeomorphism and $\chi_{-1} = \iota_G \circ \chi^{-1} \circ \iota_{\mathbb{T}^2}$.

(11) Items (g), (i) and (j) imply that $D(G(M,\omega))$ is closed by finite cartesian products and inverses of plots in $D(G(M,\omega))$.

Now let us denote $\{\mathcal{O}_i\}_{i \in I}$ the manifold atlas of $\mathbb{T}^2$. Put $\mathcal{O}$ the disjoint union $\coprod_{i \in I} \mathcal{O}_i$ and $\chi : \mathcal{O} \to G(M,\omega)$ the map whose restriction to each $\mathcal{O}_i$ is $\pi|_{\mathcal{O}_i}$. It follows that $(\chi, \mathcal{O})$ is a plot in $D(G(M,\omega))$.

LEMMA 2.2. *Let $\chi : \mathcal{O}_\chi \to G(M,\omega)$ be a plot in $D(G(M,\omega))$ and $x_0 \in \mathcal{O}_\chi$. There exists a plot $\psi : \mathcal{O}_\psi \to G(M,\omega)$ in $P(G(M,\omega))$ and submersions $p : \mathcal{O}_\psi \to \mathcal{O}$, $q : \mathcal{O}_\psi \to \mathcal{O}_\chi$ such that $x_0 \in q(\mathcal{O}_\psi)$. Moreover, p and q satisfy $\pi|_{\mathcal{O}_i} \circ p = \psi = \chi \circ q$ for every $i \in I$.*

PROOF. From the definition of the diffeology $D(G(M,\omega))$ there is a neighbourhood $\mathcal{O}_{x_0}$ of x_0 in $\mathcal{O}_\chi$, an $i \in I$ and a smooth map $h : \mathcal{O}_{x_0} \to \mathcal{O}_i$ such that $\chi|_{\mathcal{O}_{x_0}} = \pi|_{\mathcal{O}_i} \circ h$. Put $\mathcal{O}_\psi = \{(y,p) \in \mathcal{O}_{x_0} \times \mathcal{O}_i : \chi(y) = \pi(p)\}$ and define

$\psi : \mathcal{O}_\psi \to G(M,\omega)$ by $\psi(y,p) = \chi(y) = \pi(p)$. Endowing $G(M,\omega)$ with the D-topology, we find that $\mathcal{O}_\psi$ is an open subset of $\mathcal{O}_{x_0} \times \mathcal{O}_i$. The maps p and q are the respective projections. In particular, by definition the map $p : \mathcal{O}_\psi \to \mathcal{O}_i$ satisfies $\psi = \pi|_{\mathcal{O}_i} \circ p$, whence $(\psi, \mathcal{O}_\psi)$ is a plot in $D(G(M,\omega))$. The equality $\psi = \chi \circ q$ follows from the definition of ψ. □

REMARK 2.3. The construction of the plot $(\psi, \mathcal{O}_\psi)$ implies $\psi = \pi|_{\mathcal{O}_i} \circ (h \circ q)$.

2.2. Construction of $C_c^\infty(G(M,\omega))$. In this spirit, let us introduce the following spaces of functions:

(1) For every plot $\chi : \mathcal{O}_\chi \to G(M,\omega)$ in $D(G(M,\omega))$ consider the bundle of 1-densities $\Omega^1(T\mathcal{O}_\chi)$ over $\mathcal{O}_\chi$. Recall that this is a line bundle which becomes trivial once we choose one of its sections. We denote $C_c^\infty(\mathcal{O}_\chi; \Omega^1\mathcal{O}_\chi)$ the associated $C^\infty(\mathcal{O}_\chi)$-module of smooth, compactly supported sections.
(2) Let $\phi : X \to \mathcal{O}_\chi$ be a surjective submersion. The short exact sequence

$$0 \to \ker d\phi \to TX \xrightarrow{d\phi} T\mathcal{O}_\chi \to 0$$

implies $\Omega^1(TX) = \Omega^1(\ker d\phi) \otimes \phi^*(\Omega^1(T\mathcal{O}_\chi))$. Integration along the fibers is a surjective map

$$\phi_! : C_c^\infty(X; \Omega^1(TX)) \to C_c^\infty(\mathcal{O}_\chi; \Omega^1(T\mathcal{O}_\chi)) \qquad \phi_!(f) : x \mapsto \int_{\phi^{-1}(x)} f$$

Note that $C_c^\infty(X; \Omega^1 X)$ is a $C^\infty(\mathcal{O}_\chi)$-module with $f\zeta = (f \circ \phi) \cdot \zeta$ for every $f \in C^\infty(\mathcal{O}_\chi)$ and $\zeta \in C_c^\infty(X; \Omega^1 X)$.

Recall the plot $(\overline{\pi}, \mathcal{O})$ in $D(G(M,\omega))$ which arises from the disjoint union of plots $(\pi|_{\mathcal{O}_i}, \mathcal{O}_i)$ in $P(G(M,\omega))$, mentioned in item (c) of remark 2.1.

LEMMA 2.4. *Let $\chi : \mathcal{O}_\chi \to G(M,\omega)$ be a plot in $D(G(M,\omega))$ and f a density in $C_c^\infty(\mathcal{O}_\chi, \Omega^1\mathcal{O}_\chi)$. Then there exist a plot $\psi : \mathcal{O}_\psi \to G(M,\omega)$ in $D(G(M,\omega))$, submersions $p : \mathcal{O}_\psi \to \mathcal{O}$ and $q : \mathcal{O}_\psi \to \mathcal{O}_\chi$ and $g \in C^\infty(\mathcal{O}_\psi, \Omega^1\mathcal{O}_\psi)$, such that $q_!(g) = f$.*

PROOF. By lemma 2.2 and the compactness of the support of f we find a finite number of plots $(\psi_j, \mathcal{O}_{\psi_j})$ in $D(G(M,\omega))$ and submersions $p_j : \mathcal{O}_{\psi_j} \to \mathcal{O}$ and $q_j : \mathcal{O}_{\psi_j} \to \mathcal{O}_\chi$ such that $\bigcup_j q_j(\mathcal{O}_{\psi_j})$ covers the support of f. Put $\mathcal{O}_\psi$ the disjoint union $\coprod_j \mathcal{O}_{\psi_j}$ and $\psi = \coprod_j \psi_j$. It is easy to see that $(\psi, \mathcal{O}_\psi)$ is a plot in $D(G(M,\omega))$. The maps $q = \coprod_j p_j : \mathcal{O}_\psi \to \mathcal{O}_\chi$ and $p = \coprod_j p_j : \mathcal{O}_\psi \to \mathcal{O}$ are submersions. From item (b) above, we have that f is of the form $q_!(g)$. □

Using lemma 2.4 we can define our convolution algebra.

DEFINITION 2.5.

(1) We define $C_c^\infty(G(M,\omega))$ to be the quotient

$$\frac{\oplus_{i \in I} C_c^\infty(\mathcal{O}_i; \Omega^1(\mathcal{O}_i))}{\mathcal{I}}$$

where $\mathcal{I}$ is the subspace spanned by $p_!(f)$ where $p : \mathcal{O}_\psi \to \mathcal{O}$ is a submersion such that $q_!(f) = 0$. Here the data $\mathcal{O}_\psi, p, q$ is as in lemma 2.2.
(2) Let $\chi : \mathcal{O}_\chi \to G(M,\omega)$ be a plot in $D(G(M,\omega))$. With the notation of lemma 2.4 we define a linear map $Q_\chi : C_c^\infty(\mathcal{O}_\chi; \Omega^1\mathcal{O}_\chi) \to C_c^\infty(G(M,\omega))$. It is defined by putting $Q_\chi(f)$ the class of $p_!(g) \in \oplus_{i \in I} C_c^\infty(\mathcal{O}_i; \Omega^1(\mathcal{O}_i))$

in the above quotient. Again, the data $\mathcal{O}_\psi, p, q, g$ is as in Lemma 2.4, in particular $f = q_!(g)$.

REMARK 2.6.

(1) Item (b) in remark 2.1 implies that, in item (a) of definition 2.5 we can take the direct sum to be finite.
(2) The map Q_χ is well defined. Indeed, for $\ell = 1, 2$ consider plots $\mathcal{O}_{\psi_\ell}$, submersions $p_\ell : \mathcal{O}_{\psi_\ell} \to \mathcal{O}$, $q_\ell : (O)_{\psi_\ell} \to \mathcal{O}_\chi$ and $g_\ell \in C_c^\infty((O)_{\psi_\ell}; \Omega^1((O)_{\psi_\ell}))$ such that $(q_1)_!(g_1) = (q_2)_!(g_2)$. Put $\mathcal{O}_\psi = \mathcal{O}_{\psi_1} \coprod \mathcal{O}_{\psi_2}$, $q = q_1 \coprod q_2$ and $p = p_1 \coprod p_2$. The density $g = g_1 \coprod (-g_2) \in C_c^\infty(\mathcal{O}_\psi; \Omega^1(\mathcal{O}_\psi))$ obviously satisfies $q_!(g) = 0$ and $p_!(g) = (p_1)_!(g_1) - (p_2)_!(g_2)$.

The following characterization of the map Q_χ is proven exactly as in [**AS09**, Prop. 4.4].

PROPOSITION 2.7. *To every plot $\chi : \mathcal{O}_\chi \to G(M, \omega)$ in $D(G(M, \omega))$ we can associate a linear map $Q_\chi : C_c^\infty(\mathcal{O}_\chi; \Omega^1 \mathcal{O}_\chi) \to C_c^\infty(G(M, \omega))$. The maps Q_χ are characterised by the following properties:*

(1) *If $\mathcal{O}_\chi = \mathcal{O}_i$, Q_χ is the quotient map*
$$C_c^\infty(\mathcal{O}_i; \Omega^1 \mathcal{O}_i) \subset \oplus_{i \in I} C_c^\infty(\mathcal{O}_i; \Omega^1(\mathcal{O}_i)) \to \frac{\oplus_{i \in I} C_c^\infty(\mathcal{O}_i; \Omega^1(\mathcal{O}_i))}{\mathcal{I}}$$
We use the notation Q_i instead of Q_χ in this case.
(2) *For every smooth map $q : \mathcal{O}_\psi \to \mathcal{O}_\chi$ which is a submersion such that $\chi \circ q = \psi$, we have $Q_\psi = Q_\chi \circ q_!$.*

REMARK 2.8.

(1) In section 3 we provide a bijection between the function space $C_c^\infty(G(M, \omega))$ and the convolution algebra of the action groupoid $\mathbb{T}^2 \rtimes_\lambda \mathbb{R} \rightrightarrows \mathbb{T}^2$. This $*$-algebra structure is defined using the groupoid structure of the manifold $\mathbb{T}^2 \rtimes_\lambda \mathbb{R} \rightrightarrows \mathbb{T}^2$, rather than the group structure. Hence, this bijection identifies the completion of $C_c^\infty(G(M, \omega))$ with the irrational rotation algebra A_λ; This is the algebra that we want to attach to the group $G(M, \omega)$, since it is the leaf space of the foliation by irrational rotation on $\mathbb{T}^2$.
(2) Using the group structures of $\mathbb{T}^2$ and $G(M, \omega)$, it is possible to endow $C_c^\infty(G(M, \omega))$ with a different $*$-algebra structure. We provide the explicit construction of this structure in Appendix A.
(3) The bijection we define in §3 also conveys the $*$-algebra structure arising from the group structure of $\mathbb{T}^2 \rtimes_\lambda \mathbb{R}$ to the $*$-algebra structure of $C_c^\infty(G(M, \omega))$ described in Appendix A. This is explained in Appendix A, in case the reader finds it useful.

3. Isomorphism with irrational rotation algebra

The purpose of this section is to prove Theorem 1.2. To this end, in §3.1 we describe the diffeological structure of the action groupoid $\mathbb{T}^2 \rtimes_\lambda \mathbb{R} \rightrightarrows \mathbb{T}^2$ and in §3.2 we construct the associated $*$-algebra. The proof of Theorem 1.2 is given in §3.3.

3.1. The transformation groupoid of the irrational rotation. Let us recall the following:

(1) In [**AS09**, Examples 3.4, item 4] it is shown that, for any $p \in \mathbb{T}^2$, a neighborhood of the identity element $(p, 0)$ in $\mathbb{T}^2 \rtimes_\lambda \mathbb{R}$ is diffeomorphic to $\mathcal{O}_i \times (-\epsilon_{ij}, \epsilon_{ij})$, for some open subset $\mathcal{O}_i$ of $\mathbb{T}^2$.

(2) Under this diffeomorphism, the source map becomes the projection to $\mathcal{O}_i$. The range map takes the form of the exponential $r(p,t) = exp(tX)(p)$, where X is the infinitesimal generator of the irrational rotation action; it is a submersion $r : \mathcal{O}_i \times (-\epsilon_{ij}, \epsilon_{ij}) \to \mathcal{O}_j$, where $\mathcal{O}_j$ is some other open subset of $\mathbb{T}^2$.

(3) Put $\phi_{ij} : \mathcal{O}_i \times (-\epsilon_{ij}, \epsilon_{ij}) \to s^{-1}(\mathcal{O}_i) \cap r^{-1}(\mathcal{O}_j)$ this diffeomorphism. Then $(\mathcal{O}_i \times (-\epsilon_{ij}, \epsilon_{ij}), \phi_{ij}, \mathbb{T}^2 \rtimes_\lambda \mathbb{R})$ is a minimal bisubmersion[5] in the sense of [**Zam22**, Definitions 2.4, 2.6]. These are the path-holonomy bisubmersions of the Lie algebroid $A(\mathbb{T}^2 \rtimes_\lambda \mathbb{R})$.

(4) Put $\mathcal{A}$ the atlas of bisubmersions generated by the path-holonomy bisubmersions (cf. [**Zam22**, Dfn. 3.4]). The Lie groupoid $\mathbb{T}^2 \rtimes_\lambda \mathbb{R} \rightrightarrows \mathbb{T}^2$ is the quotient groupoid associated to this atlas (cf. [**Zam22**, §3.2]).

(5) It follows from the previous item that the C^*-algebra of the atlas $\mathcal{A}$ constructed in [**Zam22**, Appendix A] is isomorphic with the crossed product $C(\mathbb{T}^2) \rtimes_\lambda \mathbb{Z}$, in other words with A_λ.

Since $C_c^\infty(\mathbb{T}^2 \rtimes_\lambda \mathbb{R})$ is $*$-isomorphic with the $*$-algebra associated with the atlas $\mathcal{A}$ (cf. [**Zam22**, Appendix A]), we need to understand the relation of bisubmersions as above with plots in $D(G(M,\omega))$. We explain this relation in the following lemmas.

LEMMA 3.1. *For every $i \in I$ the path-holonomy bisubmersion $\mathcal{O}_i \times (-\epsilon_{ij}, \epsilon_{ij})$ is a plot in $D(G(M,\omega))$.*

PROOF. Consider the submersion $r : \mathcal{O}_i \times (-\epsilon_{ij}, \epsilon_{ij}) \to \mathcal{O}_j$ and define $\chi_{ij} = \pi|_{\mathcal{O}_j} \circ r$. Obviously, $(\chi_{ij}, \mathcal{O}_i \times (-\epsilon_{ij}, \epsilon_{ij}))$ is a plot in $D(G(M,\omega))$. □

LEMMA 3.2. *Every bisubmersion $(U, \phi_U, \mathbb{T}^2 \rtimes_\lambda \mathbb{R})$ in the atlas $\mathcal{A}$ is a plot in $D(G(M,\omega))$.*

PROOF. Up to shrinking U if necessary, we can assume that $\phi(U)$ is contained in $\mathcal{O}_i \times (-\epsilon_{ij}, \epsilon_{ij})$, for some $i, j \in I$. The map $\chi_U : U \to G(M,\omega)$ defined by $\chi = \pi|_{\mathcal{O}_j} \circ r \circ \phi$ makes (χ_U, U) a plot in $D(G(M,\omega))$. □

LEMMA 3.3. *Let $(\chi, \mathcal{O}_\chi)$ be a plot in $D(G(M,\omega))$ and $p : \mathcal{O}_\chi \to \mathcal{O}_i$ a submersion for some $i \in I$. Then $(\mathcal{O}_\chi \times (-\epsilon_{ij}, \epsilon_{ij}), (p \times \mathrm{id}) \circ \phi_{ij}, \mathbb{T}^2 \rtimes_\lambda \mathbb{R})$ is a bisubmersion for (the "singular subalgebroid") $\Gamma_c A(\mathbb{T}^2 \rtimes_\lambda \mathbb{R})$. Moreover, $p \times \mathrm{id}$ is a morphism of bisubmersions.*

PROOF. The composition $\mathcal{O}_\chi \times (-\epsilon_{ij}, \epsilon_{ij}) \overset{(p\times \mathrm{id})}{\longrightarrow} \mathcal{O}_i \times (-\epsilon_{ij}, \epsilon_{ij}) \overset{\phi_{ij}}{\longrightarrow} \mathbb{T}^2 \rtimes_\lambda \mathbb{R}$ is a submersion. The module of right-invariant vector fields of $\mathbb{T}^2 \rtimes_\lambda \mathbb{R}$ corresponding to the ("singular") subalgebroid $\Gamma_c A(\mathbb{T}^2 \rtimes_\lambda \mathbb{R})$ is $\Gamma_c(\ker ds)$. It is straightforward to see that [**Zam22**, definition 2.4] is satisfied. □

---

[5] Here, the ("singular") subalgebroid is the entire Lie algebroid of $\mathbb{T}^2 \rtimes_\lambda \mathbb{R}$. We put *singular* in quotes because, in fact, there are no singularities: The module of sections of any Lie algebroid is projective, thanks to the Serre-Swan theorem.

REMARK 3.4. Lemma 3.3 shows that, for every plot $(\chi, \mathcal{O}_\chi)$ in $D(G(M,\omega))$, there is $\epsilon > 0$ such that $\mathcal{O}_\chi \times (-\epsilon, \epsilon)$ is a bisubmersion of (the "singular subalgebroid") $\Gamma_c A(\mathbb{T}^2 \rtimes_\lambda \mathbb{R})$. To explain this, let us look at the dimension of $\mathcal{O}_\chi$.

(1) If $dim\mathcal{O}_\chi = 1$, there is an $\epsilon > 0$ and a diffeomorphism $\eta : \mathcal{O}_\chi \to \mathbb{R}$ such that $h = \iota \circ \eta$, where ι is the inclusion map in (1.1). Note that then the condition $\chi = \pi \circ h$ implies that the map χ is constant.

   Now define $s, r : \mathcal{O}_\chi \times (-\epsilon, \epsilon) \to \mathbb{T}^2$ by $s(x,t) = \iota(t)$ and $r(x,t) = \iota(\phi(x) + t)$. It is straightforward that $(\mathcal{O}_\chi, s, r)$ is a bisubmersion.
(2) If $dim\mathcal{O}_\chi \geq 2$, the map h is a submersion because: Identifying $\mathcal{O}_i$ with an open ball in $\mathbb{R}^2$, let $X = \partial_x + \lambda\partial_y$ be the infinitesimal generator of the irrational rotation action and put F the $C^\infty(\mathcal{O}_i)$-submodule of $\mathcal{X}(\mathcal{O}_i)$ generated by the restriction of X to $\mathcal{O}_i$. Now choose a vector field Y of $\mathcal{O}_\chi$ such that $dh(Y) = X$ and put $h^{-1}(F)$ the span of Y. This is a foliation of $\mathcal{O}_\chi$ and by definition h maps leaves of $h^{-1}(F)$ to leaves of F. The condition $\chi = \pi \circ h$ implies that h also maps (small) transversals of $h^{-1}(F)$ to (small) transversals of F. So, if L is a leaf and S a transversal of $h^{-1}(F)$ at u, then $h(L)$ is a leaf and $h(S)$ is a transversal of F at $h(u)$. The decompositions $T_u\mathcal{O}_\chi = T_uL \oplus T_uS$ and $T_{h(u)}\mathcal{O}_i = T_{h(u)}(h(L)) \oplus T_{h(u)}(h(S))$ show that dh is surjective.
(3) Note that in all cases, there is a submersion $r : \mathcal{O}_\chi \times (-\epsilon, \epsilon) \to \mathcal{O}_j$ for some $j \in I$. In the case of lemma 3.3, r is the composition of the target map of $\mathbb{T}^2 \rtimes_\lambda \mathbb{R}$ with the map $(p \times \mathrm{id}) \circ \phi_{ij}$. When $dim\mathcal{O}_\chi = 1$ we already defined r and when $dim\mathcal{O}_\chi \geq 2$, r is defined as in lemma 3.3.

3.2. The convolution algebra. Because of the previous discussion, the convolution algebra of the transformation groupoid $\mathbb{T}^2 \rtimes_\lambda \mathbb{R} \rightrightarrows \mathbb{T}^2$ can also be constructed using bisubmersions, as we explained in [**Zam22**, Appendix A]. This construction is more suitable for our purposes, so let us recall it.

(1) Since the range map $r : \mathcal{O}_i \times (-\epsilon_{ij}, \epsilon_{ij}) \to \mathcal{O}_j$ is a submersion we have the exact sequence
$$0 \to \ker dr \to T(\mathcal{O}_i \times (-\epsilon_{ij}, \epsilon_{ij})) \xrightarrow{dr} r^*(T\mathcal{O}_j) \to 0$$
   Whence $\Omega^1(T(\mathcal{O}_i \times (-\epsilon_{ij}, \epsilon_{ij})))$ is isomorphic to $\Omega^1(\ker dr) \otimes r^*\Omega^1(T\mathcal{O}_j)$.
(2) The convolution algebra is defined as
$$C_c^\infty(\mathbb{T}^2 \rtimes_\lambda \mathbb{R}) = \frac{\oplus_{i,j} C_c^\infty(\mathcal{O}_i \times (-\epsilon_{ij}, \epsilon_{ij}); \Omega^1(\ker dr) \otimes r^*\Omega^1(\mathcal{O}_j))}{\mathcal{J}}$$
   where the subspace $\mathcal{J}$ is defined in a way analogous to the definition of the subspace $\mathcal{I}$ we gave in definition 2.5. The exact definition of $\mathcal{J}$ is given right after [**Zam22**, Lemma A.4]. It is analogous to our lemmata 2.2, 2.4.
(3) Given a bisubmersion $r : \mathcal{O}_i \times (-\epsilon_{ij}, \epsilon_{ij}) \to \mathcal{O}_j$, put
$$Q_{ij}^{\mathcal{A}} : C_c^\infty(\mathcal{O}_i \times (-\epsilon_{ij}, \epsilon_{ij}); \Omega^1(\ker dr) \otimes r^*\Omega^1(\mathcal{O}_j)) \to C_c^\infty(\mathbb{T}^2 \rtimes_\lambda \mathbb{R})$$
   the obvious quotient map.
(4) Let $(U, \phi_U, \mathbb{T}^2 \rtimes \mathbb{R})$ be a bisubmersion adapted to the atlas $\mathcal{A}$. Recall from [**Zam22**, Appendix A] that there is a linear map $Q_U^{\mathcal{A}} : C_c^\infty(U; \Omega^1(U)) \to C_c^\infty(\mathbb{T}^2 \rtimes_\lambda \mathbb{R})$ such that:

- If $(U, \phi_U, \mathbb{T}^2 \rtimes \mathbb{R})$ is a bisubmersion $r : \mathcal{O}_i \times (-\epsilon_{ij}, \epsilon_{ij}) \to \mathcal{O}_j$ then $Q_U^{\mathcal{A}} = Q_{ij}^{\mathcal{A}}$.
- If $p : U' \to U$ is a morphism of bisubmersions which is a submersion, then $Q_{U'}^{\mathcal{A}} = Q_U^{\mathcal{A}} \circ p_!$.

(5) Let us recall the definition of the $*$-algebra structure of $C_c^\infty(\mathbb{T}^2 \rtimes_\lambda \mathbb{R})$ from [**Zam22**, Appendix A]. Let $(U, \phi_U, \mathbb{T}^2 \rtimes_\lambda \mathbb{R})$ and $(V, \phi_V, \mathbb{T}^2 \rtimes_\lambda \mathbb{R})$ be two bisubmersions adapted to the atlas $\mathcal{A}$ and $f_U \in C^\infty(U; \Omega^1(U))$, $f_V \in C^\infty(V; \Omega^1(V))$.

- The inverse of $(U, \phi_U, \mathbb{T}^2 \rtimes_\lambda \mathbb{R})$ is the bisubmersion $(U^{-1}, \phi_{U^{-1}}, \mathbb{T}^2 \rtimes_\lambda \mathbb{R})$ where $U^{-1} = U$ and $\phi_{U^{-1}} = \iota_{\mathbb{T}^2 \rtimes_\lambda \mathbb{R}} \circ \phi_U$, where $\iota_{\mathbb{T}^2 \rtimes_\lambda \mathbb{R}}$ is the inversion map of $\mathbb{T}^2 \rtimes_\lambda \mathbb{R}$. It is adapted to $\mathcal{A}$. Put $f_U^* = \overline{f}_U \circ \iota_{\mathbb{T}^2 \rtimes_\lambda \mathbb{R}}^{-1}$. Involution is defined by

$$(Q_U^{\mathcal{A}}(f_U))^* = Q_{U^{-1}}^{\mathcal{A}}(f_U^*)$$

- The composition $(U \circ V, \phi_U \cdot \phi_V, \mathbb{T}^2 \rtimes_\lambda \mathbb{R})$ is defined by $U \circ V = U \times_{s_U, r_V} V$ and $\phi_U \cdot \phi_V = m_{\mathbb{T}^2 \rtimes_\lambda \mathbb{R}} \circ (\phi_U, \phi_V)$, where $m_{\mathbb{T}^2 \rtimes_\lambda \mathbb{R}}$ is the groupoid multiplication. It is adapted to $\mathcal{A}$ as well. Convolution is defined by

$$Q_U^{\mathcal{A}}(f_U) * Q_V^{\mathcal{A}}(f_V) = Q_{V \circ W}^{\mathcal{A}}(f_U \otimes f_V)$$

As we saw in §3.1 every plot $(\chi, \mathcal{O}_\chi)$ in $D(G(M, \omega))$, gives rise to a bisubmersion $\mathcal{O}_\chi \times (-\epsilon, \epsilon)$ together with a submersion $r : \mathcal{O}_\chi \times (-\epsilon, \epsilon) \to \mathcal{O}_j$. The exact sequence

$$0 \to \ker dr \to T(\mathcal{O}_\chi \times (-\epsilon, \epsilon)) \xrightarrow{dr} r^*(\mathcal{O}_j) \to 0 \tag{3.1}$$

induces an isomorphism of $\Omega^1(\mathcal{O}_\chi) \otimes \Omega^1(-\epsilon, \epsilon)$ with $\Omega^1(\ker dr) \otimes r^*\Omega^1(\mathcal{O}_j)$. Now, every $g \in C_c^\infty(\mathcal{O}_\chi; \Omega^1(\mathcal{O}_\chi))$ can be identified with

$$\hat{g} = g \otimes \ell \in C^\infty(\mathcal{O}_\chi \times (-\epsilon, \epsilon)); \Omega^1(\ker dr) \otimes r^*\Omega^1(\mathcal{O}_j))$$

where ℓ is the Lebesque measure in $(-\epsilon, \epsilon)$. This defines the quotient map

$$C^\infty(\mathcal{O}_\chi; \Omega^1(\mathcal{O}_\chi)) \ni g \mapsto Q_\chi^{\mathcal{A}}(\hat{g}) \in C_c^\infty(\mathbb{T}^2 \rtimes_\lambda \mathbb{R})$$

3.3. The bijection. For every $i \in I$ consider the plot $(\chi_{ij}, \mathcal{O}_i \times (-\epsilon_{ij}, \epsilon_{ij}))$ defined in lemma 3.1 and the submersion r. We obtain a linear map

$$C_c^\infty(\mathcal{O}_i \times (-\epsilon_{ij}, \epsilon_{ij}); \Omega^1(\ker dr) \otimes r^*\Omega^1(\mathcal{O}_j)) \ni f \mapsto Q_{\chi_{ij}}(f) \in C_c^\infty(G(M, \omega))$$

It follows from proposition 2.7 that $Q_{\chi_{ij}}(f) = Q_j(r_!(f))$. Lemma 3.2 shows that the induced map

$$\Phi : C_c^\infty(\mathbb{T}^2 \rtimes_\lambda \mathbb{R}) \to C_c^\infty(G(M, \omega))$$

is well defined. The fact that r is a submersion shows that Φ is surjective. The difficulty is to prove the injectivity of Φ. We do this in Proposition 3.5 below.

PROPOSITION 3.5. *The map Φ is injective.*

PROOF. Let f be an element of $C_c^\infty(\mathcal{O}_i \times (-\epsilon_{ij}, \epsilon_{ij}); \Omega^1(\ker dr) \otimes r^*\Omega^1(\mathcal{O}_j))$ such that $\Phi([f])$ vanishes, where $[f]$ is the class of f in $C_c^\infty(\mathbb{T}^2 \rtimes_\lambda \mathbb{R}))$. This means that there are plots $(\psi, \mathcal{O}_\psi)$ and $(\chi, \mathcal{O}_\chi)$ in $D(G(M, \omega))$, a density $g \in C_c^\infty(\mathcal{O}_\psi; \Omega^1\mathcal{O}_\psi)$ and submersions $p : \mathcal{O}_\psi \to \mathcal{O}_j$, $q : \mathcal{O}_\psi \to \mathcal{O}_\chi$ such that $p_!(g) = r_!(f)$ and $q_!(g) = 0$.

Because of lemma 3.3 and remark 3.4, there is an $\epsilon > 0$ such that $\mathcal{O}_\psi \times (-\epsilon, \epsilon)$, $\mathcal{O}_\chi \times (-\epsilon, \epsilon)$ and $\mathcal{O}_j \times (-\epsilon, \epsilon)$ are bisubmersions. Put $\hat{p} = p \times \mathrm{id}$, $\hat{q} = q \times \mathrm{id}$. Let σ be a local section of the submersion $r : \mathcal{O}_i \times (-\epsilon, \epsilon) \to \mathcal{O}_j$ and put:

$$\hat{\sigma} = (\sigma \circ pr_1) \times \mathrm{pr}_2 : \mathcal{O}_j \times (-\epsilon, \epsilon) \to \mathcal{O}_i \times (-\epsilon, \epsilon)$$
$$\hat{p} = \hat{\sigma} \circ (p \times \mathrm{id}) : \mathcal{O}_\psi \times (-\epsilon, \epsilon) \to \mathcal{O}_i \times (-\epsilon, \epsilon)$$
$$\hat{q} = q \times \mathrm{id} : \mathcal{O}_\psi \times (-\epsilon, \epsilon) \to \mathcal{O}_\chi \times (-\epsilon, \epsilon)$$

It follows that $f = \hat{p}_!(\hat{g})$ and $\hat{q}_!(\hat{g}) = 0$, whence the class $[f]$ vanishes. $\square$

The bijection Φ conveys to $C_c^\infty(G(M, \omega))$ the $*$-algebra structure of $C_c^\infty(\mathbb{T}^2 \rtimes_\lambda \mathbb{R})$. To see this, recall from Lemma 3.3 that every plot $\chi : \mathcal{O}_\chi \to G(M, \omega)$ gives rise to a bisubmersion $\mathcal{O}_\chi \times (-\epsilon, \epsilon)$ of $\mathbb{T}^2 \rtimes_\lambda \mathbb{R}$. The bijection Φ also conveys the completion of $C_c^\infty(\mathbb{T}^2 \rtimes_\lambda \mathbb{R})$. This proves Theorem 1.2.

Appendix A. Another C^*-algebra attached to $G(M, \omega)$

As we mentioned in Remark 2.8, the space $C_c^\infty(G(M, \omega))$ admits another $*$-algebra structure, arising from the group structure of $\mathbb{T}^2$, rather than the groupoid structure of $\mathbb{T}^2 \rtimes_\lambda \mathbb{R}$. For the convenience of the reader, in this appendix we describe this structure. Moreover, we show that the bijection Φ conveys this structure to the $*$-algebra structure associated with the group structure of $\mathbb{T}^2 \rtimes \mathbb{R}$. We will use the notation of lemma 2.4.

A.1. Another $*$-algebra structure for $C_c^\infty(G(M, \omega))$. Let us define the convolution first. If $\chi_i : \mathcal{O}_{\chi_i} \to G(M, \omega)$ are plots in $D(G(M, \omega))$, $i = 1, 2$, then for every (x, y) in $\mathcal{O}_{\chi_1} \times \mathcal{O}_{\chi_2}$ we have $\Omega^1(\mathcal{O}_{\chi_1} \times \mathcal{O}_{\chi_2})_{(x,y)} = \Omega^1(\mathcal{O}_{\chi_1})_x \otimes \Omega^1(\mathcal{O}_{\chi_2})_y$. This identification allows us to define the tensor product of $f_1 \in C_c^\infty(\mathcal{O}_{\chi_1}; \Omega^1 \mathcal{O}_{\chi_1})$ and $f_2 \in C_c^\infty(\mathcal{O}_{\chi_2}; \Omega^1 \mathcal{O}_{\chi_2})$ by:

$$f_1 \otimes f_2 : (x, y) \mapsto f_1(x) \otimes f_2(y) \in C_c^\infty(\mathcal{O}_{\chi_1} \times \mathcal{O}_{\chi_2}; \Omega^1(\mathcal{O}_{\chi_1} \times \mathcal{O}_{\chi_2}))$$

DEFINITION A.1. Given f_i in $C_c^\infty(\mathcal{O}_{\chi_i}; \Omega^1 \mathcal{O}_{\chi_i})$, $i = 1, 2$ we define $Q_{\chi_1}(f_1) * Q_{\chi_2}(f_2)$ to be the class of $(m_{\mathbb{T}^2})_! \left((p_1)_!(g_1) \otimes (p_2)_!(g_2)\right)$.

PROPOSITION A.2. *The product in definition* A.1 *makes $C_c^\infty(G(M, \omega))$ an associative algebra over $\mathbb{C}$. Namely, for every f_i in $C_c^\infty(\mathcal{O}_{\chi_i}; \Omega^1 \mathcal{O}_{\chi_i})$, $i = 1, 2, 3$ and every $\lambda, \mu \in \mathbb{C}$ we have:*

(1) $(Q_{\chi_1}(f_1) * Q_{\chi_2}(f_2)) * Q_{\chi_3}(f_3) = Q_{\chi_1}(f_1) * (Q_{\chi_2}(f_2) * Q_{\chi_3}(f_3))$
(2) $Q_{\chi_1}(f_1) * (Q_{\chi_2}(f_2) + Q_{\chi_3}(f_3)) = Q_{\chi_1}(f_1) * Q_{\chi_2}(f_2) + Q_{\chi_1}(f_1) * Q_{\chi_2}(f_3)$
(3) $(Q_{\chi_1}(f_1) + Q_{\chi_2}(f_2)) * Q_{\chi_3}(f_3) = Q_{\chi_1}(f_1) * Q_{\chi_2}(f_3) + Q_{\chi_1}(f_2) * Q_{\chi_2}(f_3)$
(4) $(\lambda Q_{\chi_1}(f_1)) * (\mu Q_{\chi_1}(f_1)) = (\lambda\mu)(Q_{\chi_1}(f_1) * Q_{\chi_2}(f_2))$

PROOF. Associativity follows from a straightforward calculation, applying the familiar Fubini theorem. The other algebraic properties follow immediately from the definition. $\square$

Now let us define involution. Consider the inversion map $\iota_{\mathbb{T}^2} : \mathbb{T}^2 \to \mathbb{T}^2$. Let $\chi : \mathcal{O}_\chi \to G(M, \omega)$ be a plot in $D(G(M, \omega))$. In view of item (j) in remark 2.1 put $\mathcal{O}_\chi^{-1} = \iota_{\mathbb{T}^2}(\mathcal{O}_\chi)$ and $\chi^{-1} = \chi \circ \iota_{\mathbb{T}^2}^{-1}$. Then $(\chi^{-1}, \mathcal{O}_\chi^{-1})$ is obviously an element of $D(G(M, \omega))$.

DEFINITION A.3. For every $f \in C_c^\infty(\mathcal{O}_\chi; \Omega^1(\mathcal{O}_\chi))$ we define:

(1) $f^* = \overline{f} \circ \iota_{\mathbb{T}^2}^{-1}$ in $C_c^\infty(\mathcal{O}_\chi^{-1}; \Omega^1(\mathcal{O}_\chi^{-1}))$

(2) $(Q_\chi(f))^* = Q_{\chi^{-1}}(f^*)$.

PROPOSITION A.4. *The operation $(Q_\chi(f)) \mapsto (Q_\chi(f))^*$ is an involution. It makes $C_c^\infty(G(M,\omega))$ a $*$-algebra over $\mathbb{C}$. Namely, for every f_i in $C_c^\infty(\mathcal{O}_{\chi_i};\Omega^1\mathcal{O}_{\chi_i})$, $i=1,2$ and every $\lambda \in \mathbb{C}$ we have:*

(1) $((Q_\chi(f_1))^*)^* = (Q_\chi(f_1))$
(2) $(Q_{\chi_1}(f_1) * Q_{\chi_2}(f_2))^* = (Q_{\chi_2}(f_2))^* * (Q_{\chi_1}(f_1))^*$
(3) $(Q_{\chi_1}(f_1) + Q_{\chi_2}(f_2))^* = (Q_{\chi_1}(f_1))^* + (Q_{\chi_2}(f_2))^*$
(4) $(\lambda(Q_{\chi_1}(f_1)))^* = \overline{\lambda}(Q_{\chi_1}(f_1))^*$

PROOF. Items (c) and (d) are immediate consequences of definition A.3. As for items (a) and (b):

(1) This follows from the fact that $\iota_{\mathbb{T}^2} \circ \iota_{\mathbb{T}^2} = \mathrm{id}_{\mathbb{T}^2}$ and $\overline{\overline{f}} = f$.
(2) In view of lemma 2.4, definition A.3 implies that $(Q_\chi(f))^*$ is the class of $(\iota_{\mathbb{T}^2}^{-1})_!(p_!(\overline{g}))$. Whence $(Q_{\chi_1}(f_1) * Q_{\chi_2}(f_2))^*$ is the class of

$$(\iota_{\mathbb{T}^2}^{-1})_!((m_{\mathbb{T}^2})_!\,((p_1)_!(\overline{g}_1) \otimes (p_2)_!(\overline{g}_2))) = (\iota_{\mathbb{T}^2}^{-1} \circ m_{\mathbb{T}^2})_!\,((p_1)_!(\overline{g}_1) \otimes (p_2)_!(\overline{g}_2)))$$

We have $\iota_{\mathbb{T}^2}^{-1} = \iota_{T^2}$, so $\iota_{\mathbb{T}^2}^{-1} \circ m_{\mathbb{T}^2} = \iota_{\mathbb{T}^2} \circ m_{\mathbb{T}^2} = m_{\mathbb{T}^2} \circ (\iota_{\mathbb{T}^2} \times \iota_{\mathbb{T}^2}) = m_{\mathbb{T}^2} \circ (\iota_{\mathbb{T}^2}^{-1} \times \iota_{\mathbb{T}^2}^{-1})$. Whence the right-hand side is

$$(m_{\mathbb{T}^2})_!\,((\iota_{\mathbb{T}^2}^{-1})_!((p_2)_!(\overline{g}_2)) \otimes (\iota_{\mathbb{T}^2}^{-1})_!((p_1)_!(\overline{g}_1)))$$

The latter is precisely $(Q_{\chi_2}(f_2))^* * (Q_{\chi_1}(f_1))^*$.

□

A.2. Another $*$-algebra structure for $C^\infty(\mathbb{T}^2 \rtimes_\lambda \mathbb{R})$. This $*$-algebra structure arises from viewing $\mathbb{T}^2 \rtimes_\lambda \mathbb{R}$ as a group. This group structure is the cartesian group structure arising from the groups $\mathbb{T}^2 = S^1 \times S^1$ and $(\mathbb{R},+)$. Explicity, for $i=1,2$ consider $g_i = (e^{2\pi i\theta_i}, e^{2\pi i\eta_i}) \in \mathbb{T}^2$ and $t_i \in \mathbb{R}$. Then:

$$(g_1,t_1)\cdot(g_2,t_2) = ((e^{2\pi i(\theta_1+\theta_2)}, e^{2\pi i(\eta_1+\eta_2)}), t_1+t_2)$$

and

$$g_1^{\ \ 1} - ((e^{2\pi i(-\theta_i)}, c^{2\pi i(-\eta_i})), \quad t_1)$$

Let us demote $m : (\mathbb{T}^2 \rtimes_\lambda \mathbb{R}) \times (\mathbb{T}^2 \rtimes_\lambda \mathbb{R}) \to \mathbb{T}^2 \rtimes_\lambda \mathbb{R}$ and $inv : \mathbb{T}^2 \rtimes_\lambda \mathbb{R} \to \mathbb{T}^2 \rtimes_\lambda \mathbb{R}$ the above multiplication and inversion maps respectively.

Now let us define the new convolution and involution in $C_c^\infty(\mathbb{T}^2 \rtimes_\lambda \mathbb{R})$. Let $(U,\phi_U,\mathbb{T}^2 \rtimes_\lambda \mathbb{R})$ and $(V,\phi_V,\mathbb{T}^2 \rtimes_\lambda \mathbb{R})$ be two bisubmersions adapted to the atlas $\mathcal{A}$ and $f_U \in C^\infty(U;\Omega^1(U))$, $f_V \in C^\infty(V;\Omega^1(V))$.

(1) The inverse of $(U,\phi_U,\mathbb{T}^2 \rtimes_\lambda \mathbb{R})$ is the bisubmersion $(U^{-1},\phi_{U^{-1}},\mathbb{T}^2 \rtimes_\lambda \mathbb{R})$ where $U^{-1} = U$ and $\phi_{U^{-1}} = inv \circ \phi_U$. It is adapted to $\mathcal{A}$. Put $f_U^* = \overline{f}_U \circ inv^{-1}$. Involution is defined by

$$(Q_U^{\mathcal{A}}(f_U))^* = Q_{U^{-1}}^{\mathcal{A}}(f_U^*)$$

(2) The composition $(U \circ V, \phi_U \cdot \phi_V, \mathbb{T}^2 \rtimes_\lambda \mathbb{R})$ is defined by $U \circ V = U \times_{s_U,r_V} V$ and $\phi_U \cdot \phi_V = m \circ (\phi_U,\phi_V)$. It is adapted to $\mathcal{A}$ as well. Convolution is defined by

$$Q_U^{\mathcal{A}}(f_U) * Q_V^{\mathcal{A}}(f_V) = Q_{V\circ W}^{\mathcal{A}}(m_!(f_U \otimes f_V))$$

he above involution and involution make $C_c^\infty(\mathbb{T}^2 \rtimes_\lambda \mathbb{R})$ a $*$-algebra. The proof of this is as in §A.1, so we omit it.

A.3. Φ is a $*$-isomorphism.

PROPOSITION A.5. *The bijection Φ is a $*$-homomorphism for the $*$-algebra structures defined in* §A.1 *and* A.2.

PROOF. First we show that Φ respects involution. Choose a section f in $C_c^\infty(\mathcal{O}_i \times (-\epsilon_{ij}, \epsilon_{ij}); \Omega^1(\ker dr) \otimes r^*\Omega^1(\mathcal{O}_j))$. Put $U = \mathcal{O}_i \times (-\epsilon_{ij}, \epsilon_{ij})$ and $\chi_{ij} = \pi|_{\mathcal{O}_i} \circ r$. Recall that the map r is really the restriction of the target map of the groupoid $\mathbb{T}^2 \rtimes_\lambda \mathbb{R}$. The inversion map $\iota_{\mathbb{T}^2 \rtimes_\lambda \mathbb{R}}$ satisfies $r \circ \iota_{\mathbb{T}^2 \rtimes_\lambda \mathbb{R}} = s$, where s is the source map of $\mathbb{T}^2 \rtimes_\lambda \mathbb{R}$. It follows that $U^{-1} = \mathcal{O}_j \times (-\epsilon_{ij}, \epsilon_{ij})$ and $r : \mathcal{O}_j \times (-\epsilon_{ij}, \epsilon_{ij}) \to \mathcal{O}_i$. In particular, the following diagram commutes:

$$\begin{array}{ccc}
\mathcal{O}_i \times (-\epsilon_{ij}, \epsilon_{ij}) & \xrightarrow{\ r\ } & \mathcal{O}_j \\
{\scriptstyle \iota_{\mathbb{T} \rtimes_\lambda \mathbb{R}}}\big\downarrow & & \big\downarrow{\scriptstyle \iota_{\mathbb{T}}} \\
\mathcal{O}_j \times (-\epsilon_{ij}, \epsilon_{ij}) & \xrightarrow{\ r\ } & \mathcal{O}_i
\end{array}$$

Hence we have $(Q_{ij}^{\mathcal{A}}(f))^* = Q_{ji}^{\mathcal{A}}(\overline{f} \circ \iota^{-1}_{\mathbb{T}^2 \rtimes_\lambda \mathbb{R}})$ and $(Q_j(r_!(f)))^* = Q_i(\overline{r_!(f)} \circ \iota^{-1}_{\mathbb{T}^2})$. Therefore, we find:

$$\begin{aligned}
\Phi((Q_{ij}^{\mathcal{A}}(f))^*) &= \Phi(Q_{ji}^{\mathcal{A}}(\overline{f} \circ \iota^{-1}_{\mathbb{T}^2 \rtimes_\lambda \mathbb{R}})) = Q_i(r_!(\overline{f} \circ \iota^{-1}_{\mathbb{T}^2 \rtimes_\lambda \mathbb{R}})) \\
&= Q_i(r_!(\overline{f}) \circ \iota^{-1}_{\mathbb{T}^2}) = Q_i(\overline{r_!(f) \circ \iota^{-1}_{\mathbb{T}^2}}) \\
&= (Q_i(r_! f))^* = (\Phi(Q_{ij}^{\mathcal{A}}(f)))^*
\end{aligned}$$

The fact that Φ respects convolution follows from the definitions of convolution in Definition A.1 and §3.2 and the commutativity of the following diagram:

$$\begin{array}{ccc}
(\mathbb{T}^2 \rtimes_\lambda \mathbb{R}) \times (\mathbb{T}^2 \rtimes_\lambda \mathbb{R}) & \xrightarrow{\ m\ } & \mathbb{T}^2 \rtimes_\lambda \mathbb{R} \\
{\scriptstyle r \times r}\big\downarrow & & \big\downarrow{\scriptstyle r} \\
\mathbb{T}^2 \times \mathbb{T}^2 & \xrightarrow{\ m_{\mathbb{T}^2}\ } & \mathbb{T}^2
\end{array}$$

□

Acknowledgments

I would like to thank Alan Weinstein for bringing the article [**Wei89**] to my attention. I would also like to thank Collegium Helveticum and the Institute of Mathematics of the University of Zurich, for their hospitality during the revision of this article.

References

[AA18] I. Androulidakis and P. Antonini, *Integrable lifts for transitive Lie algebroids*, Internat. J. Math. **29** (2018), no. 9, 1850062, 26 pp., DOI 10.1142/S0129167X18500623. MR3845398

[AM85] R. Almeida and P. Molino, *Suites d'Atiyah et feuilletages transversalement complets* (French, with English summary), C. R. Acad. Sci. Paris Sér. I Math. **300** (1985), no. 1, 13–15. MR778785

[AS09] I. Androulidakis and G. Skandalis, *The holonomy groupoid of a singular foliation*, J. Reine Angew. Math. **626** (2009), 1–37, DOI 10.1515/CRELLE.2009.001. MR2492988

[Cra04] M. Crainic, *Prequantization and Lie brackets*, J. Symplectic Geom. **2** (2004), no. 4, 579–602. MR2197220

[Mac05] K. C. H. Mackenzie, *General theory of Lie groupoids and Lie algebroids*, London Mathematical Society Lecture Note Series, vol. 213, Cambridge University Press, Cambridge, 2005, DOI 10.1017/CBO9781107325883. MR2157566

[Sou70] J.-M. Souriau, *Structure des systèmes dynamiques* (French), Dunod, Paris, 1970. Maîtrises de mathématiques. MR0260238

[Sou80] J.-M. Souriau, *Groupes différentiels* (French), Differential geometrical methods in mathematical physics (Proc. Conf., Aix-en-Provence/Salamanca, 1979), Lecture Notes in Math., vol. 836, Springer, Berlin, 1980, pp. 91–128. MR607688

[Wei89] A. Weinstein, *Cohomology of symplectomorphism groups and critical values of Hamiltonians*, Math. Z. **201** (1989), no. 1, 75–82, DOI 10.1007/BF01161996. MR990190

[Zam22] M. Zambon, *Singular subalgebroids* (English, with English and French summaries), Ann. Inst. Fourier (Grenoble) **72** (2022), no. 6, 2109–2190, DOI 10.5802/aif.3493. With an appendix by Iakovos Androulidakis. MR4500354

National and Kapodistrian University of Athens, Department of Mathematics, Panepistimiopolis, GR-15784 Athens, Greece

Current address: Institut für Mathematik, Universität Zürich, Winterthurerstrasse 190, CH-8057 Zürich, Switzerland and Collegium Helveticum, The Institute of Advanced Studies of ETH, UZH and ZHdK, ETH-Zentrum, Schmelzbergstrasse 25, 8092 Zürich, Switzerland

Email address: `iandroul@math.uoa.gr`

Contemporary Mathematics
Volume **794**, 2024
https://doi.org/10.1090/conm/794/15942

Diffeologies and generalized Kadomtsev-Petviashvili hierarchies

Anahita Eslami-Rad, Jean-Pierre Magnot, Enrique G. Reyes, and Vladimir Rubtsov

ABSTRACT. We present a simplified, but extended proof of the well-posedness of the Kadomtsev-Petviashvili hierarchy of formal pseudo-differential symbols with coefficients in an abstract diffeological differential algebra, and with initial value a formal symbol of arbitrary order.

1. Introduction

The physicists B. Kadomtsev and V. Petviashvili obtained in 1970, during their study of ion-sound waves in the theory of collective phenomena in plasma, the famous two-dimensional integrable nonlinear partial differential equation now known as the Kadomtsev-Petviashvili (KP) equation,

$$3u_{yy} = (4u_t - 12u_x\, u - u_{xxx})_x.$$

The KP equation is a natural generalization of the Korteweg and de Vries (KdV) equation

$$4u_t - 12u_x\, u - u_{xxx} = 0,$$

a one-dimensional model equation or a laboratory for non-linear phenomena. The KP equation, like KdV, is completely integrable: it admits an infinite number of commuting flows and it may be considered an infinite-dimensional analogue of a classical Hamiltonian system of integrable differential equations that can be solved explicitly. The so-called KP hierarchy of flows that we obtain (see Subsection 3.2 below) is ubiquitous by its extremely wide spectrum of applications, ranging from pure mathematics to numerical models of the weather forecast.

Indeed, the full study of this system has created unexpected links with infinite dimensional geometry (more specifically with Sato grassmanians and loop groups), groups of diffeomorphisms, and Hodge theory, and it is intuitively natural that generalizations of the KP hierarchy will uncover new correspondences with unexpected partial differential equations, such as the ones appearing in [**13**].

In this note we review how diffeologies appear naturally in the study of the initial value problem for this non-linear, infinite dimensional integrable system, along

---

2020 *Mathematics Subject Classification.* Primary 37K10, 22E65, 35R15, 57P99.

Key words and phrases. differential algebra, formal pseudo-differential operators, integrable systems, jet bundles.

Research partially supported by the FONDECYT grant #1201894.

the lines of our previous works [**6**, **12**]. We consider a very general framework, and we show how to proceed with the analysis of the KP hierarchy in a way that really simplifies our previous efforts, see [**6**, **12**]. We take advantage of this short presentation to expose two new results, a grading in a group of (infinite order) formal pseudo-differential operators, and an extension of the statement of integrability and well-posedness for a wider class of initial conditions. The reader is assumed to be familiar with diffeologies.

2. Preliminaries on pseudo-differential symbols in a diffeological setting

First of all, we set our basic framework.

Let $\mathbb{K}$ be a field and let the quadruple $(A, +, *, \partial)$ be a differential algebra over $\mathbb{K}$; the symbols $+, *$ indicate standard binary algebra operations on A. Usually we will simply omit the symbol "$*$" for clarity reasons.

DEFINITION 2.1. The set of pseudo-differential symbols is

$$\Psi DO(A) = \left\{ \sum_{-\infty<k\leq N} a_k \, \xi^k : a_k \in A \text{ and } N \in \mathbb{Z} \right\},$$

where ξ is a formal variable.

We always assume that A is a unital algebra, that is, there exists $1 \in A$ such that $\forall a \in A, 1 * a = a * 1 = a$. The set $\Psi DO(A)$ can be equipped with an algebra structure.

DEFINITION 2.2. We define addition and multiplication on $\Psi DO(A)$ as follows.

$$\begin{array}{rccc} + : & \Psi DO(A) \times \Psi DO(A) & \longrightarrow & \Psi DO(A) \\ & (\sum a_n \xi^n, \sum b_m \xi^m) & \longmapsto & \sum (a_p + b_p)\xi^p \end{array}$$

and

$$\begin{array}{rccc} \cdot : & \Psi DO(A) \times \Psi DO(A) & \longrightarrow & \Psi DO(A) \\ & (a(\xi) = \sum a_n \xi^n, b(\xi) = \sum b_m \xi^m) & \longmapsto & \sum_{m,n} \sum_{\alpha \in \mathbb{N}} \frac{1}{\alpha!}(a_n * \partial^\alpha b_m)(D_\xi^\alpha \xi^n)\xi^m \end{array}$$

THEOREM 2.3. *$(\Psi DO(A), +, \cdot)$ is a unital (associative) $\mathbb{K}-$algebra.*

This theorem is well-known, see for example [**3**], or the more recent papers [**5**, **6**]. The following observation is at the basis of the results presented herein. In its proof we use the fact that $(\Psi DO(A), +, \cdot)$ is naturally graded: if $P = \sum_{-\infty<k\leq N} a_k \in \Psi DO(A)$, $a_N \neq 0$, we denote its degree by $deg_\xi(P) = N$.

PROPOSITION 2.1. *The decomposition*

$$\Psi DO(A) = DO(A) \oplus IO(A),$$

in which $DO(A)$ is the subalgebra of symbols of differential operators, and $IO(A)$ is the algebra of symbols of integral operators, is a vector space splitting. We note that $IO(A)$ is a non unital algebra.

PROOF. The proof relies on standard arguments (here and henceforth we use $\mathbb{N}^* = \mathbb{N} \setminus \{0\}$):

- If $(n, m) \in \mathbb{N}^2$, then for all $\alpha \in \mathbb{N}$, we have that
$$deg_\xi\left((D_\xi^\alpha \xi^n)\xi^m\right) \in \mathbb{N},$$
which shows that $DO(A)$ is closed under multiplication,

- On the other hand, if $(n,m) \in (\mathbb{N}^*)^2$, then for all $\alpha \in \mathbb{N}$, we have that
$$deg_\xi\left((D_\xi^\alpha \xi^{-n})\xi^{-m}\right) < 0,$$
which shows that $IO(A)$ is closed under multiplication.

□

This proposition implies that a symbol $a \in \Psi DO(A)$ decomposes uniquely under this direct sum as
$$a = a_+ + a_-, \quad a_+ \in DO(A) \text{ and } a_- \in IO(A).$$
In the most classical frameworks, A is an algebra of $\mathbb{R}-, \mathbb{C}-, \mathbb{H}-$valued functions most usually with domain S^1, see [**7**–**10**, **16**], equipped with a derivation ∂. In this case, we usually replace the formal variable ξ for ∂ in $\Psi DO(A)$ and speak about "formal pseudo-differential operators" or, rather informally, of "pseudo-differential operators". We have found in [**13**] that in full generality it is quite natural to stick to symbols, as in Definition 2.1, and here we will follow this convention. Of great importance for us is the fact that the algebras of functions
$$A = C^\infty(S^1, \mathbb{K}),$$
$\mathbb{K} = \mathbb{R}, \mathbb{C}$, (or $\mathbb{H}$), are Fréchet algebras, and that they are equipped with a natural notion of differentiability making addition, multiplication and differentiation smooth. Considering addition and multiplication in $\Psi DO(A)$, we can say that addition and multiplication on $\Psi DO(A)$ are *smooth* in the following sense: if
$$a = \sum_{n\in\mathbb{Z}} a_n \xi^n, \qquad b = \sum_{n\in\mathbb{Z}} b_n \xi^n,$$
then setting
$$a + b = c = \sum_{n\in\mathbb{Z}} c_n \xi^n$$
and
$$a \cdot b = d = \sum_{n\in\mathbb{Z}} d_n \xi^n,$$
the map
$$((a_n)_{n\in\mathbb{Z}}, (b_n)_{n\in\mathbb{Z}}) \mapsto ((c_n)_{n\in\mathbb{Z}}, (d_n)_{n\in\mathbb{Z}})$$
is smooth in the relevant infinite product.

These observations motivate us to consider abstract settings, and to understand smoothness on infinite products by means of the infinite product diffeology. Hereafter, smoothness will be understood in this way on objects that arise as infinite products of infinite-dimensional diffeological spaces.

We assume henceforth that the algebra A is a **diffeological algebra**. This means that addition, scalar multiplication, and multiplication are smooth operations; it also means that the inversion operation is a smooth operation on A^*, the group of the units of A, in which A^* is equipped with the subset diffeological structure of
$$A^* \subset Diff(A),$$
in which we understand A^* as acting on the left on A by multiplication. For completeness, we recall that the group of diffeomorphisms $Diff(X)$ of a diffeological

space X is not endowed with the subset diffeology of functional diffeology $\mathcal{P}_f$ of $C^\infty(X,X)$, but with the pull-back diffeology

$$\Phi^*(\mathcal{P}_f \times \mathcal{P}_f)$$

where

$$\Phi : g \in Diff(X) \mapsto (g, g^{-1}) \in C^\infty(X,X)^2.$$

The reason for this choice is the known fact that inversion may not be smooth for the subset diffeology of the functional diffeology. We also assume that ∂ is smooth[1]. It follows that, identifying a formal pseudo-differential operator $a \in \Psi DO(A)$ with its sequence of partial symbols, the space $\Psi DO(A)$ is a linear subspace of $A^{\mathbb{Z}}$ and it carries a natural diffeological structure. We obtain, see [**6**]:

PROPOSITION 2.2. *$\Psi DO(A)$ is a diffeological algebra.*

3. An algebraic resolution of the KP hierarchy

3.1. The Mulase bicross product and its subgroups. Let t be a countable family of independent variables $t_1, t_2, \ldots$. We denote by A_t the diffeological algebra of formal series

$$A_t = A[[t_1, t_2, \ldots]],$$

and we fix a valuation on A_t by setting

$$val_t(t_n) = n.$$

The interested reader may find an explicit construction of A_t in [**1**]. After [**5**], other choices of valuations can be made, but in this work we choose to concentrate on the standard one given in Mulase's original works [**14, 15**].

DEFINITION 3.1 (Mulase, [**15**]). The Mulase spaces of formal pseudo-differential and differential symbols of infinite order, $\widehat{\Psi}(A_t)$ and $\widehat{\mathcal{D}}_{A_t}$ respectively, are:
(3.1)

$$\widehat{\Psi}(A_t) = \left\{ \sum_{\alpha \in \mathbb{Z}} a_\alpha \, \xi^\alpha : a_\alpha \in A_t \text{ and } \exists (C,N) \in \mathbb{R} \times \mathbb{N} \text{ so that } val_t(a_\alpha) > C\alpha - N \ \forall \ \alpha \gg 0 \right\}$$

and

$$\widehat{\mathcal{D}}_{A_t} = \left\{ P = \sum_{\alpha \in \mathbb{Z}} a_\alpha \, \xi^\alpha : P \in \widehat{\Psi}(A_t) \text{ and } a_\alpha = 0 \text{ for } \alpha < 0 \right\}. \tag{3.2}$$

REMARK 3.2. As a set of formal series, invertibility of an element $A \in \widehat{\Psi}(A_t)$ depends on its monomial of $t-$valuation 0. This monomial is of non-positive $\xi-$order. Therefore, one can prove that $\widehat{\Psi}(A_t)^*$ is a diffeological group if and only if A^* is a diffeological group, by analyzing classical formulas of inversion of series. However, we need to deal with "more subtle" spaces of symbols in order to work with diffeological groups and to have a group splitting that reflects the vector space splitting of Proposition 2.1.

---

[1] In a diffeological setting, smoothness is disconnected from boundedness, since we do not have norms or bornologies. In the classical example of $\partial = d/dx$ acting on $C^\infty(S^1)$, the operator d/dx is unbounded for any Sobolev norm, but smooth with respect to the underlying Fréchet structure of $C^\infty(S^1)$.

In addition to (3.1) and (3.2), we define

$$\mathcal{I}_{A_t} = \left\{ P \in \widehat{\Psi}(A_t) : \forall \alpha \geq 0, a_\alpha = 0 \right\}$$

and

$$G_{A_t} = 1 + \mathcal{I}(A_t).$$

REMARK 3.3. In order to prove that G_{A_t} is a diffeological group, it is not necessary to assume that A^* is a diffeological group.

We notice that

$$A_t \subset \widehat{\Psi}(A_t)$$

as the 0-order term,

$$G_{A_t} \subset \Psi(A_t) \subset \widehat{\Psi}(A_t),$$

(in fact, $\Psi(A_t)$ is composed of elements of $\widehat{\Psi}(A_t)$ with uniformly bounded degree in ξ) and that

$$\widehat{\mathcal{D}}_{A_t} \not\subset \Psi(A_t).$$

Analysing the operations on $\widehat{\Psi}(A_t)$, see [**15**] and our later paper [**5**], we obtain the following result, as a consequence of our work in [**6**]:

LEMMA 3.4. *The spaces $\widehat{\Psi}(A_t)$ and $\widehat{\mathcal{D}}_{A_t}$ are diffeological algebras.*

The constraints appearing in Mulase's definitions of $\widehat{\Psi}(A_t)$ allow us to construct *diffeological groups* of pseudo-differential operators. We proceed thus.

DEFINITION 3.5. Let $\mathcal{K}$ be the ideal of A_t generated by $t_1, t_2, \cdots$. If $P \in \widehat{\Psi}(A_t)$, we denote by $P|_{t=0}$ the equivalence class $P \, mod \, \mathcal{K}$ (i.e. the projection on the constant coefficient of the $t-$series), and we identify it with an element of $\Psi(A)$. We also set $G_A = 1 + IO(A)$, and we define the spaces

$$G(\widehat{\Psi}(A_t)) = \{P \in \widehat{\Psi}(A_t) : P|_{t=0} \in G_A\} \tag{3.3}$$

and

$$\widehat{\mathcal{D}}^{\times}_{A_t} = \{P \in \widehat{\mathcal{D}}_A : P|_{t=0} = 1\}. \tag{3.4}$$

We now state a result established in [**12**] in the context of Frölicher spaces and, for algebras A equipped with an abstract valuation, in [**6**]. The proof we give here is inspired by [**12, 15**]; it is different to the one given in [**6**].

THEOREM 3.6.

(1) *For any $U \in G(\widehat{\Psi}(A_t))$ there exist unique $S \in G_{A_t}$ and $Y \in \widehat{\mathcal{D}}^{\times}_A$ such that*

$$U = S^{-1} Y.$$

In other words, there exists a unique global factorization of the diffeological group $G(\widehat{\Psi}(A_t))$ as a group defined by bicross product of diffeological groups,

$$G(\widehat{\Psi}(A_t)) = G_{A_t} \bowtie \widehat{\mathcal{D}}^{\times}_{A_t}.$$

(2) *The factorization $U \mapsto (S, Y)$ is (diffeologicaly) smooth.*

PROOF. Part (1) is already proved in [**14**, **15**] (see also the exposition in [**5**]), but we have to examine carefully the original proof in order to check that the map $U \mapsto S$ is smooth. Once we have this first result, we finish the proof of the theorem by remarking that $Y = SU$.

Let us follow the proof of [**14**, Lemma 3]. We solve the equation

$$(SU)_- = 0$$

for given

$$U = \sum_{\nu \in \mathbb{Z}} u_\nu \xi^\nu$$

and unknown

$$S = \sum_{\mu \in \mathbb{Z} - \mathbb{N}} s_\mu \xi^\mu .$$

This equation turns into the infinite system

$$\sum_{\nu \in \mathbb{Z}-\mathbb{N}} \left(\sum_{i \in \mathbb{N}} \binom{\nu}{i} \xi^i u_{\mu+i-\nu} \right) s_\nu = -u_\mu \quad \text{for } \mu \in \mathbb{Z} \setminus \mathbb{N}$$

with the convention

$$\binom{\nu}{i} = (-1)^{|\nu|} \binom{|\nu|}{i} \text{ if } \nu < 0.$$

We use the notation $[\,.\,]_k$ as the sum of the monomials in t with valuation equal to k, and by $[\,.\,]_{m(t)}$ the coefficient related the $t-$monomial $m(t)$.

We then solve this infinite system by induction on μ. For $\mu = -1$, we study the $t-$monomials.

$$\left[\sum_{\nu \in \mathbb{Z}-\mathbb{N}} \left(\sum_{i \in \mathbb{N}} \binom{\nu}{i} \xi^i u_{-1+i-\nu} \right) s_\nu \right]_0 = \left[\left(\sum_{i \in \mathbb{N}} \binom{-1}{i} \xi^i u_i \right) s_{-1} \right]_0 = -[s_{-1}]_0$$

thus $[s_{-1}]_0 = [u_{-1}]_0$. Then, at the monomial of valuation 1, which is only the t_1-monomial,

$$\begin{aligned} \left[\sum_{\nu \in \mathbb{Z}-\mathbb{N}} \left(\sum_{i \in \mathbb{N}} \binom{\nu}{i} \xi^i u_{-1+i-\nu} \right) s_\nu \right]_{t_1} &= \left[\left(\sum_{i \in \mathbb{N}} \binom{-1}{i} \xi^i u_i \right) \right]_{t_1} [s_{-1}]_0 \\ &\quad + \left[\left(\sum_{i \in \mathbb{N}} \binom{-1}{i} \xi^i u_i \right) \right]_0 [s_{-1}]_{t_1} \\ &= \left[\left(\sum_{i \in \mathbb{N}} \binom{-1}{i} \xi^i u_i \right) \right]_{t_1} [u_{-1}]_0 \\ &\quad -[s_{-1}]_{t_1} \end{aligned}$$

we get

$$[s_{-1}]_{t_1} = [u_{-1}]_{t_1} + \left[\left(\sum_{i \in \mathbb{N}} \binom{-1}{i} \xi^i u_i \right) \right]_{t_1} [u_{-1}]_0$$

which shows that the maps

$$u \mapsto [s_{-1}]_0$$

and

$$u \mapsto [s_{-1}]_{t_1}$$

are smooth. The same calculations can be used for any $t-$monomial for s_{-1}, and they show that

$$u \mapsto s_{-1}$$

is smooth. Let us now assume that $s_{-1}, \ldots s_{-k}$ are determined, and are smooth with respect to U, and let us determine s_{-k-1}. Then we write, for $\mu = -k-1$,

$$\begin{aligned} &\sum_{\nu \leq -k-1} \left(\sum_{i \in \mathbb{N}} \binom{\nu}{i} \xi^i u_{-k-1+i-\nu} \right) s_\nu \\ &= -u_{-k-1} - \sum_{\nu=-k}^{-1} \left(\sum_{i \in \mathbb{N}} \binom{\nu}{i} \xi^i u_{-k-1+i-\nu} \right) s_\nu \end{aligned}$$

noticing that the right side is fully determined. Then we carry out induction with respect to the valuation on t, for which we sketch the first two steps as we did before:

- for the $t = 0$ term:

$$[s_{-k-1}]_0 = (-1)^{-k-1} \left[u_{-k-1} + \sum_{\nu=-k}^{-1} \left(\sum_{i \in \mathbb{N}} \binom{\nu}{i} \xi^i u_{-k-1+i-\nu} \right) s_\nu \right]_0,$$

- for the t_1-monomial:

$$\begin{aligned} [s_{-k-1}]_{t_1} &= (-1)^{-k-1} \left(\left[u_{-k-1} + \sum_{\nu=-k}^{-1} \left(\sum_{i \in \mathbb{N}} \binom{\nu}{i} \xi^i u_{-k-1+i-\nu} \right) s_\nu \right]_{t_1} + \right. \\ &\left. \left[\left(\sum_{i \in \mathbb{N}} \binom{-1}{i} \xi^i u_i \right) \right]_{t_1} [s_{-k-1}]_0 \right) \end{aligned}$$

Thus we have here a construction of S made of smooth operations, so that the map

$$U \mapsto S$$

is smooth. Hence the map

$$U \mapsto Y = SU$$

is also smooth. This completes the proof. □

In order to solve the KP hierarchy, we need the following groups (see [**6**, **12**] for the $n = 1$ case):

DEFINITION 3.7.

$$\overline{\Psi}_n = \left\{ P = \sum_{\alpha \in \mathbb{Z}} a_\alpha \xi^\alpha \ : a_\alpha \in A_t \text{ and } val_T(a_\alpha) \geq n\alpha \text{ and } P|_{t=0} \in 1 + IO(A) \right\}$$

and

$$\overline{\mathcal{D}_n} = \left\{ P = \sum_{\alpha \in \mathbb{Z}} a_\alpha \xi^\alpha : P \in \overline{\Psi}_n(A_t) \text{ and } a_\alpha = 0 \text{ for } \alpha < 0 \right\}.$$

PROPOSITION 3.1. *Let us assume that the algebra A, understood as a diffeological group with respect to its addition, has its internal tangent space at 1 that is diffeologically isomorphic to A.*

(1) *The groups* $\overline{\Psi}_n$, G_{A_t} *and* $\overline{\mathcal{D}_n}$ *are diffeological Lie groups, that is, they have an internal tangent space at their neutral element which carries a smooth Lie bracket.*

(2) *Moreover, we have a diffeological matched pair (or bicross product)*

$$\overline{\Psi}_n = G_{A_t} \bowtie \overline{\mathcal{D}}_n.$$

REMARK 3.8. The condition that a diffeological vector space is isomorphic with its tangent space at a given point seems natural in a Fréchet setting but becomes not obvious for a general diffeology. A recent work that develops a counter-example is [**2**].

PROOF. If the internal tangent space at 1 of A is diffeologically isomorphic to A, then it is trivial that we have an inclusion of diffeological vector spaces

$$T_1\overline{\mathcal{D}_n} \subset T_1\overline{\Psi}_n \subset \widehat{\Psi}(A_t)$$

and therefore that the Ad-action of $\overline{\mathcal{D}_n}$ (resp. $\overline{\Psi_n}$) on $T_1\overline{\mathcal{D}_n}$ (resp. $T_1\overline{\Psi_n}$) differentiates as in the usual case of a Fréchet algebra. Therefore, the Lie bracket is defined in a classical way by commutators $[A, B] = AB - BA$. The same holds for G_{A_t}, for which it is trivial that

$$T_1 G_{A_t} = IO(A_t).$$

The claim (2) follows from the inclusion of $\overline{\Psi}_n$ into $G(\widehat{\Psi}(A_t))$ and Theorem 3.4. □

We conclude that we have two diffeological inclusion sequences of diffeological groups:

$$\overline{\Psi}_1 \subset \cdots \subset \overline{\Psi}_n \subset \cdots \subset G(\widehat{\Psi}(A_t))$$

and

$$\overline{\mathcal{D}_1} \subset \cdots \subset \overline{\mathcal{D}_n} \subset \cdots \subset \widehat{\mathcal{D}}^{\times}_{A_t}.$$

These sequences will allow us to solve the initial value problem for the KP hierarchy, as we explain in Section 4. It is shown in [**12**] that $\overline{\Psi}_1$ has much better regularity properties than $G(\widehat{\Psi}(A_t))$. We discuss one of them in the next remark.

REMARK 3.9. In all the present constructions, we do not need to define a tangent space at the identity on any of our groups. This is a good point since actually the notion of tangent space of a manifold has multiple generalizations in the setting of diffeologies, and it often happens that, in one or another definition of the tangent space at one point of a diffeological vector space V, the tangent space is not (diffeologically) isomorphic to V. Even more, for similar reasons, considering a diffeological group, the tangent space at the identity element may not carry a smooth Lie algebra structure. This is the difference between general diffeological groups and diffeological *Lie* groups: for the latter we consider only such diffeological groups which have a diffeological Lie algebra structure on their tangent space at the identity, see [**11**]. It seems to be difficult to prove that a diffeological group is not a diffeological Lie group: actually, we do not know of any rigorous example of this statement. Now, assuming that, $\forall a \in A$, the kinematic or internal tangent space at a is (diffeologically) isomorphic to A, we conjecture that the group $G(\widehat{\Psi}(A_t))$ may well be an example of a diffeological group which is not a diffeological Lie group (that is, for which the Lie bracket is ill-defined). On the other hand, it is proven in [**12**] that $\overline{\Psi}_1$ is a diffeological Lie group if A is a Fréchet algebra (with tangent space isomorphic to A, as stated before).

3.2. The algebraic initial value problem (IVP). First of all we state the solution to the IVP for the KP hierarchy after [**15**] and the more recent re-workings of Mulase's result appearing in our papers [**5**] and [**6**].

THEOREM 3.10. *Consider the KP system of equations*

$$\frac{dL}{dt_k} = [(L^k)_+, L] \tag{3.5}$$

with initial condition $L(0) = L_0 \in \Psi DO(A)$*, and let* $Y \in \overline{\mathcal{D}_n}$ *and* $S \in G_{A_t}$ *be the unique solution to the factorization problem* $\exp(t_k\, L_0^k) = S^{-1}(t_k)\, Y(t_k)$*. The unique solution to Equation* (3.5) *with* $L(0) = L_0$ *is*

$$L(t_k) = Y\, L_0\, Y^{-1} = S\, L_0\, S^{-1}.$$

What is important for us at this point is that at this level of generality *the initial condition is an arbitrarily given formal pseudo-differential operator.* We postpone the proof of Theorem 3.5 until next section where we consider the well-posedness of (3.5). Before that, we stress that a somewhat similar statement in a more general setting is given in [**6**], and that the very general setting considered in that reference may include the case of coefficients being Laurent series in t under mild assumptions on the control of the order, and also sets of coefficients with arbitrary valuations. However, the proof given in [**6**] is drastically more complicated than the one that we present here, and "concrete" examples in such general settings are still to be constructed.

4. Proof of well-posedness in a diffeological setting

The following result, stated in [**6**] for more specific initial values L_0, gives a synthesized statement of our main results on the KP hierarchy (3.5), and it states smooth dependence on the initial conditions.

THEOREM 4.1. *Consider the KP hierarchy* 3.5 *with initial condition* $L(0) = L_0$ *of order* $n \in \mathbb{N}$*. Then,*

(1) *There exists a pair* $(S, Y) \in G_{A_t} \times \overline{\mathcal{D}_n}$ *such that a solution to Equation* (3.5) *with* $L|_{t=0} = L_0$ *is*
$$L(t_1, t_2, \cdots) = Y\, L_0\, Y^{-1} = S L_0 S^{-1}.$$
and $L \in \Psi DO^n(A_t)$*.*

(2) *The pair* (S, Y) *is uniquely determined by the smooth decomposition problem*
$$exp\left(\sum_{k\in\mathbb{N}} t_k L_0^k\right) = S^{-1} Y$$
and the solution L *depends smoothly on the initial condition* L_0*.*

(3) *The solution operator* L *is smoothly dependent on the initial value* L_0*.*

PROOF: THEOREM 3.10 AND THEOREM 4.1. We model our proof after Reyman and Semenov-Tian-Shansky's [**18**], see also [**6, 12, 17**]. The proof is based on r-matrix theory.

We set

$$U = \exp\left(\sum_{k\in\mathbb{N}} t_k L_0^k\right).$$

Then,

$$U \in \overline{\Psi}_n.$$

We consider the smooth decomposition

$$U \mapsto (S, Y) \in G_{A_t} \times \overline{\mathcal{D}}_n$$

following Proposition 3.1, and we set

$$L = Y\, L_0\, Y^{-1}.$$

We make two obvious observations:

(1) $L^k = Y L_0^k Y^{-1}$

(2) $U\, L_0^k U^{-1} = L_0^k$ since L_0 commutes with $U = \exp(\sum_k t_k\, L_0^k)$.

It follows that

$$L^k = Y L_0^k Y^{-1} = S S^{-1} Y L_0^k Y^{-1} S S^{-1} = S L_0^k S^{-1}.$$

We take t_k-derivative of U for each $k \geq 1$. We get the equation

$$L_0^k U = -S^{-1} S_{t_k} S^{-1} Y + S^{-1} Y_{t_k}$$

and so, using $U = S^{-1}\, Y$, we obtain the decomposition

$$S L_0^k S^{-1} = -S_{t_k} S^{-1} + Y_{t_k} Y^{-1}.$$

Since $S_{t_k} S^{-1} \in \mathcal{I}_A$ and $Y_{t_k} Y^{-1} \in \mathcal{D}_A$, we conclude that

$$(L^k)_+ = Y_{t_k} Y^{-1} \quad \text{and} \quad (L^k)_- = -S_{t_k} S^{-1}.$$

Now we take t_k-derivative of L:

$$\begin{aligned}\frac{dL}{dt_k} &= Y_{t_k} L_0 Y^{-1} - Y L_0 Y^{-1} Y_{t_k} Y^{-1} \\ &= Y_{t_k} Y^{-1} Y L_0 Y^{-1} - Y L_0 Y^{-1} Y_{t_k} Y^{-1} \\ &= (L^k)_+ L - L\,(L^k)_+ \\ &= [(L^k)_+, L].\end{aligned}$$

It remains to check the initial condition: We have

$$L(0) = Y(0) L_0 Y(0)^{-1},$$

but $Y(0) = 1$ since $Y \in \overline{\mathcal{D}}_n$. Therefore

$$L(0) = L_0.$$

Smoothness with respect to t is already proved by construction, and we have established smoothness of the map $L_0 \mapsto Y$ at the beginning of the proof. Thus, the map

$$L_0 \mapsto L(t) = Y(t) L_0 Y^{-1}(t)$$

is smooth in $\overline{\Psi}(A_t)$.

In order to finish the proof, we need to prove that the solution L belongs to $\Psi DO^n(A_t)$. For this, we consider above mentioned relation

$$L^k = S L_0^k S^{-1},$$

which for $k = 1$, implies that

$$L = L_0 + S[L_0, S^{-1}].$$

$[L_0, S^{-1}] \in \Psi DO^{n-1}(A_t)$, which ends the proof. □

REMARK 4.2. We note that at this level of generality we do not assert that the solution to the initial value problem for the KP hierarchy that we just found is unique. There seem to be difficulties in the algebraic proofs of uniqueness if we do not assume some additional conditions, see [**15**]. On the other hand, it is possible to prove the uniqueness of smooth solutions rigorously if we restrict to *e.g.* commutative Fréchet algebras and to a specific kind of initial conditions, see [**12**] and [**15**].

5. Outlook

There are several examples we can give. Our approach can be applied to $A = C^\infty(S^1, \mathbb{R})$, and therefore we obtain a solution to the Cauchy problem for the KP hierarchy in this classical case treated by algebraic and geometric means in, for instance, [**7**–**9**]. However, there is no reason why A should be a commutative algebra. We can therefore prove well-posedness of the KP hierarchy posed on the algebra of quaternions that appears in [**16**] if we start from a good class of initial conditions, see [**14, 15**]. Very general non-commutative algebras have been considered in [**4**], [**15**] and [**13**]. Our main theorems imply existence, regularity and smooth dependence on initial conditions also in these cases, as long as we start with diffeological algebras A.

Let us finish with some questions: first, what about the Kadomtsev-Petviashvili *equations*? Our theory, in fact, implies smooth existence theorems for them. We have presented results on this issue in [**12**] and in more depth, in [**13**]. Second, what about other hierarchies such as the Gelfand-Dickey flows? We are investigating this issue and we will present our results elsewhere. Finally, we wonder whether the algebraic deformation described in [**4**] has a pending diffeological realization.

References

[1] N. Bourbaki, *Elements of Mathematics.* Springer-Verlag, Berlin. (1998)

[2] J. D. Christensen and E. Wu, *Exterior bundles in diffeology*, Israel J. Math. **253** (2023), no. 2, 673–713, DOI 10.1007/s11856-022-2372-9. MR4575422

[3] L. A. Dickey, *Soliton equations and Hamiltonian systems*, 2nd ed., Advanced Series in Mathematical Physics, vol. 26, World Scientific Publishing Co., Inc., River Edge, NJ, 2003, DOI 10.1142/5108. MR1964513

[4] A. Eslami Rad, *Lie bialgebroid of pseudo-differential operators*, J. Nonlinear Math. Phys. **29** (2022), no. 4, 869–895, DOI 10.1007/s44198-022-00067-4. MR4514143

[5] A. Eslami Rad and E. G. Reyes, *The Kadomtsev-Petviashvili hierarchy and the Mulase factorization of formal Lie groups*, J. Geom. Mech. **5** (2013), no. 3, 345–364, DOI 10.3934/jgm.2013.5.345. MR3110105

[6] A. Eslami Rad, J.-P. Magnot, and E. G. Reyes, *The Cauchy problem of the Kadomtsev-Petviashvili hierarchy with arbitrary coefficient algebra*, J. Nonlinear Math. Phys. **24** (2017), no. suppl. 1, 103–120, DOI 10.1080/14029251.2017.1418057. MR3750840

[7] B. Enriquez, S. Khoroshkin, A. Radul, A. Rosly, and V. Rubtsov, *Poisson-Lie aspects of classical W-algebras*, The interplay between differential geometry and differential equations, Amer. Math. Soc. Transl. Ser. 2, vol. 167, Amer. Math. Soc., Providence, RI, 1995, pp. 37–59, DOI 10.1090/trans2/167/03. MR1343983

[8] L. Guieu and C. Roger, *L'algèbre et le groupe de Virasoro: Aspects géométriques et algébriques, généralisations* (French), Les Publications CRM, Montreal, QC, 2007. MR2362888

[9] B. Khesin and I. Zakharevich, *Poisson-Lie group of pseudodifferential symbols*, Comm. Math. Phys. **171** (1995), no. 3, 475–530. MR1346170

[10] B. A. Kuperschmidt, *KP or mKP: Mathematics of Lagrangian, Hamiltonian and Integrable Systems.* Math. Surv. Monographs **78** (2000)

[11] J. Leslie, On a Diffeological Group Realization of certain Generalized symmetrizable Kac-Moody Lie Algebras *J. Lie Theory* **13** , 427-442 (2003).

[12] J.-P. Magnot and E. G. Reyes, *Well-posedness of the Kadomtsev-Petviashvili hierarchy, Mulase factorization, and Frölicher Lie groups*, Ann. Henri Poincaré **21** (2020), no. 6, 1893–1945, DOI 10.1007/s00023-020-00896-3. MR4100916

[13] J.-P. Magnot, E. G. Reyes, and V. Rubtsov, *On (t_2, t_3)-Zakharov-Shabat equations of generalized Kadomtsev-Petviashvili hierarchies*, J. Math. Phys. **63** (2022), no. 9, Paper No. 093501, 11, DOI 10.1063/5.0093238. MR4482715

[14] M. Mulase, *Complete integrability of the Kadomtsev-Petviashvili equation*, Adv. in Math. **54** (1984), no. 1, 57–66, DOI 10.1016/0001-8708(84)90036-7. MR761762

[15] M. Mulase, *Solvability of the super KP equation and a generalization of the Birkhoff decomposition*, Invent. Math. **92** (1988), no. 1, 1–46, DOI 10.1007/BF01393991. MR931203

[16] I. McIntosh, *The quaternionic KP hierarchy and conformally immersed 2-tori in the 4-sphere*, Tohoku Math. J. (2) **63** (2011), no. 2, 183–215, DOI 10.2748/tmj/1309952086. MR2812451

[17] A. M. Perelomov, *Integrable systems of classical mechanics and Lie algebras. Vol. I*, Birkhäuser Verlag, Basel, 1990. Translated from the Russian by A. G. Reyman [A. G. Reĭman], DOI 10.1007/978-3-0348-9257-5. MR1048350

[18] A. G. Reyman and M. A. Semenov-Tian-Shansky, *Reduction of Hamiltonian systems, affine Lie algebras and Lax equations. II*, Invent. Math. **63** (1981), no. 3, 423–432, DOI 10.1007/BF01389063. MR620678

Department of Mathematics, FaMAF, Universidad Nacional de Cordoba, Medina Allende, Ciudad Universitaria , 5000-Cordoba, Argentina
Email address: aeslami@famaf.unc.edu.ar

Univ. Angers, CNRS, LAREMA, SFR MATHSTIC, F-49000 Angers, France, and Lycée Jeanne d'Arc, Avenue de Grande Bretagne, 63000 Clermont-Ferrand, France
Email address: magnot@math.cnrs.fr

Departamento de Matemática y Ciencia de la Computación, Universidad de Santiago de Chile (USACH), Casilla 307 Correo 2, Santiago, Chile
Email address: enrique.reyes@usach.cl ; e_g_reyes@yahoo.ca

Univ. Angers, CNRS, LAREMA, SFR MATHSTIC, F-49000 Angers, France
Email address: volodya@univ-angers.fr

Contemporary Mathematics
Volume **794**, 2024
https://doi.org/10.1090/conm/794/15926

Smooth A_∞-form on a diffeological loop space

Norio Iwase

ABSTRACT. To construct an A_∞-form for a loop space in the category of diffeological spaces, we have two minor problems. Firstly, the concatenation of paths in the category of diffeological spaces needs a small technical trick, which apparently restricts the number of iterations of concatenations. Secondly, we do not know a *smooth* decomposition of an associahedron as a simplicial or a cubical complex. To resolve these difficulties, we introduce a notion of a q-cubic set which enjoys good properties on dimensions and representabilities, and show, using it, that the smooth loop space of a reflexive diffeological space is a h-unital smooth A_∞-space. In appendix, we show an alternative solution by modifying the concatenation to be stable without assuming reflexivity on spaces nor stability on paths.

Introduction

A site is a concrete category with a 'coverage' assigning a 'covering family' to each object. For a site C, we denote by $\mathsf{Obj}(\mathsf{C})$ the class of objects, by $\mathsf{Mor}_{\mathsf{C}}(A,B)$ the set of morphisms from A to B, and by $\mathsf{Cov}_{\mathsf{C}}(U)$ the set of covering families on $U \in \mathsf{Obj}(\mathsf{C})$. We denote by Set the category of sets and maps between sets. For a given set X, we have two contravariant functors $\mathcal{M}_X$, $\mathcal{K}_X : \mathsf{C} \to \mathsf{Set}$ defined by

(1) $\mathcal{M}_X(U) = \mathsf{Map}(U,X)$ the set of maps from U to X and
(2) $\mathcal{K}_X(U) = \{P \in \mathcal{M}_X(U) \mid P : U \to X \text{ is locally constant}\}$,

where we say $P : U \to X$ is locally constant, if there exists a covering family $\{g_\alpha : V_\alpha \to U\}_{\alpha\in\Lambda}$ of U such that $P\circ g_\alpha$ is constant for any $\alpha \in \Lambda$.

In [**Che73, Che75, Che77, Che86**], K. T. Chen introduced a site C^r-Convex which is a category of convex sets with non-void interiors in $\mathbb{R}^n$ for some $n \geq 0$, and smooth functions between them in the ordinary sense (see [**KM97**]), with a 'coverage' assigning a 'covering family' to each convex set with non-void interior, which is the set of open coverings by interiors of convex sets.

In [**Sou80**], J. M. Souriau introduced a similar but a slightly more sophisticated site C-Domain which is a category of open sets in $\mathbb{R}^n$ for some $n \geq 0$, and smooth functions between them in the ordinary sense, with a 'coverage' assigning a 'covering family' to each open set, which is the set of open coverings in the usual sense.

2020 *Mathematics Subject Classification.* Primary 58A40; Secondary 58A03, 58A05, 57N60.
Key words and phrases. Diffeology, homotopy associativity, A_∞ structure.
This research was partially supported by Grant-in-Aid for Scientific Research (C) #23K03093 and Exploratory Research #18K18713 from Japan Society for the Promotion of Science.

We call a pair $(X, \mathcal{D}_X)$ a diffeological space, if it satisfies the following conditions.

(D1) X is a set and $\mathcal{D}_X : \mathsf{C\text{-}Domain} \to \mathsf{Set}$ is a contravariant functor.

(D2) For any $U \in \mathrm{Obj}(\mathsf{C\text{-}Domain})$, $\mathcal{K}_X(U) \subset \mathcal{D}_X(U) \subset \mathcal{M}_X(U)$.

(D3) For any $U \in \mathrm{Obj}(\mathsf{C\text{-}Domain})$ and any $P \in \mathcal{M}_X(U)$, $P \in \mathcal{D}_X(U)$ if there exists $\{U_\alpha\}_{\alpha\in\Lambda} \in \mathrm{Cov}_{\mathsf{C\text{-}Domain}}(U)$ such that $P|_{U_\alpha} \in \mathcal{D}_X(U_\alpha)$ for all $\alpha \in \Lambda$.

A map $f : X \to Y$ is said to be smooth, if the natural transformation $f_* : \mathcal{M}_X \to \mathcal{M}_Y$ satisfies $f_*(\mathcal{D}_X(U)) \subset \mathcal{D}_Y(U)$ for any $U \in \mathrm{Obj}(\mathsf{C\text{-}Domain})$. We denote by $\mathsf{Diffeology}$, the category of diffeological spaces and smooth maps between diffeological spaces. An element of $\mathcal{D}_X(U)$ is called a plot of X on U, and $\mathcal{D} = \bigcup_{U\in\mathrm{Obj}(\mathsf{C\text{-}Domain})} \mathcal{D}_X(U)$ is called a 'diffeology' on X. If we replace the site $\mathsf{C\text{-}Domain}$ by the site $\mathsf{C^r\text{-}Convex}$, we obtain Chen's smooth category denoted by Chen. From now on, we discuss in the smooth category $\mathsf{Diffeology}$, rather than Chen, while we believe that entirely similar arguments can be performed also in Chen. Let $\mathbb{N}$ be the set of non-negative integers.

REMARK 0.1. For any set X, both $\mathcal{K}_X$ and $\mathcal{M}_X$ give diffeologies on X. In fact, $\mathcal{K}_X$ gives the finest diffeology on X and $\mathcal{M}_X$ gives the coarsest diffeology on X (see [**IZ13**, 1.18]).

REMARK 0.2. Let $\mathsf{C\text{-}Euclidean}$ be the full-subcategory of $\mathsf{C\text{-}Domain}$ consisting of all Euclidean spaces of some dimension $\in \mathbb{N}$. Even if we replace $\mathsf{C\text{-}Domain}$ with $\mathsf{C\text{-}Euclidean}$ in the definition of $\mathsf{Diffeology}$, we recover $\mathsf{Diffeology}$ itself (see [**IZ13**, Exercise 3]).

REMARK 0.3. For $n \in \mathbb{N}$, let $\mathsf{C\text{-}Euclidean}_n$ be the full-subcategory of $\mathsf{C\text{-}Euclidean}$ consisting of all Euclidean spaces of dimension up to n. If we replace $\mathsf{C\text{-}Domain}$ with $\mathsf{C\text{-}Euclidean}_0$ in the definition of $\mathsf{Diffeology}$, we must obtain Set the category of sets. If we replace $\mathsf{C\text{-}Domain}$ with $\mathsf{C\text{-}Euclidean}_1$ in the definition of $\mathsf{Diffeology}$, we must obtain the category of diffeological spaces with 'wire-diffeology' (see [**IZ13**, art. 1.10]).

In this paper, a manifold is assumed to be paracompact. We denote by $\mathsf{Manifold}$ the category of smooth manifolds and smooth maps between them which can be embedded into $\mathsf{Diffeology}$ as a full subcategory (see [**IZ13**]). One of the advantage to expand our playground to $\mathsf{Diffeology}$ than to restrict ourselves in $\mathsf{Manifold}$ is that the category $\mathsf{Diffeology}$ is cartesian-closed, complete and cocomplete (see [**IZ13**]).

The path space in $\mathsf{Diffeology}$ is defined using the real line $\mathbb{R}$ in place of the closed interval $[0, 1]$ (see [**IZ13**, Chapter 5]). This definition gives a nice diffeology on a path space, while it causes a technical issue on concatenation:

$$\mathrm{Paths}(X) = C^\infty(\mathbb{R}, X)\ (= \mathcal{D}_X(\mathbb{R}) \text{ as a set})$$

A work-around can easily be found as in [**IZ13**, art. 5.4] by compressing the moving part into an open subinterval $(\varepsilon, 1-\varepsilon) \subset (0, 1) \subset \mathbb{R}$, where $0 < \varepsilon \ll 1$:

$$\mathrm{stPaths}_\varepsilon(X) = \{\, u \in \mathrm{Paths}(X) \mid \forall\, t \le \varepsilon\ u(t) = u(0)\ \&\ \forall\, t \ge 1-\varepsilon\ u(t) = u(1) \,\}.$$

On the other hand, if we consider A_∞-form of concatenations using $\mathrm{stPaths}_\varepsilon(X)$, we need some more tricks to concatenate many paths. In this paper, we adopt slightly different ways to consider a smooth (h-unital) A_∞-form for a concatenation.

Let $\mathcal{P}(X) = \{\, u \in \mid u \circ \pi_{set} = u \,\}$, where $\pi_{set} : \mathbb{R} \to \mathbb{R}$ is a continuous idempotent, i.e, $\pi_{set} \circ \pi_{set} = \pi_{set}$, which is defined as follows:

$$\pi_{set}(t) = \max\{0, \min\{1, t\}\} = \min\{1, \max\{0, t\}\}.$$

Then by definition, π_{set} enjoys the following properties.

(1) $\pi_{set}(t) = 0$, $t \leq 0$, (2) $\pi_{set}(t) + \pi_{set}(1-t) = 1$, (3) $\pi_{set}(t) = t$, $0 < t < 1$. Then by (1) and (2) above, we have $\pi_{set}(t) = 1$, $t \geq 1$ as well.

1. Basic properties on subductions

Let us recall basic properties on subductions in Diffeology used in this paper.

LEMMA 1.1. *Let $\varpi_1 : K \to X$ and $\varpi_2 : L \to Y$ be two subductions. Then, $\varpi = \varpi_1 \times \varpi_2 : K \times L \to X \times Y$ is also a subduction.*

PROOF. This can be obtained using [**Kih19**, Lemma 2.5]. But we give here a direct proof: clearly, ϖ is a smooth surjection, and so we are left to show that a plot on $X{\times}Y$ can be pulled back to $K{\times}L$ locally. Let $P : V \to X{\times}Y$ be a plot. We denote by pr_k the canonical projection from a product to its k-th factor, $k = 1, 2$. Then, $P_1 = \mathrm{pr}_1{\circ}P : V \to X$ and $P_2 = \mathrm{pr}_2{\circ}P : V \to Y$ are plots, for each pr_k is smooth. Since $\varpi_1 : K \to X$ and $\varpi_2 : L \to Y$ are subductions, there is an open covering $\{V_\alpha\}$ of V such that there are plots $Q_{1\alpha} : V_\alpha \to K$ and $Q_{2\alpha} : V_\alpha \to L$ satisfying $P_k|_{V_\alpha} = \varpi_k{\circ}Q_{k\alpha}$, $k = 1, 2$. Using the data $Q_{1\alpha}$ and $Q_{2\alpha}$, we obtain a smooth map $Q_\alpha : V_\alpha \to K \times L$ satisfying $Q_{k\alpha} = \mathrm{pr}_k{\circ}Q_\alpha$, $k = 1, 2$. Thus we obtain $\varpi{\circ}Q_\alpha = P|_{V_\alpha}$, and hence $X \times Y$ has the push-forward diffeology by $\varpi : K \times L \to X \times Y$. □

From now on, we assume that L, Y and X are diffeological spaces, and that $\varpi : L \twoheadrightarrow Y$ is a subduction. Here, we remark that $\dim Y \leq \dim L$ diffeologically.

LEMMA 1.2. *For a map $g : Y \to X$, g is smooth iff $g{\circ}\varpi : L \to X$ is smooth.*

PROOF. It is sufficient to show that g is smooth if $g{\circ}\varpi : L \to X$ is smooth: let $P : U \to Y$ be a plot. Since ϖ is a subduction, there is an open covering $\{V_\alpha\}$ of U and plots $\{P_\alpha : V_\alpha \to L\}$ such that $P|_{V_\alpha} = \varpi{\circ}P_\alpha$ for all α. Since $g{\circ}P|_{V_\alpha} = (g{\circ}\varpi){\circ}P_\alpha : V_\alpha \to X$ is smooth for all α, so is $g{\circ}P : U \to X$. Thus g is smooth. □

PROPOSITION 1.3. *ϖ induces an induction $\varpi^* : C^\infty(Y,X) \to C^\infty(L,X)$.*

PROOF. Since $\varpi : L \to Y$ is a smooth surjection, $\varpi^* : C^\infty(Y,X) \to C^\infty(L,X)$ is a smooth injection. Now we are left to show that a plot in $\mathrm{Im}\,\varpi^* \subset C^\infty(L,X)$ can be pulled back to $C^\infty(Y,X)$: let $P : U \to \mathrm{Im}\,\varpi^* \subset C^\infty(L,X)$ be a plot. Then, for any $\mathbb{u} \in U$, there is $F_{\mathbb{u}} \in C^\infty(Y,X)$ such that $P(\mathbb{u}) = F_{\mathbb{u}}{\circ}\varpi$. For $t, s \in L$ with $\varpi(t) = \varpi(s)$ and $\mathbb{u} \in U$, the adjoint $\widehat{P} : U{\times}L \to X$ of P is a smooth map satisfying $\widehat{P}(\mathbb{u},t) = P(\mathbb{u})(t) = F_{\mathbb{u}}{\circ}\varpi(t) = F_{\mathbb{u}}{\circ}\varpi(s) = P(\mathbb{u})(s) = \widehat{P}(\mathbb{u},s)$. Thus the smooth map $\widehat{P} : U \times L \to X$ induces a map $\widehat{Q} : U \times Y \to X$ such that $\widehat{Q}{\circ}(1 \times \varpi) = \widehat{P}$, where $1 \times \varpi : U \times L \to U \times Y$ is a subduction by Lemma 1.1. By Lemma 1.2, $\widehat{Q}$ is smooth and hence its adjoint $Q : U \to C^\infty(Y,X)$ is a plot satisfying $(\varpi^*{\circ}Q(\mathbb{u}))(t) = Q(\mathbb{u})(\varpi(t)) = \widehat{Q}(\mathbb{u},\varpi(t)) = \widehat{Q}{\circ}(1 \times \varpi)(\mathbb{u},t) = \widehat{P}(\mathbb{u},t) = P(\mathbb{u})(t)$, $(\mathbb{u},t) \in U \times L$, which implies $\varpi^*{\circ}Q = P$, and we have done. □

We further assume that Y is a diffeological quotient L/ϖ_{set} by a relation ϖ_{set} on L, i.e, $Y = \{\,\widehat{\varpi}_{set}(t) \mid t \in L\}$ and $\varpi(t) = \widehat{\varpi}_{set}(t)$, where $\widehat{\varpi}_{set}$ is an equivalence relation on L generated by ϖ_{set}. Hence $\varpi{\circ}\varpi_{set} = \varpi$ as relations from L to Y.

PROPOSITION 1.4. $\mathrm{Im}\,\varpi^* = \{f \in C^\infty(L,X) \mid f{\circ}\varpi_{set} = f|_{\mathrm{Dom}\,\varpi_{set}}$ *as relations*$\}$.

PROOF. We show that $\{f \in C^\infty(L,X) \mid f\circ\varpi_{set} = f|_{\mathrm{Dom}\,\varpi_{set}}\} \subset \mathrm{Im}\,\varpi^*$: let $f \in C^\infty(L,X)$. If $f\circ\varpi_{set} = f|_{\mathrm{Dom}\,\varpi_{set}}$ as relations, then we get $f\circ\widehat{\varpi}_{set} = f$ as relations, and hence f induces a map $g : Y \to X$ such that $f = g\circ\varpi$. By Lemma 1.2, g is smooth, and hence $f \in \mathrm{Im}\,\varpi^*$. The converse is clear by $\varpi\circ\varpi_{set} = \varpi|_{\mathrm{Dom}\,\varpi_{set}}$. □

If the relation ϖ_{set} is a continuous idempotent on L, then Y is topologically the same as $\mathrm{Im}\,\varpi_{set} \subset L$, while Y might not be a diffeological subspace of L.

2. Cubic complex in Topology

In Topology, we use the symbol $\mathbb{I}_{top}$ for the topological subspace $[0,1]$ of $\mathbb{R}$. We remark that the topology of $\mathbb{I}_{top} = [0,1]$ is the same as the quotient topology induced by a continuous map $\pi_{top} : \mathbb{R} \to \mathbb{I}_{top}$ given by $\pi_{top}(t) = \pi_{set}(t) \in [0,1]$.

Now we introduce a generalised notion of a simplicial or cubical complex using an idea of a cubic set: a q-cubic set σ in $\mathbb{R}^n$ is defined as a convex body in some affine subspace L_σ in $\mathbb{R}^n$, inductively on q, $-1 \le q \le n$ (see also [**II19**]).

(1) The -1-cubic set in $\mathbb{R}^n$ is the empty set $\emptyset \subset \mathbb{R}^n$. In this case, $L_\emptyset = \emptyset$.
(2) A 0-cubic set in $\mathbb{R}^n$ is a point $\mathbb{p} \in \mathbb{R}^n$. In this case, $L_{\mathbb{p}} = \{\mathbb{p}\}$.
(3) Let σ_1 and σ_2 be respectively q_1-cubic and q_2-cubic sets in $\mathbb{R}^n$ with $-1 \le q{-}1 \le q_1{+}q_2 \le q \le n$, where σ_1 and σ_2 are convex bodies in affine subspaces L_1 and L_2, respectively. Let V_1 and V_2 be vector subspaces of $\mathbb{R}^n$ such that $V_1 \cap V_2 = \{\mathbb{0}\}$, $L_1 = \mathbb{a}_1 + V_1$ and $L_2 = \mathbb{a}_2 + V_2$ for some $\mathbb{a}_1 \in \sigma_1$ and $\mathbb{a}_2 \in \sigma_2$.

($q_1 + q_2 = q{-}1$ and $L_1 \cap L_2 = \emptyset$ (or $\mathbb{a}_2 - \mathbb{a}_1 \notin V_1 + V_2$))] The subset $\sigma_1 * \sigma_2 = \{(1{-}t)\cdot\mathbb{x} + t\cdot\mathbb{y}\,;\ \mathbb{x} \in \sigma_1,\ \mathbb{y} \in \sigma_2,\ t \in I_{top}\} \subset \mathbb{R}^n$ is a q-cubic set in $\mathbb{R}^n$. In this case, we have a relative homeomorphism $\phi_{\sigma_1,\sigma_2} : (\sigma_1 \times I_{top} \times \sigma_2, \sigma_1 \times \{0,1\} \times \sigma_2) \to (\sigma_1 * \sigma_2, \sigma_1 \amalg \sigma_2)$ given by $\phi_{\sigma_1,\sigma_2}(\mathbb{x}, t, \mathbb{y}) = (1{-}t)\cdot\mathbb{x} + t\cdot\mathbb{y}$.

($q_1 + q_2 = q$ and $L_1 \cap L_2 \neq \emptyset$ (or $\mathbb{a}_2 - \mathbb{a}_1 \in V_1 + V_2$))] Let $L_1 \cap L_2 = \{\mathbb{a}\}$, $\mathbb{a} \in \mathbb{R}^n$. Then the subset $\sigma_1\, {}_{L_1}\!\!\times_{L_2} \sigma_2 = \{\mathbb{x} + \mathbb{y} - \mathbb{a}\,;\ \mathbb{x} \in \sigma_1,\ \mathbb{y} \in \sigma_2\}$ is a q-cubic set in $\mathbb{R}^n$. In this case, we have a homeomorphism $\psi_{\sigma_1,\sigma_2} : \sigma_1 \times \sigma_2 \to \sigma_1\, {}_{L_1}\!\!\times_{L_2} \sigma_2$ given by $\psi_{\sigma_1,\sigma_2}(\mathbb{x}, \mathbb{y}) = \mathbb{x} + \mathbb{y} - \mathbb{a}$.

For each $n \ge 0$ and q with $-1 \le q \le n$, we denote by $C(n)^q$ the set of all q-cubic sets in $\mathbb{R}^n$ and $C(n) = \{\emptyset\} \cup \bigcup_{q \ge 0} C(n)^q$. Then the above construction yields two natural products: the join $* : C(n)^q \times C(n')^{q'} \to C(n{+}n'{+}1)^{q+q'+1}$ induced by (2) above using $\mathbb{R}^n \approx \mathbb{R}^n \times \{\mathbb{0}\} \times \{\mathbb{0}\} = V_1 \subset \mathbb{R}^n \times \mathbb{R} \times \mathbb{R}^{n'} \supset V_2 = \{\mathbb{0}\} \times \{0\} \times \mathbb{R}^{n'} \approx \mathbb{R}^{n'}$ with $\mathbb{a}_t = (\mathbb{0}, t, \mathbb{0})$ for $t = 1, 2$, and the product $\times : C(n)^q \times C(n')^{q'} \to C(n{+}n')^{q+q'}$ induced by (2) above using $\mathbb{R}^n \approx \mathbb{R}^n \times \{\mathbb{0}\} = V_1 \subset \mathbb{R}^n \times \mathbb{R}^{n'} \supset V_2 = \{\mathbb{0}\} \times \mathbb{R}^{n'} \approx \mathbb{R}^{n'}$ with $\mathbb{a} = \mathbb{a}_1 = \mathbb{a}_2 = (\mathbb{0}, \mathbb{0})$.

The notion of a face of a cubic set is inductively given as follows.

(1) Let σ be a cubic set. Then the emptyset $\emptyset$ and σ itself are faces of σ.
(2) Let σ_1 and σ_2 be two cubic set. Then we have the following.
    (a) A face of $\sigma_1 * \sigma_2$ is expressed as $\tau_1 * \tau_2$ for some faces τ_1 and τ_2 of σ_1 and σ_2, respectively. Therefore $\sigma_1 = \sigma_1 * \emptyset$ and $\sigma_2 = \emptyset * \sigma_2$ are faces of $\sigma_1 * \sigma_2$.
    (b) A face of $\sigma_1\, {}_{L_1}\!\!\times_{L_2} \sigma_2$ is expressed as $\tau_1\, {}_{L_1}\!\!\times_{L_2} \tau_2$ for some faces τ_1 and τ_2 of σ_1 and σ_2, respectively.

We denote $\tau \prec \sigma$ if $\tau \in C(n)$ is a face of $\sigma \in C(n)$.

An ordered subset $\mathbb{K} \subset C(n)$ is called a cubic complex, if the followings hold.

(0) $\forall \tau, \sigma \in \mathbb{K}\ \ \tau \cap \sigma \in C(n)$, $\tau \cap \sigma \prec \tau$ and $\tau \cap \sigma \prec \sigma$.

(1) $\emptyset \in \mathbb{K}$, (2) $\forall \tau \in C(n)\ \forall \sigma \in \mathbb{K}\ \ \tau \prec \sigma \implies \tau \in \mathbb{K}$,

A subset $\mathbb{L} \subset \mathbb{K}$ with the following properties is called a cubic subcomplex of $\mathbb{K}$.

(1) $\emptyset \in \mathbb{L}$, (2) $\forall \tau \in \mathbb{K}\ \forall \sigma \in \mathbb{L}\ \ \tau \prec \sigma \implies \tau \in \mathbb{L}$.

Then we denote $\dim \mathbb{K} = \max\{\dim \sigma \mid \sigma \in \mathbb{K}\}$, where $\dim \sigma = q$ if $\sigma \in C(n)^q$.

For any q-cubic set $\sigma \in K$, $\mathbb{K}(\sigma) = \{\tau \in C(n) \mid \tau \leqq \sigma\}$ for $q \geq -1$ and $\mathbb{K}(\dot{\sigma}) = \{\tau \in C(n) \mid \tau \lneqq \sigma\}$ for $q \geq 0$ are cubic subcomplexes of $\mathbb{K}$.

Proposition 2.1. *For any q-cubic set $\sigma \in C(n)$, $q \geq 0$, we have $\partial\sigma = |\mathbb{K}(\dot{\sigma})|$.*

For any two cubic complexes $\mathbb{K} \subset C(n)$ and $\mathbb{L} \subset C(m)$, we obtain

(1) $\mathbb{K} * \mathbb{L} := \{\sigma * \tau \mid \sigma \in \mathbb{K}, \tau \in \mathbb{L}\} \subset C(n+m+1)$

(2) $\mathbb{K} \times \mathbb{L} := \{\sigma \times \tau \mid \sigma \in \mathbb{K}, \tau \in \mathbb{L}\} \subset C(n+m)$

Proposition 2.2. *For any two cubic sets $\sigma, \tau \in C(n)$ and 0-cubic sets a, b, we have $(\sigma * a) \times (\tau * b) \approx \mathbb{L} * c$, where $\mathbb{L} = (\sigma * a) \times \tau \cup \sigma \times (\tau * b)$ and $c = (a, b)$.*

For any cubic complexes $\mathbb{K} \subset C(n)$ and $\mathbb{K}' \subset C(m)$, an order-preserving map $\varphi : \mathbb{K} \to \mathbb{K}'$ is called a cubic map, if the following conditions are satisfied.

(1) $\varphi^{-1}(\emptyset) = \{\emptyset\}$, (2) $\forall \tau' \in \mathbb{K}'\ \forall \sigma \in \mathbb{K}\ \ \tau' \prec \varphi(\sigma) \implies \exists \tau \prec \sigma\ \varphi(\tau) = \tau'$. In particular, the image of a cubic map $\varphi : \mathbb{K} \to \mathbb{K}'$ is a cubic subcomplex of $\mathbb{K}'$.

Proposition 2.3. *Let $*$ be a 0-cubic set. The following maps are cubic maps.*

(1) *The trivial map $\phi : \mathbb{K} \to \mathbb{K}(*)$ given by $\phi(\emptyset) = \emptyset$ and $\phi(\tau) = \{*\}$, $\tau \in \mathbb{K} \smallsetminus \{\emptyset\}$.*

(2) *The natural inclusion $\phi : \mathbb{L} \hookrightarrow \mathbb{K}$ of cubic subcomplex $\mathbb{L}$ of $\mathbb{K}$.*

(3) *For two cubic maps $\phi_1 : \mathbb{K}(1) \to \mathbb{K}'_1$ and $\phi_2 : \mathbb{K}(2) \to \mathbb{K}'_2$, maps*

 (a) *$\phi : \mathbb{K}(1) * \mathbb{K}(2) \to \mathbb{K}'_1 * \mathbb{K}'_2$ given by $\phi(\tau_1 * \tau_2) = \phi_1(\tau_1) * \phi_2(\tau_2)$ and*

 (b) *$\psi : \mathbb{K}(1) \times \mathbb{K}(2) \to \mathbb{K}'_1 \times \mathbb{K}'_2$ given by $\psi(\tau_1 \times \tau_2) = \phi_1(\tau_1) \times \phi_2(\tau_2)$.*

For a cubic complex $\mathbb{K} \subset C(n)$, $n \geq 0$, we denote $\mathbb{K}^q = \{\sigma \in K\,;\, \sigma \text{ is } q\text{-cubic}\}$, $q \geq -1$ and by $|\mathbb{K}| = \bigcup_{\sigma \in \mathbb{K}} \sigma$ the polyhedron in $\mathbb{R}^n$ associated to $\mathbb{K}$. For a cubic complexes $\mathbb{K}$ and $\mathbb{L}$, a continuous map $f : |\mathbb{L}| \to |\mathbb{K}|$ is called cubic, if there exists a map $\varphi : \mathbb{L} \to \mathbb{K}$ such that $f|_\tau : \tau \to \varphi(\tau) \subset |\mathbb{K}|$ for any $\tau \in \mathbb{L}$. Such a map f is often denoted by $|\varphi| : |\mathbb{L}| \to |\mathbb{K}|$.

3. Smooth cubic complex

In Diffeology, we use the symbol $\mathbb{I}$ for a special diffeological space: let $\mathbb{I} = \mathbb{R}/\pi_{set}$ be the diffeological quotient of $\mathbb{R}$, where $\mathbb{I} = [0,1]$ as a set, with a subduction $\pi : \mathbb{R} \to \mathbb{I}$ given by $\pi(t) = \pi_{set}(t) \in [0,1]$, so that we obtain $\dim \mathbb{I} = 1$ diffeologically and $\pi \circ \pi_{set} = \pi$. The underlying topology (D-Topology in [**IZ13**, 2.8]) of $\mathbb{I}$ is the same as $\mathbb{I}_{top} \subset \mathbb{R}$, while $\mathbb{I} \hookrightarrow \mathbb{R}$ can not be an induction.

Theorem 3.1. *By Propositions 1.3 and 1.4, $\pi^* : C^\infty(\mathbb{I}, X) \to \mathcal{P}(X)$ is a diffeomorphism, and so $\mathbb{I}$ represents the functor $\mathcal{P}()$. Moreover we have $\dim \mathbb{I} = 1$.*

As is well-known, there is a smooth function $\lambda : \mathbb{R} \to \mathbb{R}$ enjoying

(1) $\lambda(t) = 0$, $t \leq 0$, (2) $\lambda(t) + \lambda(1{-}t) = 1$, (3) $\lambda'(t) > 0$, $0 < t < 1$. Then by (1) and (2) above, we have $\lambda(t) = 1$, $t \geq 1$ as well and hence $\lambda \circ \pi_{set} = \lambda$. Thus λ induces a smooth injection $\hat{\lambda} : \mathbb{I} \to \mathbb{R}$ satisfying $\lambda = \hat{\lambda} \circ \pi$.

DEFNINITION 3.2. We give a diffeology on a q-cubic set $\sigma \in C(n)$ by a subduction $\pi_\sigma : \mathbb{R}^q \to \sigma$, equipped with a smooth injection $\hat{\lambda}_\sigma : \sigma \to \mathbb{R}^n$ induced from some smooth map $\lambda_\sigma : \mathbb{R}^q \to \mathbb{R}^n$ such that $\lambda_\sigma = \hat{\lambda}_\sigma \circ \pi_\sigma$, by induction on q.

(1) 0-cubic set is a one point space which has the trivial diffeology.
(2) Let σ_1 and σ_2 be cubic set with subductions $\pi_i : \mathbb{R}^{q_i} \to \sigma_i$, $i = 1, 2$, equipped with smooth injections $\hat{\lambda}_i : \sigma_i \to \mathbb{R}^n$ induced from some smooth maps $\lambda_i : \mathbb{R}^{q_i} \to \mathbb{R}^n$ such that $\lambda_i = \hat{\lambda}_i \circ \pi_i$, $i = 1, 2$.
  (a) If $\sigma = \sigma_1 * \sigma_2 \subset \mathbb{R}^n$, we have a subduction $\pi_\sigma = \phi_{\sigma_1,\sigma_2} \circ (\pi_1 \times \pi \times \pi_2) : \mathbb{R}^{q_1} \times \mathbb{R} \times \mathbb{R}^{q_2} \to \sigma_1 * \sigma_2$ and a smooth map $\lambda_\sigma = \phi_{\sigma_1,\sigma_2} \circ (\lambda_1 \times \lambda \times \lambda_2) : \mathbb{R}^{q_1} \times \mathbb{R} \times \mathbb{R}^{q_2} \to \mathbb{R}^n$, which induces a smooth injection $\hat{\lambda}_\sigma : \sigma \to \mathbb{R}^n$.
  (b) If $\sigma = \sigma_1 {}_{L_1}{\times}_{L_2} \sigma_2 \subset \mathbb{R}^n$, we have a subduction $\pi_\sigma = \psi_{\sigma_1,\sigma_2} \circ (\pi_1 \times \pi_2)$ and a smooth map $\lambda_\sigma = \psi_{\sigma_1,\sigma_2} \circ (\lambda_1 \times \lambda_2) : \mathbb{R}^{q_1} \times \mathbb{R}^{q_2} \to \mathbb{R}^n$, which induces a smooth injection $\hat{\lambda}_\sigma : \sigma \to \mathbb{R}^n$.

For any q-cubic set σ, we clearly have $\mathbf{dim}\,\sigma = q$ diffeologically, and by Proposition 1.3, we obtain that $\pi_\sigma^* : C^\infty(\sigma, X) \to C^\infty(\mathbb{R}^q, X)$ is an induction.

PROPOSITION 3.3. *σ is a smooth neighbourhood deformation retract of $\sigma * \tau$.*

PROOF. Since there is a subduction $\pi_\sigma = \phi_{\sigma_1,\sigma_2} \circ (\pi_1 \times \pi \times \pi_2) : \mathbb{R}^{q_1} \times \mathbb{R} \times \mathbb{R}^{q_2} \to \sigma_1 * \sigma_2$, we have a deformation $h_s : \mathbb{R}^{q_1} \times (-\infty, 1) \times \mathbb{R}^{q_2} \to \mathbb{R}^{q_1} \times \mathbb{R} \times \mathbb{R}^{q_2}$ given by

$$h_s(\mathbb{x}, t, \mathbb{y}) = (\mathbb{x}, t{-}\lambda(s), \mathbb{y}), \quad s \in \mathbb{R},$$

which is clearly smooth and thus inducing a smooth deformation $\hat{h}_s : O \to \sigma * \tau$, $s \in \mathbb{R}$, where $O = \sigma * \tau \smallsetminus \tau$. By definition, it induces a smooth deformation of O relative to σ, and σ is a smooth deformation retract of O a neighbourhood of σ. □

For a cubic complex $\mathbb{K} \subset C(n)$, we introduce a smooth structure on the polyhedron $|\mathbb{K}|$ as $|\mathbb{K}| = \operatorname{\mathbf{colim}}_{\sigma \in \mathbb{K}} \sigma$, which is called a smooth cubic polyhedron. Then by definition, we obtain a smooth injection $\hat{\lambda}_{\mathbb{K}} : |\mathbb{K}| \to \mathbb{R}^n$ by collecting smooth maps $\hat{\lambda}_\sigma : \sigma \to \mathbb{R}^n$, $\sigma \in \mathbb{K}$. Then we instantly see that $\mathbf{dim}\,|\mathbb{K}| = \mathbf{dim}\,\mathbb{K}$ diffeologically.

REMARK 3.4. Following the above definition, we obtain a cube $\mathbb{I}^n = (\mathbb{R}/\pi_{set})^n$ as a diffeological quotient of $\mathbb{R}^n$ which instantly implies $\mathbf{dim}\,\mathbb{I}^n = n$ diffeologically. The cube $\mathbb{I}^n$ is set-theoretically the same as $[0, 1]^n$, while its diffeology is different from the induced diffeology from $\mathbb{R}^n$ which is used in [**HS20**].

4. Associahedra as cubic complexes

Let us introduce associahedra $K_n \subset \mathbb{R}^n$, $n \geq 1$ in $\mathsf{Topology}$ as follows, which is slightly modified from the definition by Stasheff (see [**Sta63**], [**IM89**] or [**Iwa12**]):

$$K_n = \{(t_1, \dots, t_n) \,\big|\, t_1 = 0 \leq t_k \leq k{-}1{-}\textstyle\sum_{i=1}^{k-1} t_i \ (1 < k < n),\ t_n = n{-}1 - \sum_{i=1}^{n-1} t_i\},$$

or equivalently, we can describe the associahedron as follows.

$$K_n' = \{(u_1, \dots, u_n) \,|\, 0 = u_1 \leq u_2 \leq \cdots \leq u_{n-1} \leq u_n = n{-}1,\ u_k \leq k{-}1 \ (1 < k < n)\}.$$

Let H_1^{n-2}: $x_1 + \cdots + x_n = n{-}1, x_1 = 0$ be an affine space where K_n is a convex body.

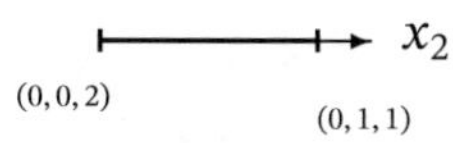

$(K_3 \subset H_1^1 \approx \mathbb{R})$

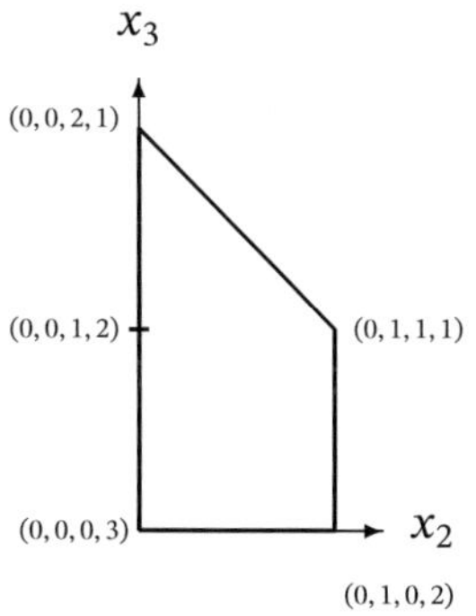

$(K_4 \subset H_1^2 \approx \mathbb{R}^2)$

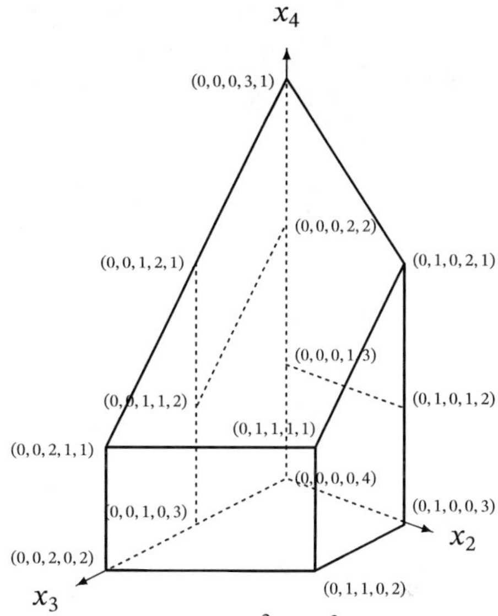

$(K_5 \subset H_1^3 \approx \mathbb{R}^3)$

Let $A(n) = \{(k,r,s) \in \mathbb{N} \mid 1 \leq k \leq r,\ 2 \leq s = n-r+1 \leq n-1\}$. Then the boundary of K_n is the union of faces corresponding to elements in $A(n)$, given as follows.

$$L_k(r,s) = \left\{ (t_1, \dots, t_n) \in K_n \,\middle|\, (t_k, \dots, t_{k+s-2}, t) \in K_s,\ t_{k+s-1} \geq t = s-1 - \textstyle\sum_{i=k}^{k+s-2} t_{k+i-1} \right\}.$$

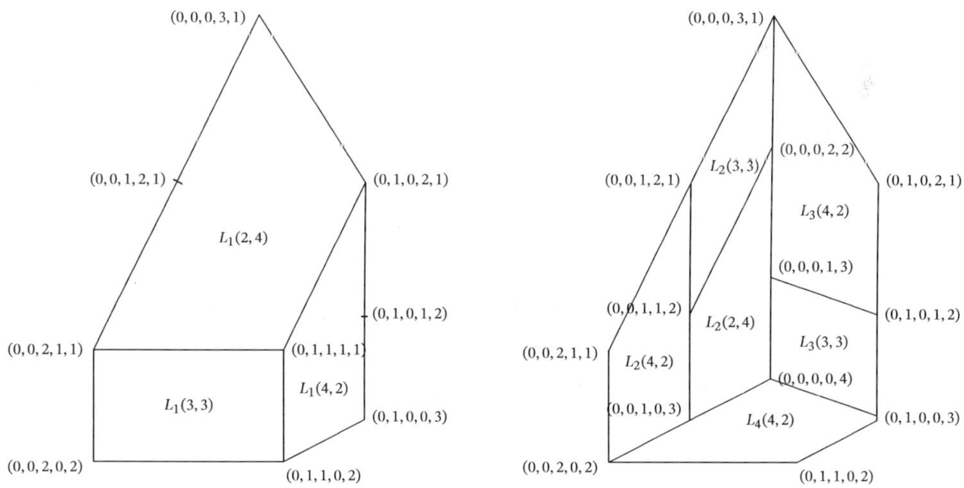

$(\partial K_5 \approx S^2)$

Following Stasheff [**Sta63**] (see also [**IM89**] or [**Iwa12**]), we introduce face operators $\partial_k : \mathbb{R}^r \times \mathbb{R}^s \to \mathbb{R}^n$, $r+s=n+1$, $1 \leq k \leq r$, as the following linear maps.

$$\partial_k((t_1, \dots, t_r), (u_1, \dots, u_s)) = \begin{cases} (u_1, \dots, u_{s-1}, u_s{+}t_1, t_2, \dots, t_r), & k = 1, \\ (t_1, \dots, t_{k-1}, u_1, \dots, u_{s-1}, u_s{+}t_k, t_{k+1}, \dots, t_r), & 2 \leq k \leq r, \end{cases}$$

If we restrict ∂_k to $K_r \times K_s$, then we obtain $\partial_k : K_r \times K_s \approx L_k(r,s) \subset K_n \subset \mathbb{R}^n$. Now we choose an interior point of K_n, as $\mathbb{b}_n = (0, {}^1\!/_2, \dots, {}^1\!/_2, {}^n\!/_2) \in K_n$. Then we see that K_n and $L_k(r,s)$ are characterised by the following two conditions.

(1) $\partial_k : K_r \times K_s \xrightarrow{\approx} L_k(r,s)$, $(k,r,s) \in A(n)$, (2) $K_n = \bigcup_{(k,r,s)\in A(n)} L_k(r,s) * \mathbb{b}_n$.

EXAMPLE 4.1. $K_2 = \{\mathbb{b}_2\}$, $K_3 = L_2(2,2) * \{\mathbb{b}_3\} * L_1(2,2)$, and $K_4 = L_2(2,3) * \mathbb{b}_4 \cup L_2(3,2) * \mathbb{b}_4 \cup L_3(3,2) * \mathbb{b}_4 \cup L_1(2,3) * \mathbb{b}_4 \cup L_1(3,2) * \mathbb{b}_4$.

Firstly, K_n is the realisation of a cubic set, namely $\mathbb{K}(n)$, which begins with

(1) $\mathbb{K}(1) = \{\emptyset\}$.

Secondly, we define cubic complexes $\mathbb{L}_k(r,s)$, $(k,r,s) \in A(n)$, assuming that $\mathbb{K}(r)$ and $\mathbb{K}(s)$ are given: let us denote $\mathbb{0}_n = (0,\dots,0) \in \mathbb{R}^n$ and $H_n = \{(0, x_2, \dots, x_n) \in \mathbb{R}^n \mid x_2 + \cdots + x_n = 0\}$. Then we have the following two linear subspaces of $\mathbb{R}^n$:

$$V_1 = \partial_k(H_r \times \{\mathbb{0}_s\}) \quad \text{and}$$
$$V_2 = \partial_k(\{\mathbb{0}_r\} \times H_s) \quad \text{with} \quad V_1 \cap V_2 = \{\mathbb{0}_n\}.$$

Let $\mathbb{a} = \partial_k(\mathbb{b}_r, \mathbb{b}_s) \in \mathbb{R}^n$. Since $K_n \subset H_n + \mathbb{b}_n$, we obtain

$$L_1 = V_1 + \mathbb{a} = \partial_k((H_r + \mathbb{b}_r) \times \{\mathbb{b}_s\}) \supset \partial_k(K_r \times \{\mathbb{b}_s\}) \approx K_r,$$
$$L_2 = V_2 + \mathbb{a} = \partial_k(\{\mathbb{b}_r\} \times (H_s + \mathbb{b}_s)) \supset \partial_k(\{\mathbb{b}_r\} \times K_s) \approx K_s.$$

Then we clearly have that $L_1 \cap L_2 = \{\mathbb{a}\}$. Hence, when $\mathbb{K}(r)$ and $\mathbb{K}(s)$ have already been defined, we must obtain the following.

(2) $\mathbb{L}_k(r,s) = \partial_k(\mathbb{K}(r) \times \{\mathbb{b}_s\}) \;{}_{L_1}\!\times_{L_2} \partial_k(\{\mathbb{b}_r\} \times \mathbb{K}(s))$, $(k,r,s) \in A(n)$.

Thirdly, we define a cubic complex $\mathbb{K}(n)$: let $V_1' = V_1 + V_2$, $V_2' = \{\mathbb{0}_n\}$, $\mathbb{a}_1' = \mathbb{a}$ and $\mathbb{a}_2' = \mathbb{b}_n$. Then $L_k(r,s) \subset V_1' + \mathbb{a}_1'$, $\mathbb{b}_n \in V_2' + \mathbb{a}_2'$ and $V_1' \cap V_2' = \emptyset$. Hence, when $\mathbb{L}_k(r,s)$, $(k,r,s) \in A(n)$, have already been defined, we must obtain the following.

(3) $\mathbb{K}(n) = \bigcup_{(k,r,s)\in A(n)} \mathbb{L}_k(r,s) * \mathbb{b}_n$.

Thus $\mathbb{K}(n)$, $n \geq 1$ can inductively be defined by the above formulas (1), (2) and (3). Then we can easily see that there is a natural homeomorphism from $|\mathbb{K}(n)|$ to K_n.

By slightly modifying the definition in Stasheff [**Sta63**], we obtain the following degeneracy operators $s_j : K_n \to K_{n-1}$, $1 \leq j \leq n$, $n \geq 2$ (see [**IM89**] or [**Iwa12**]).

$$s_j((1-t)\cdot\partial_k(\rho,\sigma) + t\cdot\mathbb{b}_n) = (1-t)\cdot s_j \circ \partial_k(\rho,\sigma) + t\cdot\mathbb{b}_{n-1},$$

where $s_j \circ \partial_k : K_r \times K_s \to K_{n-1}$ is given by

$$s_j \circ \partial_k(\rho,\sigma) = \begin{cases} \partial_{k-1}(r-1,s)\circ(s_j\rho \times \sigma), & j < k,\ r > 2, \\ \sigma, & j = 1,\ k = 2,\ r = 2, \\ \partial_k(r,s-1)\circ(\rho \times s_{j-k+1}\sigma), & k \leq j < k+s,\ r < n-1, \\ \rho, & k \leq j \leq k+1,\ r = n-1, \\ \partial_{k-1}(r-1,s)\circ(s_{j-s+1}\rho \times \sigma), & k+s \leq j \leq n,\ r > 2, \\ \sigma, & j = n,\ k = 1,\ r = 2. \end{cases}$$

Thus we may suppose that $s_j : K_n \to K_{n-1}$, $1 \leq j \leq n$, $n \geq 2$ is a realisation of a cubic map, which is denoted again by $s_j : \mathbb{K}(n) \to \mathbb{K}(n-1)$.

REMARK 4.2. J. L. Loday [**Lod07**, **Lod12**] gave a nice realisation of an associahedron and its triangulation using Tamari ordering on vertices, which could give a natural smooth structure on such triangulation in terms of trees.

5. A_∞-form in Diffeology

In this section, let us concentrate on our cubic complex of assosiahedra. The following is obtained by induction on n.

THEOREM 5.1. *$\partial_k : \mathbb{K}(r) \times \mathbb{K}(s) \to \mathbb{K}(n)$, $r + s = n + 1$, $1 \leq k \leq r$ and $s_j : \mathbb{K}(n) \to \mathbb{K}(n-1)$, $1 \leq j \leq n \geq 2$ are smooth cubic maps.*

We now state the smooth version of a strict unital A_∞-space: let G be a diffeological space with a base point $e \in G$, which is called a unit of G.

DEFNINITION 5.2 (Stasheff [**Sta63**]). *G is called a smooth (strict unital) A_∞-space, if there is a series of smooth maps $\{M(n) : |\mathbb{K}(n)| \times G^n \to G\}_{n\geq 2}$ (A_∞-form) satisfying*

(1) $M(n)(\partial_k(\rho, \sigma); g_1, \dots, g_n) = M(r)(\rho; g_1, \dots, M(s)(\sigma; g_k, \dots, g_{k+s-1}), \dots, g_n)$.
(2) $M(n)(\tau; g_1, \dots, g_{j-1}, e, g_{j+1}, \dots, g_n) = M(n-1)(s_j(\tau); g_1, \dots, g_{j-1}, g_{j+1}, \dots, g_n)$.

An A_∞-space allowing homotopy unit was introduced in [**Sta70**] as in the following definition, which shall be referred as h-unital A_∞-space in this article.

DEFNINITION 5.3 (Stasheff [**Sta70**]). *G is called a smooth h-unital A_∞-space, if there is a series of smooth maps $\{M(n) : |\mathbb{K}(n)| \times G^n \to G\}_{n\geq 2}$ (A_∞-form) satisfying*

(1) $M(n)(\partial_k(\rho, \sigma); g_1, \dots, g_n) = M(r)(\rho; g_1, \dots, M(s)(\sigma; g_k, \dots, g_{k+s-1}), \dots, g_n)$.
(2') $M_2(e, g) \simeq M_2(g, e) \simeq g$ the identity.

By [**Iwa12**, Theorem 1.4], these definitions are not the same but equivalent up to homotopy. In view of [**Iwa12**], we give two notions of internal precategories:

DEFNINITION 5.4. A pair of diffeological spaces $G = (G, X)$ is called an internal precategory, if it is equipped with three smooth (structure) maps

$$\sigma : G \to X, \quad \tau : G \to X, \quad \iota : X \to G$$

with relations $\sigma \circ \iota = \tau \circ \iota = 1$.

For an internal precategory $G = (G, X)$, we define $\{G^n; \sigma_n, \tau_n : G^n \to X\}_{n\geq 1}$ by induction:

(1) $G^1 = G$, $\sigma_1 = \sigma$ and $\tau_1 = \tau$ (n+1) $G^{n+1} = G^n \times_X G$, $\sigma_{n+1}(\chi, h) = \sigma_n(\chi)$ and $\tau_{n+1}(\chi, h) = \tau(h)$, where $G^n \times_X G$ is the pullback of τ_n and σ, that is, $G^n \times_X G = \{(\chi, h) \in G^n \times G \mid \tau_n(\chi) = \sigma(h)\}$ is a subspace of $G^n \times G$.

DEFNINITION 5.5. An internal precategory $G = (G, X)$ in Diffeology is called an internal (strict unital) smooth A_∞-category, if there is a series of smooth maps $\{M(n) : |\mathbb{K}(n)| \times G^n \to G\}_{n\geq 2}$ (A_∞-form) satisfying the conditions (0), (1) and (2):

(0) $\sigma \circ M(n)(\tau; g_1, \dots, g_n) = \sigma(g_1)$ and $\tau \circ M(n)(\tau; g_1, \dots, g_n) = \tau(g_n)$.

(1) $M(n)\circ(\partial_k{\times}1^n) = M(r)\circ(1{\times}1^{k-1}{\times_X}M(s){\times_X}1^{r-k})\circ(1{\times}T_k) : |\mathbb{K}(r)|{\times}|\mathbb{K}(s)|{\times} G^n \to G$,

$$\begin{array}{ccc}
|\mathbb{K}(r)| \times |\mathbb{K}(s)| \times G^n & \xrightarrow[\partial_k\times 1^n]{} & |\mathbb{K}(n)| \times G^n \\
\downarrow{\scriptstyle 1\times T_k} & & \\
|\mathbb{K}(r)| \times G^{k-1} \times_X (|\mathbb{K}(s)| \times G^s) \times_X G^{r-k} & & \downarrow{\scriptstyle M(n)} \\
\downarrow{\scriptstyle 1\times 1\times_X M(s)\times_X 1} & & \\
|\mathbb{K}(r)| \times G^r & \xrightarrow{M(r)} & G,
\end{array}$$

where $T_k : |\mathbb{K}(s)|{\times}G^n = |\mathbb{K}(s)|{\times}G^{k-1}\times_X G^s \times_X G^{r-k} \to G^{k-1}\times_X(|\mathbb{K}(s)|\times G^s)\times_X G^{r-k}$ is given by $T_k(a, \mathbb{x}, \mathbb{y}, \mathbb{z}) = (\mathbb{x}, a, \mathbb{y}, \mathbb{z})$.

(2) $M(n)\circ(1{\times}1^{j-1}{\times_X}\iota{\times_X}1^{n-j}) = M(n{-}1)\circ(s_j{\times}1^{j-1}{\times_X}1^{n-j}) : |\mathbb{K}(n)|{\times}G^{n-1} \to G$.

$$\begin{array}{ccc}
|\mathbb{K}(n)| \times G^{j-1} \times_X G^{n-j} & \longrightarrow & |\mathbb{K}(n)| \times G^n \\
\downarrow{\scriptstyle s_j\times 1\times_X 1} & & \downarrow{\scriptstyle M(n)} \\
|\mathbb{K}(n{-}1)| \times G^{j-1} \times_X G^{n-j} & \xrightarrow[M(n-1)]{} & G,
\end{array}$$

DEFNINITION 5.6. We call $G = (G, X)$ an internal smooth h-unital A_∞-category in Diffeology, if there is a series of smooth maps $\{M(n) : |\mathbb{K}(n)|{\times}G^n \to G\}_{n\geq 2}$ (A_∞-form) satisfying the conditions (0) and (1) in Definition 5.5 and (2') below.

(2') $M(2)|_{\{\mathbb{b}_2\}\times\{e\}\times G} \simeq M(1)\circ(s_1 \times 1)$ and $M(2)|_{\{\mathbb{b}_2\}\times G\times\{e\}} \simeq M(1)\circ(s_2 \times 1)$ the identity.

6. Path space with usual concatenation

In this section, we work in Diffeology making some additional assumptions on the diffeology $\mathcal{D}(X)$ of a diffeological space X. Let us denote by $\mathcal{F}(X) = C^\infty(X, \mathbb{R})$ the set of smooth functions on X to $\mathbb{R}$. Then we define a superset $\mathcal{D}'(X)$ of $\mathcal{D}(X)$ as the set of parametrisations P on X satisfying that $\varphi\circ P$ is smooth for any $\varphi \in \mathcal{F}(X)$.

DEFNINITION 6.1 (J. Watts [**Wat12**] (see also [**IZ13**])). A diffeological space X is said to be reflexive, if $\mathcal{D}(X) = \mathcal{D}'(X)$.

THEOREM 6.2 ([**IZ13**, Exercise 79]). *A manifold is reflexive in* Diffeology.

Let us introduce the following notion for a point $\mathbb{a}$ in X.

DEFNINITION 6.3. A diffeological space X is said to be reflexive at $\mathbb{a} \in X$, if there is a D-open neighbourhood U of $\mathbb{a}$, which is reflexive as a diffeological space.

EXAMPLE 6.4. A manifold is reflexive at any point, and hence a smooth CW complex of finite dimension is reflexive at an interior point of a top cell.

EXAMPLE 6.5. Let $f : \mathbb{R} \to \mathbb{I}$ be a map given by

$$f(t) = \pi(\sqrt{\max\{0, t\}}),$$

which is not smooth at $t = 0$. In fact, if f is smooth, f can be expressed as $f = \pi \circ \phi$ near $t = 0$ by a smooth map $\phi : \mathbb{R} \to \mathbb{R}$. Then we have $\phi(t) = \sqrt{t}$ for small $t > 0$, and hence $\phi'(0) = \lim_{t \to +0} \phi'(t) = +\infty$. It contradicts to the smoothness of ϕ at $t = 0$. On the other hand, for any smooth map $g : \mathbb{I} \to \mathbb{R}$, the composition $\psi = g \circ \pi : \mathbb{R} \to \mathbb{R}$ is smooth on $\mathbb{R}$, and constant on $(-\infty, 0]$. Thus we have

$$\lim_{t \to +0} \psi^{(r)}(t) = \psi^{(r)}(0) = \lim_{t \to -0} \psi^{(r)}(t) = 0 \quad \text{for all } r \geq 1.$$

By applying L'Hôpital's rule many times, we obtain that $\lim_{t \to +0} \psi^{(r)}(t)/t^n = 0$ for all $r, n \geq 1$. Then by induction, we can express $(\psi \circ f)^{(r)}(t)$ as the following form:

$$(\psi \circ f)^{(r)}(t) = \sum_{j=0}^{r} P_{r,j}(1/\sqrt{t}) \cdot \psi^{(j)}(\sqrt{t}), \; t > 0, \quad \text{for all } r > 1,$$

where $P_{r,j}(x)$ is a polynomial on x. Again by applying L'Hôpital's rule, we obtain that $(\psi \circ f)^{(r)}(0)$ exists and equals to $\lim_{t \to 0} (\psi \circ f)^{(r)}(t) = 0$ for all $r \geq 1$, and hence $\psi \circ f$ is smooth at $t = 0$. Thus $f \in \mathcal{D}'(\mathbb{I})$ while $f \notin \mathcal{D}(\mathbb{I})$. So, $\mathbb{I}$ is not reflexive (at $t = 0$). In contrast, $\mathbb{I}$ is reflexive at any point $t \in (0, 1) \subset \mathbb{I}$.

Let us recall the following diffeological subspace of $\mathsf{Paths}(X)$.

$$\mathcal{P}(X) = \{ u \in \mathsf{Paths}(X) \mid u \circ \pi_{set} = u \} \cong C^\infty(\mathbb{I}, X).$$

From now on, we often identify $\mathcal{P}(X)$ with $C^\infty(\mathbb{I}, X)$ without mentioning π^*.

PROPOSITION 6.6. *For $u \in \mathcal{P}(X) = C^\infty(\mathbb{I}, X)$ and $\varphi \in C^\infty(X, \mathbb{R})$, we obtain*

$$\frac{d^n(\varphi \circ u)}{dt^n} = 0 \quad \textit{on} \quad (-\infty, 0] \cup [1, \infty) \quad \textit{for all} \quad n > 0.$$

PROOF. Since u is constant on $(-\infty, 0)$ and on $(1, \infty)$, so is $\varphi \circ u$ and we obtain

$$\frac{d(\varphi \circ u)}{dt} = 0 \quad \text{on} \quad (-\infty, 0) \cup (1, \infty).$$

Hence by iterating differentiations, we obtain

$$\frac{d^n(\varphi \circ u)}{dt^n} = 0 \quad \text{on} \quad (-\infty, 0) \cup (1, \infty) \quad \text{for all} \quad n > 0.$$

Because all the derivatives are continuous, we obtain the propostion. □

Since $\mathcal{P}(X) = C^\infty(\mathbb{I}, X)$ is the left adjoint of the product functor with $\mathbb{I}$, $\mathcal{P}(X)$ is an internal precategory with the following smooth structure maps.

$$\begin{aligned}
\sigma = \sigma_X &: \mathcal{P}(X) \to X \iff \sigma_X(u) = u(0), \quad u \in \mathcal{P}(X),\\
\tau = \tau_X &: \mathcal{P}(X) \to X \iff \tau_X(u) = u(1), \quad u \in \mathcal{P}(X),\\
\iota = \iota_X &: X \to \mathcal{P}(X) \iff \iota_X(\mathbb{x})(t) = \mathbb{x}, \quad \mathbb{x} \in X \ \&\ t \in \mathbb{R}.
\end{aligned}$$

The internal precategory $\mathcal{P}(X)$ is equipped with the following structure map $\mu = \mu_X$ called a concatenation, which could fail to be well-defined in general:

$$\begin{aligned}
\mu = \mu_X &: \mathcal{P}(X) \times_X \mathcal{P}(X) \dashrightarrow \mathcal{P}(X)\\
&\iff \mu_X(u, v)(t) = \begin{cases} u(2t), & t \leq 1/2,\\ v(2t-1), & t \geq 1/2, \end{cases} \quad t \in \mathbb{R},
\end{aligned}$$

where $\mathcal{P}(X)\times_X \mathcal{P}(X)$ denotes the pullback of $\tau : \mathcal{P}(X)\to X$ and $\sigma : \mathcal{P}(X)\to X$:

$$\mathcal{P}(X)\times_X \mathcal{P}(X) = \{(u,v)\in \mathcal{P}(X)\times\mathcal{P}(X)\mid \tau(u)=\sigma(v)\}.$$

DEFINITION 6.7. We define the following subspaces of $\mathcal{P}(X)\subset C^\infty(\mathbb{R},X)$.

(1) $\mathcal{P}(X;\mathbb{a},\mathbb{b}) = \{u\in\mathcal{P}(X)\mid u(0)=\mathbb{a}\ \&\ u(1)=\mathbb{b}\}$.
(2) $\mathcal{L}(X,\mathbb{a}) = \mathcal{P}(X;\mathbb{a},\mathbb{a})$ and $\mathcal{L}(X) = \mathcal{P}(X;*,*)$.

Let $\mathbb{a},\mathbb{b},\mathbb{c}\in X$. In the remainder of this section, we assume that X is reflexive at $\mathbb{b}\in X$. For any two paths $u\in\mathcal{P}(X;\mathbb{a},\mathbb{b})$ and $v\in\mathcal{P}(X;\mathbb{b},\mathbb{c})$, and any function $\varphi\in\mathcal{F}(X)$, both u and v are smooth, and $\varphi\circ u$ and $\varphi\circ v$ are smooth functions in the ordinary sense. The function $\varphi\circ\mu(u,v)$ can be described as follows:

$$\text{(A)}\qquad \varphi\circ\mu(u,v)(t) = \mu_{\mathbb{R}}(\varphi\circ u,\varphi\circ v)(t) = \begin{cases}\varphi\circ u(2t), & t\le 1/2,\\ \varphi\circ v(2t{-}1), & t\ge 1/2.\end{cases}\quad t\in\mathbb{R}.$$

Then, $\varphi\circ\mu(u,v) = \mu_{\mathbb{R}}(\varphi\circ u,\varphi\circ v)$ is smooth: by Proposition 6.6, we obtain

$$\frac{d^n(\varphi\circ u)}{dt^n} = 0 \ \text{ and } \ \frac{d^n(\varphi\circ v)}{dt^n} = 0 \ \text{ on } \ (-\infty,0]\cup[1,\infty) \ \text{ for all } \ n>0.$$

So we define $\varphi_1,\varphi_2\in\mathcal{P}(X)$ by $\varphi_1(t)=\varphi\circ u(2t)$ and $\varphi_2(t)=\varphi\circ v(2t{-}1)$, $t\in\mathbb{R}$ to obtain $\varphi_1 = \varphi\circ\mu(u,k(\mathbb{b})) = \mu_{\mathbb{R}}(\varphi\circ u,k(\varphi(\mathbb{b})))$ and $\varphi_2 = \varphi\circ\mu(k(\mathbb{b}),v) = \mu_{\mathbb{R}}(k(\varphi(\mathbb{b})),\varphi\circ v)$, where $k(\mathbb{b})$ and $k(\varphi(\mathbb{b}))$ are the constant maps at $\mathbb{b}\in X$ and $\varphi(\mathbb{b})\in\mathbb{R}$, respectively. Since φ_1 and φ_2 are compositions of smooth maps, they are smooth satisfying

$$\frac{d^n\varphi_1}{dt^n} = 0 \ \text{ on } \ (-\infty,0]\cup[1/2,\infty) \ \text{ for all } \ n>0, \ \text{ and}$$

$$\frac{d^n\varphi_2}{dt^n} = 0 \ \text{ on } \ (-\infty,1/2]\cup[1,\infty) \ \text{ for all } \ n>0.$$

Since $\varphi\circ\mu(u,v)(t) = \mu_{\mathbb{R}}(\varphi\circ u,\varphi\circ v)(t) = \varphi_1(t)+\varphi_2(t)-\varphi(\mathbb{b})$, $t\in\mathbb{R}$ by the equation (A), the composition $\varphi\circ\mu(u,v) = \mu_{\mathbb{R}}(\varphi\circ u,\varphi\circ v)$ is also a smooth function satisfying

$$\frac{d^n(\varphi\circ\mu(u,v))}{dt^n} = 0 \ \text{ on } \ (-\infty,0]\cup\{1/2\}\cup[1,\infty) \ \text{ for all } \ n>0.$$

Here, let us recall that $\mu(u,v) : \mathbb{R}\to X$ is smooth on $\mathbb{R}\smallsetminus\{1/2\}$ and that there is a D-open neighbourhood $U\subset X$ of $\mathbb{b}$ such that U is reflexive, since X is reflexive at the point $\mathbb{b}$. Since $\varphi\circ\mu(u,v)$ is smooth for any smooth function $\varphi : X\to\mathbb{R}$, we obtain that $\mu(u,v)$ is smooth on the open set $\mu(u,v)^{-1}(U)\ni 1/2$. It means that $\mu(u,v)$ is a plot, and μ is well-defined.

THEOREM 6.8. *Let X be a diffeological space and reflexive at $\mathbb{b}$ with $\mathbb{a},\mathbb{b},\mathbb{c}\in X$. Then the concatenation $\mu=\mu_X : \mathcal{P}(X;\mathbb{a},\mathbb{b})\times\mathcal{P}(X;\mathbb{b},\mathbb{c})\to\mathcal{P}(X;\mathbb{a},\mathbb{c})$ is well-defined and smooth.*

PROOF. For any two plots $P : \mathbb{R}^k\supset U\to\mathcal{P}(X;\mathbb{a},\mathbb{b})$, $Q : \mathbb{R}^\ell\supset V\to\mathcal{P}(X;\mathbb{b},\mathbb{c})$ and any $\varphi\in\mathcal{F}(X)$, the adjoints $\widehat{P} : \mathbb{R}\times U\to X$ and $\widehat{Q} : \mathbb{R}\times V\to X$ of P and Q are smooth, and hence $\varphi\circ\widehat{P} : \mathbb{R}\times U\to\mathbb{R}$ and $\varphi\circ\widehat{Q} : \mathbb{R}\times V\to\mathbb{R}$ are smooth functions. Let $\widehat{(P{\cdot}Q)} : \mathbb{R}\times U\times V\to X$ be the adjoint map of $P{\cdot}Q := \mu\circ(P\times Q) : U\times V\to\mathcal{P}(X;\mathbb{a},\mathbb{b})\times\mathcal{P}(X;\mathbb{b},\mathbb{c})\to\mathcal{P}(X;\mathbb{a},\mathbb{c})$. The function $\varphi\circ\widehat{(P{\cdot}Q)}$ can be described as

$$\text{(B)}\qquad \varphi\circ\widehat{(P{\cdot}Q)}(t,\mathbb{x},\mathbb{y}) = \begin{cases}\varphi\circ\widehat{P}(2t,\mathbb{x}), & t\le 1/2,\\ \varphi\circ\widehat{Q}(2t{-}1,\mathbb{y}), & t\ge 1/2,\end{cases}\quad (t,\mathbb{x},\mathbb{y})\in\mathbb{R}\times U\times V.$$

Since the adjoint map $\widehat{P}$ of P is constant on $(-\infty,0]\times U$ and on $[1,\infty)\times U$, so is $\varphi\circ\widehat{P}$, and we obtain the following equation.

$$\frac{\partial(\varphi\circ\widehat{P})}{\partial t}=\frac{\partial(\varphi\circ\widehat{P})}{\partial x_i}=0 \ \text{ on } \ ((-\infty,0]\cup[1,\infty))\times U \text{ for } 1\le i\le k.$$

Similarly for the adjoint map $\widehat{Q}$ of Q, we obtain the following equation.

$$\frac{\partial(\varphi\circ\widehat{Q})}{\partial t}=\frac{\partial(\varphi\circ\widehat{Q})}{\partial y_j}=0 \ \text{ on } \ ((-\infty,0]\cup[1,\infty))\times V \text{ for } 1\le j\le \ell.$$

Hence by iterating partial differentials, we obtain

$$\frac{\partial^{n+|I|}(\varphi\circ\widehat{P})}{\partial t^n\,\partial \mathbb{x}^I}=0 \text{ on } ((-\infty,0]\cup[1,\infty))\times U \text{ for } (n,I)\in\mathbb{N}\times\mathbb{N}^k,\ n+|I|>0,$$

$$\frac{\partial^{n+|J|}(\varphi\circ\widehat{Q})}{\partial t^n\,\partial \mathbb{y}^J}=0 \text{ on } ((-\infty,0]\cup[1,\infty))\times V \text{ for } (n,J)\in\mathbb{N}\times\mathbb{N}^\ell,\ \text{if } n+|J|>0,$$

where $|I|=i_1+\cdots+i_k$ and $\partial\mathbb{x}^I=\partial x_1^{i_1}\cdots\partial x_k^{i_k}$ for $I=(i_1,\dots,i_k)$, and $|J|=j_1+\cdots+j_\ell$ and $\partial\mathbb{y}^J=\partial y_1^{j_1}\cdots\partial y_\ell^{j_\ell}$ for $J=(j_1,\dots,j_\ell)$. Let $\Phi_1=\varphi\circ\mu(\widehat{P},k(\mathbb{b}))=\mu_{\mathbb{R}}(\varphi\circ\widehat{P},k(\varphi(\mathbb{b})))$ and $\Phi_2=\mu_{\mathbb{R}}(k(\varphi(\mathbb{b})),\varphi\circ\widehat{Q})$. Then we obtain the following equations.

$$\frac{\partial^{n+|I|+|J|}\Phi_1}{\partial t^n\,\partial\mathbb{x}^I\,\partial\mathbb{y}^J}=0 \text{ on } ((-\infty,0]\cup[1/2,\infty))\times U \quad\text{and}$$

$$\frac{\partial^{n+|I|+|J|}\Phi_2}{\partial t^n\,\partial\mathbb{x}^I\,\partial\mathbb{y}^J}=0 \text{ on } ((-\infty,1/2]\cup[1,\infty))\times V$$

for $(n,I,J)\in\mathbb{N}\times\mathbb{N}^k\times\mathbb{N}^\ell$, $n+|I|+|J|>0$. Since $\varphi\circ\widehat{(P\cdot Q)}(t,\mathbb{x},\mathbb{y})=\mu_{\mathbb{R}}(\varphi\circ\widehat{P},\varphi\circ\widehat{Q})(t,\mathbb{x},\mathbb{y})$ $=\Phi_1(t,\mathbb{x},\mathbb{y})+\Phi_2(t,\mathbb{x},\mathbb{y})-\varphi\circ k(\mathbb{b})$, $(t,\mathbb{x},\mathbb{y})\in\mathbb{R}\times U\times V$ by the equation (B), the composition $\varphi\circ\widehat{(P\cdot Q)}$ is also a smooth function satisfying

$$\frac{\partial^{n+|I|+|J|}(\varphi\circ\widehat{(P\cdot Q)})}{\partial t^n\,\partial\mathbb{x}^I\,\partial\mathbb{y}^J}=0 \text{ on } ((-\infty,0]\cup\{1/2\}\cup[1,\infty))\times U\times V$$

for $(n,I,J)\in\mathbb{N}\times\mathbb{N}^k\times\mathbb{N}^\ell$, $n+|I|+|J|>0$. Therefore, by the reflexivity at a point $\mathbb{b}\in X$, we can deduce, using a similar argument given to show the well-definedness of $\mu(u,v)$, that the map $\widehat{(P\cdot Q)}:\mathbb{R}\times U\times V\to X$ is a plot, which means $\mu\circ(P\times Q):U\times V\to\mathcal{P}(X)$ is a plot, and hence μ is smooth. □

There also is an obvious smooth homotopy $\mu\circ(\iota\times_X 1)\sim 1\sim\mu\circ(1\times_X\iota)$.

COROLLARY 6.9. *If X is reflexive, then $\mu:\mathcal{P}(X)\times_X\mathcal{P}(X)\to\mathcal{P}(X)$ is smooth, and hence $(\mathcal{P}(X),X)$ is an internal H-category in* Diffeology. *If X is reflexive at the base point $*$, then the concatenation of $\mathcal{L}(X)=\mathcal{P}(X;*,*)$ is smooth.*

EXAMPLE 6.10. Let X be a manifold, or a smooth CW complex of finite dimension whose base point is in the top cell. Then the concatenation of $\mathcal{L}(X)$ is smooth.

THEOREM 6.11. *If X is a reflexive diffeological space, then $(\mathcal{P}(X),X)$ is an internal smooth h-unital A_∞-category in* Diffeology.

PROOF. Firstly, we introduce the space of concatenations as follows:

$$E_n = \{(r_1, r_2, \dots, r_n) \in \mathbb{R}^{n-1} \mid 0 < r_i \ (1 \le i \le n),\ r_1 + r_2 + \cdots + r_n = 1\}$$

which is a convex open set in an affine space of dimension $n{-}1$:

$$H^{n-1} : x_1 + x_2 + \cdots + x_n = 1.$$

Let $\mathcal{P}(X)^n = \mathcal{P}(X) \times_X \cdots \times_X \mathcal{P}(X)$. Then we define $\widehat{\beta}_n : E_n \times \mathcal{P}(X)^n \hookrightarrow \mathcal{P}(X)$ as

$$\widehat{\beta}_n(r_1, \dots, r_n; u_1, \dots, u_{n-1}) = v \in \mathcal{P}(X),$$

which is defined by

$$v(t) = \begin{cases} u_1(\frac{t}{r_1}), & t \le v_1, \\ u_i(\frac{t-v_{i-1}}{r_i}), & v_{i-1} \le t \le v_i,\ 1 < i < n, \\ u_n(\frac{t-v_{n-1}}{r_n}), & v_{n-1} \le t, \end{cases} \quad t \in \mathbb{R},$$

where $v_i = r_1 + \cdots + r_i$, $1 \le i \le n$. It is not very hard to show that $\widehat{\beta}_n$ is well-defined and also smooth by using a similar arguments to the proof of Theorem 6.8, and we leave it to the reader. Thus by taking adjoint, we obtain a smooth map $\beta_n : E_n \to C^\infty(\mathcal{P}(X)^n, \mathcal{P}(X))$.

Secondly, let $\partial^E_k : E_r \times E_s \to E_n$, $r{+}s = n{+}1$, be the smooth map defined by

$$\partial^E_k(x_1, \dots, x_r; y_1, \dots, y_s) = (x_1, \dots, x_{k-1}, x_k{\cdot}y_1, \dots, x_k{\cdot}y_s, x_{k+1}, \dots, x_r).$$

Let $\mathbb{e}_n = \frac{1}{n}(1, 1, \dots, 1) \in E_n$. Then we define $\phi_n : K_n \to E_n$ inductively by

(1) $\phi_n : K_n \ni \mathbb{b}_n \mapsto \mathbb{e}_n \in E_n$, and (2) $\phi_n{\circ}\partial_k = \partial^E_k{\circ}(\phi_r \times \phi_s)$.

By (1), $\phi_2 : K_2 = \{(0,1)\} \to \{\mathbb{e}_2\} \subset E_2$ is the trivial map, since $\mathbb{b}_2 = (0,1)$. (2) determines a smooth map $\widehat{\phi}_{k,r,s} : L_k(r,s) \to E_n$, since $\partial_k : K_r \times K_s \to L_k(r,s)$ is a diffeomorphism. There is a smooth extension $\phi_{k,r,s} : L_k(r,s) * \{\mathbb{b}_n\} \to E_n$ of $\widehat{\phi}_{k,r,s}$, $(k,r,s) \in A(n)$, such that $\phi_{k,r,s}(\mathbb{b}_n) = \mathbb{e}_n$, since E_n is star-shaped w.r.t. $\mathbb{e}_n \in E_n$. As $|\mathbb{K}(n)|$ is the colimit of $|\mathbb{L}_k(r,s) * \{\mathbb{b}_n\}|$ and the identity map $|\mathbb{L}_k(r,s) * \{\mathbb{b}_n\}| \to L_k(r,s) * \{\mathbb{b}_n\}$ is a smooth bijection, smooth maps $\phi_{k,r,s}$, $(k,r,s) \in A(n)$, give a smooth map $\phi_n : |\mathbb{K}(n)| \to E_n$.

Finally, smooth maps $M_n = \beta_n{\circ}\phi_n$ satisfy (0), (1) and (2') in Definition 5.5, which determines a smooth A_∞-form on $(\mathcal{P}(X), X)$, and hence $(\mathcal{P}(X), X)$ is an internal smooth A_∞-category in Diffeology. □

COROLLARY 6.12. *If a diffeological space X is reflexive at the base point, then the diffeological loop space $\mathcal{L}(X)$ is a smooth h-unital A_∞-space in* Diffeology.

COROLLARY 6.13. *Let X be a manifold, or a smooth CW complex of finite dimension with base point in the interior of the top cell. Then $\mathcal{L}(X)$ is a smooth h-unital A_∞-space in* Diffeology.

Appendix A. Path space with stable concatenation

In this section, we modify a concatenation to obtain a smooth A_∞-form on $\mathcal{P}(X)$ without assuming reflexivity on X. The internal precategory $\mathcal{P}(X)$ is equipped with the following stable concatenation for a fixed small $\varepsilon > 0$.

$$\mu_\varepsilon : \mathcal{P}(X) \times_X \mathcal{P}(X) \to \mathcal{P}(X)$$
$$\iff \mu_\varepsilon(u, v)(t) = \begin{cases} u(\frac{2t}{1-\varepsilon}), & t < \frac{1+\varepsilon}{2}, \\ v(\frac{2t-1-\varepsilon}{1-\varepsilon}), & t > \frac{1-\varepsilon}{2}, \end{cases} \quad t \in \mathbb{R}.$$

Then μ_ε is stable when $\frac{1-\varepsilon}{2} \leq t \leq \frac{1+\varepsilon}{2}$, and is smooth for any diffeological space X.

THEOREM A.1. *$(\mathcal{P}(X), X)$ is an internal h-unital A_∞-category in* Diffeology.

PROOF. Firstly, we introduce the space of stable concatenations as follows:

$$D_n = \{(r_1, \dots, r_n; \varepsilon_1, \dots, \varepsilon_{n-1}) \in (0,1)^{2n-1} \mid r_1 + \cdots + r_n + \varepsilon_1 + \cdots + \varepsilon_{n-1} = 1\}$$

which is a convex open set in the following affine space of dimension $2n-2$:

$$H^{2n-2} : x_1 + \cdots + x_n + x_{n+1} + \cdots + x_{2n-1} = 1.$$

We define $\widehat{\alpha}_n : D_n \times \mathcal{P}(X)^n \to \mathcal{P}(X)$ as

$$\widehat{\alpha}_n(r_1, \dots, r_n; \varepsilon_1, \dots, \varepsilon_{n-1}, r_n; u_1, \dots, u_n) = v \in \mathcal{P}(X),$$

which is defined by

$$v(r) = \begin{cases} u_1(\frac{r}{r_1}), & r < s_1 + \varepsilon_1, \\ u_i(\frac{r - s_{i-1} - \varepsilon_{i-1}}{r_i}), & s_{i-1} < r < s_i + \varepsilon_i,\ 1 < i < n, \\ u_n(\frac{r - s_{n-1} - \varepsilon_{n-1}}{r_n}), & s_{n-1} < r, \end{cases} \quad r \in \mathbb{R},$$

where $s_i = r_1 + \cdots + r_i + \varepsilon_1 + \cdots + \varepsilon_{i-1}$, $1 \leq i \leq n$, and hence $s_i - s_{i-1} - \varepsilon_{i-1} = r_i$, $1 \leq i \leq n$. Since the open sets $(-\infty, s_1 + \varepsilon_1) \supset (-\infty, s_1]$, $(s_{i-1}, s_i + \varepsilon_i) \supset (s_{i-1}, s_i]$, $1 < i < n$ and (s_{n-1}, ∞) cover entire $\mathbb{R}$, $\widehat{\alpha}_n$ is well-defined by definition and is also smooth. Thus by taking adjoint, we obtain a smooth map $\alpha_n : D_n \to C^\infty(\mathcal{P}(X)^n, \mathcal{P}(X))$.

Secondly, let $\partial^D_k : D_r \times D_s \to D_n$, $r+s = n+1$, be the smooth map defined by

$$\partial^D_k(x_1, \dots, x_r; \varepsilon_1, \dots, \varepsilon_{r-1}; y_1, \dots, y_s; \eta_1, \dots, \eta_{r-1})$$
$$= (x_1, \dots, x_{k-1}, x_k \cdot y_1, \dots, x_k \cdot y_s, x_{k+1}, \dots, x_r; \varepsilon_1, \dots, \varepsilon_{k-1}; x_k \cdot \eta_1, \dots, x_k \cdot \eta_{s-1}; \varepsilon_k, \dots, \varepsilon_{r-1}).$$

Let $\mathbb{d}_n = \frac{1}{3n-1}(2, \dots, 2; 1, \dots, 1) \in D_n$. Then we define $\psi_n : K_n \to D_n$ inductively by

(1) $\psi_n : K_n \ni \mathbb{b}_n \mapsto \mathbb{d}_n \in D_n$ and (2) $\psi_n \circ \partial_k = \partial^D_k \circ (\psi_r \times \psi_s)$.

By (1), $\psi_2 : K_2 = \{(0,1)\} \to \{\mathbb{d}_2\} \subset D_2$ is the trivial map, since $\mathbb{b}_2 = (0,1)$. (2) determines a smooth map $\widehat{\psi}_{k,r,s} : L_k(r,s) \to D_n$, since $\partial_k : K_r \times K_s \to L_k(r,s)$ is a diffeomorphism. There is a smooth extension $\psi_{k,r,s} : L_k(r,s) * \{\mathbb{b}_n\} \to D_n$ of $\widehat{\psi}_{k,r,s}$, $(k,r,s) \in A(n)$, such that $\psi_{k,r,s}(\mathbb{b}_n) = \mathbb{d}_n$, since D_n is star-shaped w.r.t. $\mathbb{d}_n \in D_n$. Since $|\mathbb{K}(n)|$ is the colimit of $|\mathbb{L}_k(r,s) * \{\mathbb{b}_n\}|$ and the identity map $|\mathbb{L}_k(r,s) * \{\mathbb{b}_n\}| \to L_k(r,s) * \{\mathbb{b}_n\}$ is a smooth bijection, smooth maps $\psi_{k,r,s}$, $(k,r,s) \in A(n)$, give rise to a smooth map $\psi_n : |\mathbb{K}(n)| \to D_n$.

Finally, smooth maps $M_n = \alpha_n \circ \psi_n$ satisfy (0), (1) and (2') in Definition 5.5, which determines a smooth A_∞-form on $(\mathcal{P}(X), X)$, and hence $(\mathcal{P}(X), X)$ is an internal smooth A_∞-category in Diffeology. □

COROLLARY A.2. *$\mathcal{L}(X)$ is a smooth h-unital A_∞-space in* Diffeology.

Acknowledgments

The author thanks Jim Stasheff, Bruno Vallette, Jordan Watts, Patrick I-Zemmour and the editor for their valuable comments, suggestions and encouragements, without which this work could not be completed for publication.

Particularly, the author wants to mention Jim Stacheff's talk "How I became a so-called *cohomological physicist*" given just after his reading the referee's final comments, which encouraged him to back from emotional words thrown on ▯. So he wants to give special thanks to Jim.

References

[Che73] K.-t. Chen, *Iterated integrals of differential forms and loop space homology*, Ann. of Math. (2) **97** (1973), 217–246, DOI 10.2307/1970846. MR380859

[Che75] K. T. Chen, *Iterated integrals, fundamental groups and covering spaces*, Trans. Amer. Math. Soc. **206** (1975), 83–98, DOI 10.2307/1997148. MR377960

[Che77] K. T. Chen, *Iterated path integrals*, Bull. Amer. Math. Soc. **83** (1977), no. 5, 831–879, DOI 10.1090/S0002-9904-1977-14320-6. MR454968

[Che86] K. T. Chen, *On differentiable spaces*, Categories in continuum physics (Buffalo, N.Y., 1982), Lecture Notes in Math., vol. 1174, Springer, Berlin, 1986, pp. 38–42, DOI 10.1007/BFb0076932. MR842915

[HS20] Tadayuki Haraguchi and Kazuhisa Shimakawa, *A model structure on the category of diffeological spaces, i*, 11 2020, Preprint, `arXiv:2011.12842`.

[II19] N. Iwase and N. Izumida, *Mayer-Vietoris sequence for differentiable/diffeological spaces*, Algebraic topology and related topics, Trends Math., Birkhäuser/Springer, Singapore, 2019, pp. 123–151, DOI 10.1007/978-981-13-5742-8_8. MR3991180

[IM89] N. Iwase and M. Mimura, *Higher homotopy associativity*, Algebraic topology (Arcata, CA, 1986), Lecture Notes in Math., vol. 1370, Springer, Berlin, 1989, pp. 193–220, DOI 10.1007/BFb0085229. MR1000378

[Iwa12] Norio Iwase, *Associahedra, multiplihedra and units in A_∞ form*, 11 2012, Preprint, `arXiv:1211.5741`.

[IZ13] P. Iglesias-Zemmour, *Diffeology*, Mathematical Surveys and Monographs, vol. 185, American Mathematical Society, Providence, RI, 2013, DOI 10.1090/surv/185. MR3025051

[Kih19] H. Kihara, *Model category of diffeological spaces*, J. Homotopy Relat. Struct. **14** (2019), no. 1, 51–90, DOI 10.1007/s40062-018-0209-3. MR3913971

[KM97] A. Kriegl and P. W. Michor, *The convenient setting of global analysis*, Mathematical Surveys and Monographs, vol. 53, American Mathematical Society, Providence, RI, 1997, DOI 10.1090/surv/053. MR1471480

[Lod07] J.-L. Loday, *Parking functions and triangulation of the associahedron*, Categories in algebra, geometry and mathematical physics, Contemp. Math., vol. 431, Amer. Math. Soc., Providence, RI, 2007, pp. 327–340, DOI 10.1090/conm/431/08279. MR2342835

[Lod12] J.-L. Loday, *Dichotomy of the addition of natural numbers*, Associahedra, Tamari lattices and related structures, Progr. Math., vol. 299, Birkhäuser/Springer, Basel, 2012, pp. 65–79, DOI 10.1007/978-3-0348-0405-9_4. MR3221534

[Sou80] J.-M. Souriau, *Groupes différentiels* (French), Differential geometrical methods in mathematical physics (Proc. Conf., Aix-en-Provence/Salamanca, 1979), Lecture Notes in Math., vol. 836, Springer, Berlin, 1980, pp. 91–128. MR607688

[Sta63] J. D. Stasheff, *Homotopy associativity of H-spaces. I, II*, Trans. Amer. Math. Soc. **108** (1963), 293–312, DOI 10.1090/s0002-9947-1963-0158400-5. **108** (1963), 275-292; ibid. MR158400

[Sta70] ———, *H-spaces from a homotopy point of view*, Lecture Notes in Mathematics, Vol. 161, Springer-Verlag, Berlin-New York, 1970. MR0270372

[Wat12] J. Watts, *Diffeologies, Differential Spaces, and Symplectic Geometry*, ProQuest LLC, Ann Arbor, MI, 2012. Thesis (Ph.D.)–University of Toronto (Canada). MR3153238

Faculty of Mathematics, Kyushu University, Fukuoka 819-0395, Japan
Email address: `iwase@math.kyushu-u.ac.jp`

Contemporary Mathematics
Volume **794**, 2024
https://doi.org/10.1090/conm/794/15939

Smooth homotopy of diffeological spaces: theory and applications to infinite-dimensional C^∞-manifolds

Hiroshi Kihara

ABSTRACT. We present the basic notions and results of our smooth homotopy theory of diffeological spaces and its applications to infinite-dimensional C^∞-manifolds, focusing on the fundamental ideas. In particular, smoothing problems for maps, sections, principal bundles, and gauge transformations are discussed.

1. Introduction

The notion of a diffeological space was introduced by Souriau [**27**, **28**], and the foundation of the theory of diffeological spaces was laid mainly by Iglesias-Zemmour [**13**]. Their notions and results have been applied to various fields of mathematics and mathematical physics.

In this paper, we use abstract homotopical methods to study smooth maps of diffeological spaces (especially, infinite-dimensional C^∞-manifolds). Since the study of continuous maps is done by topological homotopy theory, we construct a framework for a smooth homotopy theory and compare smooth and topological homotopy theories. In particular, we discuss the smoothing of continuous maps, sections, principal bundles, and gauge transformations. This paper is written with the intention of giving an introductory presentation to [**14**] and [**15**].

Let $\mathcal{D}$ denote the category of diffeological spaces. Recall that a topological space X is called *arc-generated* if its topology is final for the continuous curves from $\mathbb{R}$ to X. We adopt the category $\mathcal{C}^0$ of arc-generated spaces as a convenient category of topological spaces. (The notion of an arc-generated space is equivalent to that of a Δ-generated space; see [**14**, p. 59].)

Throughout this paper, $\mathcal{M}$ denotes one of the categories $\mathcal{D}$ and $\mathcal{C}^0$. The following subjects are dealt with in this paper:

- Model category structure and simplicial category structure on $\mathcal{M}$ ($= \mathcal{D}, \mathcal{C}^0$) (Sections 3 and 4).
- Smoothing results of continuous maps, sections, principal bundles, and gauge transformations (Sections 5-7).
- Applications to (infinite-dimensional) C^∞-manifolds (Section 8).

2020 *Mathematics Subject Classification.* Primary 18F15, 18N40; Secondary 58A40, 58B10.
Key words and phrases. Smooth homotopy, model category, convenient calculus, smoothing theorem.

Our smoothing problem for maps is formulated using the simplicial category structures of $\mathcal{D}$ and $\mathcal{C}^0$ (Problem in Section 4). On the other hand, we would like to address our smoothing problem using various results in model category theory. Thus, we study the relation between simplicial and model category structures on $\mathcal{M}$ $(= \mathcal{D}, \mathcal{C}^0)$ to establish a smoothing theorem for continuous maps (Theorem 5.5). Similarly, we study the relation between simplicial and model category structures on $\mathcal{M}/X$ $(= \mathcal{D}/X, \mathcal{C}^0/X)$ to establish a smoothing theorem for continuous sections (Theorem 6.4).

We use the smoothing theorems for maps and sections to derive smoothing results for principles bundles and gauge transformations (Propositions 5.10 and 7.4), which are unified using the notion of a Dwyer-Kan equivalence (Theorem 7.5).

In Section 8, we make a brief review of convenient calculus, which is regarded as a prime candidate for the final theory of infinite-dimensional calculus. Then, we see that the category C^∞ of C^∞-manifolds in convenient calculus can be fully faithfully embedded into the category $\mathcal{D}$ and study the smooth homotopical properties of C^∞-manifolds to apply the smoothing theorems to C^∞-manifolds (see Theorems 8.9-8.11).

2. Categories $\mathcal{S}$, $\mathcal{D}$, and $\mathcal{C}^0$

In this section, we make a brief review of simplicial sets, diffeological spaces, and arc-generated spaces. The category of simplicial sets is denoted by $\mathcal{S}$.

Categories $\mathcal{S}$ and $\mathcal{C}^0$. The category $\mathcal{C}^0$ of arc-generated spaces is a bicomplete cartesian closed category; see [**14**, Section 2.2] and [**15**, Remark 2.4] for the basic properties and the reason why we use the category $\mathcal{C}^0$. The usual topological homotopy theory is developed in the category $\mathcal{C}^0$.

On the other hand, simplicial homotopy theory is a homotopy theory developed in the category $\mathcal{S}$, which is equivalent (in an appropriate sense) to topological homotopy theory via the adjoint pair of the topological realization and singular functors

$$| \ | : \mathcal{S} \rightleftarrows \mathcal{C}^0 : S$$

(see [**20**, Section 16] and Remark 3.3(2)).

Categories $\mathcal{D}$ and $\mathcal{C}^0$. We begin with the definition of a diffeological space. A *parametrization* of a set X is a (set-theoretic) map $p : U \longrightarrow X$, where U is an open subset of $\mathbb{R}^n$ for some n.

DEFINITION 2.1.
(1) A *diffeological space* is a set X together with a specified set D_X of parametrizations of X satisfying the following conditions:
  (i) (Covering) Every constant parametrization $p : U \longrightarrow X$ is in D_X.
  (ii) (Locality) Let $p : U \longrightarrow X$ be a parametrization such that there exists an open cover $\{U_i\}$ of U satisfying $p|_{U_i} \in D_X$. Then, p is in D_X.
  (iii) (Smooth compatibility) Let $p : U \longrightarrow X$ be in D_X. Then, for every $n \geq 0$, every open set V of $\mathbb{R}^n$ and every smooth map $F : V \longrightarrow U$, $p \circ F$ is in D_X.

The set D_X is called the *diffeology* of X, and its elements are called *plots*.

(2) Let $X = (X, D_X)$ and $Y = (Y, D_Y)$ be diffeological spaces, and let $f : X \longrightarrow Y$ be a (set-theoretic) map. We say that f is *smooth* if for any $p \in D_X$, $f \circ p \in D_Y$.

The convenient properties of $\mathcal{D}$ are summarized in the following proposition. See [**7**, pp. 230-233] for initial and final structures with respect to the underlying set functor.

PROPOSITION 2.2.

(1) *The category $\mathcal{D}$ has initial and final structures with respect to the underlying set functor. In particular, $\mathcal{D}$ is complete and cocomplete.*
(2) *The category $\mathcal{D}$ is cartesian closed.*

REMARK 2.3. Let $\mathfrak{X}$ be a concrete category (i.e., a category equipped with a faithful functor to *Set*); the faithful functor $\mathfrak{X} \longrightarrow Set$ is called the underlying set functor. See [**7**, Section 8.8] for the notions of an $\mathfrak{X}$-embedding, an $\mathfrak{X}$-subspace, an $\mathfrak{X}$-quotient map, and an $\mathfrak{X}$-quotient space. $\mathcal{D}$-subspaces and $\mathcal{D}$-quotient spaces are usually called diffeological subspaces and quotient diffeological spaces, respectively.

The *underlying topological space* $\widetilde{A}$ of a diffeological space $A = (A, D_A)$ is defined to be the set A endowed with the final topology for D_A, which is often called the D-topology. Since open sets of $\mathbb{R}^n$ are arc-generated, we have the underlying topological space functor $\widetilde{\cdot} : \mathcal{D} \longrightarrow \mathcal{C}^0$.

Define the functor $R : \mathcal{C}^0 \longrightarrow \mathcal{D}$ by assigning to an arc-generated space X the set X endowed with the diffeology

$$D_{RX} = \{\text{continuous parametrizations of } X\}.$$

Then, we can easily see that

$$\widetilde{\cdot} : \mathcal{D} \rightleftarrows \mathcal{C}^0 : R$$

is an adjoint pair.

3. Model structure on $\mathcal{M}$

It is a natural idea to develop a smooth homotopy theory using the notions and results of model category theory. In this section, we thus recall the model structure on the category $\mathcal{C}^0$, and then introduce the model structure on $\mathcal{D}$.

At the end of this section, we see that the model categories $\mathcal{S}$, $\mathcal{D}$, and $\mathcal{C}^0$ are Quillen equivalent. (The reader is assumed to be familiar with the model structure on the category $\mathcal{S}$ [**8**, Chapter I]. Refer to [**12**] for basic notions and results in model category theory.)

Model structure on $\mathcal{C}^0$. First, we recall the model structure on $\mathcal{C}^0$. The standard p-simplex Δ^p and its k-th horn Λ^p_k are defined by

$$\Delta^p = \{(x_0, \ldots, x_p) \in \mathbb{R}^{p+1} \mid \sum_i x_i = 1,\ x_i \geq 0\},$$

$$\Lambda^p_k = \{(x_0, \ldots, x_p) \in \Delta^p \mid x_i = 0 \text{ for some } i \neq k\}.$$

Let Δ^p_{top} and $\Lambda^p_{k\ \text{top}}$ denote the standard p-simplex and its k-th horn, respectively, endowed with the standard topologies.

DEFINITION 3.1. Define a map $f : X \longrightarrow Y$ in $\mathcal{C}^0$ to be

(1) a *weak equivalence* if $Sf : SX \longrightarrow SY$ is a weak equivalence in the category of simplicial sets,
(2) a *fibration* if the map f has the right lifting property with respect to the inclusions $\Lambda^p_{k\ \mathrm{top}} \hookrightarrow \Delta^p_{\mathrm{top}}$ for all $p > 0$ and $0 \leq k \leq p$, and
(3) a *cofibration* if the map f has the left lifting property with respect to all maps that are both fibrations and weak equivalences.

Then, the following theorem holds.

THEOREM 3.2. *$\mathcal{C}^0$ is a cofibrantly generated model category whose object is always fibrant. $\mathcal{I}_{\mathrm{top}} = \{\dot{\Delta}^p_{\mathrm{top}} \hookrightarrow \Delta^p_{\mathrm{top}} \,|\, p \geq 0\}$ and $\mathcal{J}_{\mathrm{top}} = \{\Lambda^p_{k\ \mathrm{top}} \hookrightarrow \Delta^p_{\mathrm{top}} \,|\, p > 0, 0 \leq k \leq p\}$ are sets of generating cofibrations and generating trivial cofibrations respectively.*

REMARK 3.3.

(1) We can easily observe the following:
   - $f : X \longrightarrow Y$ is a weak equivalence in $\mathcal{C}^0 \Leftrightarrow f : X \longrightarrow Y$ is a weak homotopy equivalence.
   - $f : X \longrightarrow Y$ is a fibration in $\mathcal{C}^0 \Leftrightarrow f : X \longrightarrow Y$ is a Serre fibration.

   Therefore, the model structure on $\mathcal{C}^0$ is just the usual one (see [**12**, Section 7.10] and [**6**, Section 8]).
(2) The equivalence in an appropriate sense between simplicial and topological homotopy theories referred to in Section 2 can be precisely formulated as follows: The adjoint pair
$$| \ | : \mathcal{S} \rightleftarrows \mathcal{C}^0 : S$$
is a pair of Quillen equivalences (see [**8**, Theorem 11.4 in Chapter I]).

Model structure on $\mathcal{D}$. Next we discuss a model structure on $\mathcal{D}$. The principal part of our construction of a model structure on $\mathcal{D}$ is the construction of good diffeologies on the sets Δ^p $(p \geq 0)$. We begin by summarizing in axioms the desired properties of difffeologies on Δ^p $(p \geq 0)$, which enable us to define weak equivalences, fibrations, and cofibrations and to verify the model axioms as in the case of $\mathcal{C}^0$.

AXIOM 1. *The underlying topological space of Δ^p is the topological standard p-simplex for $p \geq 0$.*

Recall that $f : \Delta^p \longrightarrow \Delta^q$ is an *affine map* if f preserves convex combinations.

AXIOM 2. *Any affine map $f : \Delta^p \longrightarrow \Delta^q$ is smooth.*

As in the topological case, use Δ^p $(p \geq 0)$ to define the *realization functor*
$$| \ |_{\mathcal{D}} : \mathcal{S} \longrightarrow \mathcal{D}$$
and consider the smooth map $|\dot{\Delta}[p]|_{\mathcal{D}} \hookrightarrow |\Delta[p]|_{\mathcal{D}} = \Delta^p$ induced by the inclusion of the boundary $\dot{\Delta}[p]$ into $\Delta[p]$.

AXIOM 3. *The canonical smooth injection*
$$\left|\dot{\Delta}[p]\right|_{\mathcal{D}} \hookrightarrow \Delta^p$$
is a $\mathcal{D}$-embedding.

The $\mathcal{D}$-homotopical notions, especially the notion of a $\mathcal{D}$-deformation retract, are defined in the same manner as in the category of topological spaces by using the unit interval $I = [0, 1]$ endowed with a diffeology via the canonical bijection with Δ^1 ([**14**, Section 2.4]).

AXIOM 4. *The kth horn Λ^p_k is a $\mathcal{D}$-deformation retract of Δ^p for $p \geq 1$ and $0 \leq k \leq p$.*

For a subset A of the affine p-space $\mathbb{A}^p = \{(x_0, \ldots, x_p) \in \mathbb{R}^{p+1} \mid \sum x_i = 1\}$, A_{sub} denotes the set A endowed with the sub-diffeology of $\mathbb{A}^p$ ($\cong \mathbb{R}^p$). The diffeological space Δ^p_{sub}, used in [**10**] to study diffeological spaces by homotopical means, satisfies neither Axiom 3 nor 4 for $p \geq 2$ ([**14**, Proposition A.2]). Thus, we must construct a new diffeology on Δ^p, at least for $p \geq 2$.

Let (i) denote the vertex $(0, \ldots, \underset{(i)}{1}, \ldots, 0)$ of Δ^p, and let d^i denote the affine map from Δ^{p-1} to Δ^p, defined by

$$d^i((k)) = \begin{cases} (k) & \text{for } k < i, \\ (k+1) & \text{for } k \geq i. \end{cases}$$

DEFINITION 3.4. We define the *standard p-simplices* Δ^p ($p \geq 0$) inductively. Set $\Delta^p = \Delta^p_{\text{sub}}$ for $p \leq 1$. Suppose that the diffeologies on Δ^k ($k < p$) are defined. We define the map

$$\varphi_i : \Delta^{p-1} \times [0, 1) \longrightarrow \Delta^p$$

by $\varphi_i(x, t) = (1-t)(i) + td^i(x)$, and endow Δ^p with the final structure for the maps $\varphi_0, \ldots, \varphi_p$.

The following result is established in [**14**, Propositions 3.2, 5.1, 7.1, and 8.1].

PROPOSITION 3.5. *The standard p-simplices Δ^p ($p \geq 0$) in Definition 3.4 satisfy Axioms 1-4.*

Without explicit mention, the symbol Δ^p denotes the standard p-simplex defined in Definition 3.4 and a subset of Δ^p is endowed with the sub-diffeology of Δ^p. Since the diffeology of Δ^p is the sub-diffeology of $\mathbb{A}^p$ for $p \leq 1$, the $\mathcal{D}$-homotopical notions, especially the notion of a $\mathcal{D}$-deformation retract, coincide with the ordinary smooth homotopical notions in the theory of diffeological spaces (see [**13**, p. 108] and [**14**, Remark 2.14]).

By Axiom 2, we can define the singular complex $S^{\mathcal{D}}X$ of a diffeological space X to have smooth maps $\sigma : \Delta^p \longrightarrow X$ as p-simplices, thereby defining the *singular functor* $S^{\mathcal{D}} : \mathcal{D} \longrightarrow \mathcal{S}$. As in the topological case, the realization functor $|\ |_{\mathcal{D}}$ and the singular functor $S^{\mathcal{D}}$ form an adjoint pair

$$|\ |_{\mathcal{D}} : \mathcal{S} \rightleftarrows \mathcal{D} : S^{\mathcal{D}}.$$

Similarly to the arc-generated case, we define weak equivalences, fibrations, and cofibrations as follows.

DEFINITION 3.6. Define a map $f : X \longrightarrow Y$ in $\mathcal{D}$ to be

(1) a *weak equivalence* if $S^{\mathcal{D}}f : S^{\mathcal{D}}X \longrightarrow S^{\mathcal{D}}Y$ is a weak equivalence in the category of simplicial sets,
(2) a *fibration* if the map f has the right lifting property with respect to the inclusions $\Lambda^p_k \hookrightarrow \Delta^p$ for all $p > 0$ and $0 \leq k \leq p$, and
(3) a *cofibration* if the map f has the left lifting property with respect to all maps that are both fibrations and weak equivalences.

Then, we have the following theorem, which is analogous to Theorem 3.2.

THEOREM 3.7 ([**14**, Theorem 1.3]). *$\mathcal{D}$ is a cofibrantly generated model category whose object is always fibrant. $\mathcal{I} = \{\dot{\Delta}^p \hookrightarrow \Delta^p \,|\, p \geq 0\}$ and $\mathcal{J} = \{\Lambda^p_k \hookrightarrow \Delta^p \,|\, p > 0, 0 \leq k \leq p\}$ are sets of generating cofibrations and generating trivial cofibrations respectively.*

Further, we can prove the following result; see [**3**, Section 3.1] or [**13**, Chapter 5] for the smooth homotopy group $\pi^{\mathcal{D}}_p(X, x)$ for a pointed diffeological space (X, x).

THEOREM 3.8 ([**14**, Theorem 1.4]). *Let (X, x) be a pointed diffeological space. Then, there exists a natural bijection*

$$\Theta_X : \pi^{\mathcal{D}}_p(X, x) \longrightarrow \pi_p(S^{\mathcal{D}}X, x) \quad \textit{for} \quad p \geq 0,$$

that is an isomorphism of groups for $p > 0$.

The following result is immediate from Theorem 3.8

COROLLARY 3.9. *$f : X \longrightarrow Y$ is a weak equivalence in $\mathcal{D} \Leftrightarrow f : X \longrightarrow Y$ induces isomorphisms on the smooth homotopy groups.*

Quillen equivalences between $\mathcal{S}$, $\mathcal{D}$, and $\mathcal{C}^0$. Now we have the three adjoint pairs

$$\mathcal{S} \underset{S^{\mathcal{D}}}{\overset{|\ |_{\mathcal{D}}}{\rightleftarrows}} \mathcal{D} \underset{R}{\overset{\tilde{\cdot}}{\rightleftarrows}} \mathcal{C}^0, \qquad | \ | : \mathcal{S} \longrightarrow \mathcal{C}^0, \quad S : \mathcal{C}^0 \longrightarrow \mathcal{S}.$$

Since $\widetilde{\Delta^p} = \Delta^p_{\mathrm{top}}$, the composition of $(|\ |_{\mathcal{D}}, S^{\mathcal{D}})$ and $(\tilde{\cdot}, R)$ is just the adjoint pair $(|\ |, S)$. We can easily see that these adjoint pairs are Quillen pairs. Since $(|\ |, S)$ is a pair of Quillen equivalences (Remark 3.3(2)), we obtain the following result by proving that $(|\ |_{\mathcal{D}}, S^{\mathcal{D}})$ is a pair of Quillen equivalences.

THEOREM 3.10 ([**15**, Theorem 1.5]).

$$\mathcal{S} \underset{S^{\mathcal{D}}}{\overset{|\ |_{\mathcal{D}}}{\rightleftarrows}} \mathcal{D} \textit{ and } \mathcal{D} \underset{R}{\overset{\tilde{\cdot}}{\rightleftarrows}} \mathcal{C}^0$$

are pairs of Quillen equivalences, whose composite is just the pair of Quillen equivalences

$$|\ | : \mathcal{S} \rightleftarrows \mathcal{C}^0 : S.$$

REMARK 3.11.

(1) Hector [**10**] and Christensen-Wu [**3**] introduced the singular complexes $S^{\mathcal{D}}_{\mathrm{sub}}(X)$ and $S^{\mathcal{D}}_{\mathrm{aff}}(X)$, which are defined using Δ^p_{sub} and $\mathbb{A}^p$, respectively. Since the set-theoretic identity map $id : \Delta^p \longrightarrow \Delta^p_{\mathrm{sub}}$ is smooth for $p \geq 0$, we have the natural maps of simplicial sets

$$S^{\mathcal{D}}_{\mathrm{aff}}(X) \longrightarrow S^{\mathcal{D}}_{\mathrm{sub}}(X) \longrightarrow S^{\mathcal{D}}(X).$$

These maps are weak equivalences in $\mathcal{S}$, and hence $S^{\mathcal{D}}(X)$ is a fibrant approximation both for $S^{\mathcal{D}}_{\mathrm{aff}}(X)$ and $S^{\mathcal{D}}_{\mathrm{sub}}(X)$ (see [**16**]).

(2) The smooth map $id : \Delta^p \longrightarrow \Delta^p_{\mathrm{sub}}$ restricts to the diffeomorphism on the complement of the $(p-2)$-skeleton $\mathrm{sk}_{p-2}\, \Delta^p$ ([**14**, Lemma 4.2]).

(3) Construction of a model structure on the category $\mathcal{D}$ has been attempted by several authors. Christensen and Wu [**3**] defined fibrations, cofibrations, and weak equivalences, using affine p-spaces instead of the standard p-simplices ([**3**, Definition 4.8]), and conjectured that with their definitions, $\mathcal{D}$ is a model category. Recently, Pavlov [**26**] however resolved this conjecture in the negative. In the latest version of [**9**], Haraguchi and Shimakawa use the notion of a tame map ([**9**, Definition 3.8]) to define fibrations, and claim that $\mathcal{D}$ is a model category which is not cofibrantly generated ([**9**, Theorem 5.6]).

See [**26**, Section 1.4] for model structures on larger categories such as the category of smooth sets (i.e., sheaves of sets on the site of smooth manifolds and open covers).

4. $\mathcal{S}$-category structure on $\mathcal{M}$

The basic notions of topological homotopy theory such as $\mathcal{C}^0$-homotopy relation $\simeq_{\mathcal{C}^0}$, $\mathcal{C}^0$-homotpy sets $[A, X]_{\mathcal{C}^0}$, and $\mathcal{C}^0$-homotopy types are defined using the standard 1-simplex Δ^1_{top}, which is isomorphic to the unit interval I. Similarly, the basic notions of smooth homotopy theory such as $\mathcal{D}$-homotopy relation $\simeq_{\mathcal{D}}$, $\mathcal{D}$-homotopy sets $[A, X]_{\mathcal{D}}$, and $\mathcal{D}$-homotopy types are defined using the standard 1-simplex Δ^1.

These $\mathcal{M}$-homotopy notions are incorporated into the framework of $\mathcal{S}$-category theory as follows.

Since $\mathcal{M}$ $(= \mathcal{D}, \mathcal{C}^0)$ is cartesian closed, $\mathcal{M}$ is itself an $\mathcal{M}$-category with hom-object $\mathcal{M}(X, Y)$. Further, we have the adjoint pair

$$| \ |_{\mathcal{M}} : \mathcal{S} \rightleftarrows \mathcal{M} : S^{\mathcal{M}} = \begin{cases} | \ |_{\mathcal{D}} : \mathcal{S} \rightleftarrows \mathcal{D} : S^{\mathcal{D}} & \text{for } \mathcal{M} = \mathcal{D}, \\ | \ | : \mathcal{S} \rightleftarrows \mathcal{C}^0 : S & \text{for } \mathcal{M} = \mathcal{C}^0. \end{cases}$$

Thus, $\mathcal{M}$ is also an $\mathcal{S}$-category via the singular functor $S^{\mathcal{M}}$ (in other words, $\mathcal{M}$ is an $\mathcal{S}$-category with function complex $S^{\mathcal{M}}\mathcal{M}(X, Y)$).

Set

$$\Delta^p_{\mathcal{M}} = \begin{cases} \Delta^p & \text{for } \mathcal{M} = \mathcal{D}, \\ \Delta^p_{\mathrm{top}} & \text{for } \mathcal{M} = \mathcal{C}^0. \end{cases}$$

Since the natural bijections

$$\mathcal{S}(\Delta[p], S^{\mathcal{M}}\mathcal{M}(A, X)) \cong \mathcal{M}(\Delta^p_{\mathcal{M}}, \mathcal{M}(A, X)) \cong \mathcal{M}(\Delta^p_{\mathcal{M}} \times A, X)$$

exist, the p-simplices of $S^{\mathcal{M}}\mathcal{M}(A, X)$ are identified with the maps from $\Delta^p_{\mathcal{M}} \times A$ to X. In particular, the 0- and 1-simplices of $S^{\mathcal{M}}\mathcal{M}(A, X)$ are just the maps and $\mathcal{M}$-homotopies from A to X. Hence, $\pi_0(S^{\mathcal{M}}\mathcal{M}(A, X))$ is nothing but the $\mathcal{M}$-homotopy set $[A, X]_{\mathcal{M}}(= \mathcal{M}(A, X)/ \simeq_{\mathcal{M}})$.

Since $\widetilde{\cdot} : \mathcal{D} \longrightarrow \mathcal{C}^0$ preserves finite products ([**14**, Proposition 2.13]), it induces a natural map

$$\widetilde{\cdot} : S^{\mathcal{D}}\mathcal{D}(A, X) \longrightarrow S\mathcal{C}^0(\widetilde{A}, \widetilde{X})$$

for $A, X \in \mathcal{D}$ (in other words, $\widetilde{\cdot} : \mathcal{D} \longrightarrow \mathcal{C}^0$ is an $\mathcal{S}$-functor). Thus, we consider the following problem.

PROBLEM. *Under what condition is the map*

$$\widetilde{\cdot} : S^{\mathcal{D}}\mathcal{D}(A, X) \longrightarrow S\mathcal{C}^0(\widetilde{A}, \widetilde{X})$$

a weak equivalence in $\mathcal{S}$*?*

The π_0-part of the map $\widetilde{\cdot}$ is just the obvious map

$$[A, X]_{\mathcal{D}} := \mathcal{D}(A, X)/\simeq_{\mathcal{D}} \longrightarrow \mathcal{C}^0(\widetilde{A}, \widetilde{X})/\simeq_{\mathcal{C}^0} = [\widetilde{A}, \widetilde{X}]_{\mathcal{C}^0}.$$

Hence, the above problem is a higher homotopical version of the smoothing problem for maps.

REMARK 4.1.

(1) The map $\widetilde{\cdot}$ need not be a weak equivalence (see [**3**, Examples 3.12 and 3.20]).
(2) Similarly, we consider (higher homotopical versions of) smoothing problems for sections, principal bundles, and gauge transformations (see Sections 5-7).

5. Smoothing of continuous maps

In this section, we discuss the smoothing problem of continuous maps.

$\mathcal{S}$-category structure and model structure on $\mathcal{M}$. We are mainly concerned with function complexes $S^{\mathcal{D}}\mathcal{D}(A, X)$ and $S\mathcal{C}^0(A, X)$, and their relation, which come from $\mathcal{S}$-category structures on $\mathcal{D}$ and $\mathcal{C}^0$ (see Section 4). On the other hand, since $\mathcal{D}$ and $\mathcal{C}^0$ have model structures, homotopy function complexes $\mathrm{map}_{\mathcal{D}}(A, X)$ and $\mathrm{map}_{\mathcal{C}^0}(A, X)$ are defined and model category theory provides tools to investigate their relation. Thus, we must study the relation between function complexes and homotopy function complexes for the categories $\mathcal{D}$ and $\mathcal{C}^0$ (see Figure 5.1).

$$\begin{array}{ccccc} & \mathcal{D} & & \mathcal{C}^0 \\ \text{simplicial} & S^{\mathcal{D}}\mathcal{D}(A, X) & \longleftrightarrow & S\mathcal{C}^0(A, X) \\ & \updownarrow & & \updownarrow \\ \text{model} & \mathrm{map}_{\mathcal{D}}(A, X) & \longleftrightarrow & \mathrm{map}_{\mathcal{C}^0}(A, X) \end{array}$$

FIGURE 5.1

Since $\mathcal{C}^0$ is a simplicial model category, the following result follows from [**12**, Proposition 16.1.3] (see [**15**, the discussion leading up to Proposition 4.7]).

LEMMA 5.1. *Let A be a cofibrant arc-generated space and X an arc-generated space. Then, the equality*

$$\mathrm{map}_{\mathcal{C}^0}(A, X) = S\mathcal{C}^0(A, X)$$

holds.

On the other hand, $\mathcal{D}$ is not a simplicial model category, but we can prove the following result (see [**15**, Section 4.3]).

PROPOSITION 5.2. *Let A be a cofibrant diffeological space and X a diffeological space. Then, there exists a homotopy equivalence*

$$\mathrm{map}_{\mathcal{D}}(A, X) \simeq S^{\mathcal{D}}\mathcal{D}(A, X).$$

REMARK 5.3.

(1) The significant difference that $\mathcal{C}^0$ is a simplicial model category but $\mathcal{D}$ is not is due to the fact that $|\ | : \mathcal{S} \longrightarrow \mathcal{C}^0$ preserves finite products but $|\ |_{\mathcal{D}} : \mathcal{S} \longrightarrow \mathcal{D}$ does not.
(2) It seems to be an inevitable feature of homotopical algebra of smooth spaces that one must deal with a category of smooth spaces which is not a simplicial model category (see [**15**, Remark 4.5]).

Classes $\mathcal{W}_{\mathcal{D}}$ and $\mathcal{V}_{\mathcal{D}}$. To describe a sufficient condition under which

$$\widetilde{\cdot} : S^{\mathcal{D}}\mathcal{D}(X, Y) \hookrightarrow S\mathcal{C}^0(\widetilde{X}, \widetilde{Y})$$

is a weak equivalence in $\mathcal{S}$, we introduce and study the following subclasses $\mathcal{W}_{\mathcal{D}}$ and $\mathcal{V}_{\mathcal{D}}$ of $\mathcal{D}$.

The subclass $\mathcal{W}_{\mathcal{D}}$ is defined by

$$\mathcal{W}_{\mathcal{D}} = \{A \in \mathcal{D} \mid A \text{ has the } \mathcal{D}\text{-homotopy type of a cofibrant diffeological space}\}.$$

This class is obviously a diffeological version of Milnor's class $\mathcal{W}$ of topological spaces that have the homotopy type of a CW-complex [**21**].

The subclass $\mathcal{V}_{\mathcal{D}}$ is defined by

$$\mathcal{V}_{\mathcal{D}} = \left\{A \in \mathcal{D} \mid id : A \longrightarrow R\widetilde{A} \text{ is a weak equivalence in } \mathcal{D}\right\};$$

where the map $id : A \longrightarrow R\widetilde{A}$ is the unit of the adjoint pair $(\widetilde{\cdot}, R)$, which is set-theoretically the identity map (see Section 2).

Parts 1 and 2 of the following proposition follow from Theorem 3.10. Noting that $S^{\mathcal{D}}R = S$, we can easily prove Part 3.

PROPOSITION 5.4 ([**15**, Corollary 1.6]).

(1) $A \in \mathcal{W}_{\mathcal{D}} \Rightarrow \widetilde{A}$ *has the homotopy type of a CW complex.*
(2) $A \in \mathcal{W}_{\mathcal{D}} \Rightarrow A \in \mathcal{V}_{\mathcal{D}}$.
(3) *The following conditions are equivalent:*
    (i) $A \in \mathcal{V}_{\mathcal{D}}$.
    (ii) $S^{\mathcal{D}}A \hookrightarrow S\widetilde{A}$ *is a weak equivalence in* $\mathcal{S}$.
    (iii) $\pi_*^{\mathcal{D}}(A, a) \underset{\cong}{\longrightarrow} \pi_*(\widetilde{A}, a)$ *for any* $a \in A$.

Smoothing theorem for maps. We can use Lemma 5.1 and Proposition 5.2 to establish the following theorem.

THEOREM 5.5 ([**15**, Theorem 1.7]). *If A is in $\mathcal{W}_{\mathcal{D}}$ and X is in $\mathcal{V}_{\mathcal{D}}$, then the natural inclusion*

$$S^{\mathcal{D}}\mathcal{D}(A, X) \hookrightarrow S\mathcal{C}^0(\widetilde{A}, \widetilde{X})$$

is a weak equivalence in $\mathcal{S}$.

PROOF. We may assume that A is a cofibrant diffeological space. Note that the left Quillen functor $\widetilde{\cdot} : \mathcal{D} \longrightarrow \mathcal{C}^0$ defines the map

$$\widetilde{\cdot} : \mathrm{map}_{\mathcal{D}}(A, X) \longrightarrow \mathrm{map}_{\mathcal{C}^0}(\widetilde{A}, \widetilde{X})$$

([**12**, Proposition 16.2.1(1)]). By Lemma 5.1 and Proposition 5.2, we have only to show that this map is a weak equivalence in $\mathcal{S}$.

Since X is in $\mathcal{V}_{\mathcal{D}}$, $id_\sharp : \mathrm{map}_{\mathcal{D}}(A, X) \longrightarrow \mathrm{map}_{\mathcal{D}}(A, R\widetilde{X})$ is a weak equivalence in $\mathcal{S}$ (see [**12**, Theorem 17.6.3(1)]). Since $\widetilde{\cdot} : \mathcal{D} \rightleftarrows \mathcal{C}^0 : R$ is a Quillen pair, $\mathrm{map}_{\mathcal{D}}(A, R\widetilde{X}) \cong \mathrm{map}_{\mathcal{C}^0}(\widetilde{A}, \widetilde{X})$ holds (see [**12**, Proposition 17.4.16]). Considering the commutative diagram

$$\begin{array}{ccc} \mathrm{map}_{\mathcal{D}}(A, X) & \xrightarrow{\ \widetilde{\cdot}\ } & \mathrm{map}_{\mathcal{C}^0}(\widetilde{A}, \widetilde{X}) \\ {\scriptstyle id_\sharp}\downarrow & \nearrow & \\ \mathrm{map}_{\mathcal{D}}(A, R\widetilde{X}), & & \end{array}$$

we thus see that $\widetilde{\cdot} : \mathrm{map}_{\mathcal{D}}(A, X) \longrightarrow \mathrm{map}_{\mathcal{C}^0}(\widetilde{A}, \widetilde{X})$ is a weak equivalence in $\mathcal{S}$. □

Smoothing result for principal bundles. We use Theorem 5.5 to prove a smoothing result of continuous principal bundles. We begin by introducing the basic notions.

DEFINITION 5.6. Let X be an object of $\mathcal{M}$.

(1) A covering $\{U_i\}$ of X is $\mathcal{M}$-numerable if there exists an $\mathcal{M}$-partition of unity $\{\varphi_i : X \longrightarrow \mathbb{R}\}$ subordinate to $\{U_i\}$ (i.e., $\mathrm{supp}\, \varphi_i \subset U_i$).
(2) X is $\mathcal{M}$-paracompact if every open covering of X is $\mathcal{M}$-numerable.

DEFINITION 5.7. Let B be an object of $\mathcal{M}$ and G a group in $\mathcal{M}$.

(1) Let $\mathcal{M}G$ denote the category of right G-objects of $\mathcal{M}$ (i.e., objects of $\mathcal{M}$ endowed with a right G-action). Regard B as an object of $\mathcal{M}G$ with trivial G-action.

    An object $\pi : E \longrightarrow B$ of the overcategory $\mathcal{M}G/B$ is called a principal G-bundle if there exists an open cover $\{U_i\}$ of B such that $E|_{U_i} := \pi^{-1}(U_i) \cong U_i \times G$ in $\mathcal{M}G/U_i$.

    Let $\mathsf{P}\mathcal{M}G/B$ denote the full subcategory of $\mathcal{M}G/B$ consisting of principal G-bundles.
(2) A principal G-bundle $\pi : E \longrightarrow B$ is called $\mathcal{M}$-numerable if it has a trivialization open cover $\{U_i\}$ which is $\mathcal{M}$-numerable.

    Let $(\mathsf{P}\mathcal{M}G/B)_{\mathrm{num}}$ denote the full subcategory of $\mathsf{P}\mathcal{M}G/B$ consisting of $\mathcal{M}$-numerable principal G-bundles.

A group in $\mathcal{D}$ (resp. $\mathcal{C}^0$) is called a *diffeological group* (resp. *arc-generated group*). Since $\widetilde{\cdot} : \mathcal{D} \longrightarrow \mathcal{C}^0$ preserves finite products, we see that

$$G \text{ is a diffeological group } \Rightarrow \widetilde{G} \text{ is an arc-generated group.}$$

Further, we can prove the following.

LEMMA 5.8. *Let B be a diffeological space and G a diffeological group. Then, the functor $\widetilde{\cdot} : \mathcal{D} \longrightarrow \mathcal{C}^0$ induces a functor*

$$\mathsf{P}\mathcal{D}G/B \longrightarrow \mathsf{P}\mathcal{C}^0\widetilde{G}/\widetilde{B},$$

which restricts to a functor

$$(\mathsf{P}\mathcal{D}G/B)_{\mathrm{num}} \longrightarrow (\mathsf{P}\mathcal{C}^0\widetilde{G}/\widetilde{B})_{\mathrm{num}}.$$

REMARK 5.9. Let $\mathcal{T}$ be the category of topological spaces. Since the functor $\widetilde{\cdot} : \mathcal{D} \longrightarrow \mathcal{T}$ does not preserve finite products, the underlying topological space of a diffeological group G need not be a topological group (see [**15**, Remark 5.8]).

For an essentially small category $\mathcal{A}$, $K\mathcal{A}$ denotes the set of isomorphism classes of objects of $\mathcal{A}$. We can derive the following result from Theorem 5.5

PROPOSITION 5.10 ([**15**, Theorem 5.17]). *Let X be a diffeological space and G a diffeological group. If X is in $\mathcal{W}_{\mathcal{D}}$ and G is in $\mathcal{V}_{\mathcal{D}}$, then the functor $\widetilde{\cdot} : \mathcal{D} \longrightarrow \mathcal{C}^0$ induces the bijection*

$$K(\mathsf{P}\mathcal{D}G/X)_{\mathrm{num}} \xrightarrow[\cong]{} K(\mathsf{P}\mathcal{C}^0\widetilde{G}/\widetilde{X})_{\mathrm{num}}.$$

PROOF (sketch). Let $\pi : EG \longrightarrow BG$ denote the universal principal G-bundle constructed by Christensen and Wu. Then, we have the natural bijection

$$K(\mathsf{P}\mathcal{D}G/X)_{\mathrm{num}} \xrightarrow[\cong]{} [X, BG]_{\mathcal{D}}$$

([**4**, Theorem 5.10]).

By Lemma 5.8, $\widetilde{\pi} : \widetilde{EG} \longrightarrow \widetilde{BG}$ is a $\mathcal{C}^0$-numerable principal $\widetilde{G}$-bundle. Since EG is $\mathcal{D}$-contractible [**4**, Corollary 5.5], $\widetilde{EG}$ is contractible, which implies the universality of $\widetilde{\pi}$ by the arc-generated version of [**5**, Theorem 7.5]. Hence, we can identify the map $K(\mathsf{P}\mathcal{D}G/X)_{\mathrm{num}} \longrightarrow K(\mathsf{P}\mathcal{C}^0\widetilde{G}/\widetilde{X})_{\mathrm{num}}$ with the natural map

$$[X, BG]_{\mathcal{D}} \longrightarrow [\widetilde{X}, \widetilde{BG}]_{\mathcal{C}^0}.$$

Further, we can show that G is in $\mathcal{V}_{\mathcal{D}}$ if and only if BG is in $\mathcal{V}_{\mathcal{D}}$. Therefore, this map is bijective by Theorem 5.5. □

6. Smoothing of continuous sections

In this section, we discuss a smoothing problem of continuous sections of fiber bundles.

Quillen equivalences between the overcategories of $\mathcal{S}, \mathcal{D}$, and $\mathcal{C}^0$. We have the Quillen pairs

$$|\ |_{\mathcal{D}} : \mathcal{S} \rightleftarrows \mathcal{D} : S^{\mathcal{D}} \text{ and } \widetilde{\cdot} : \mathcal{D} \rightleftarrows \mathcal{C}^0 : R,$$

which are pairs of Quillen equivalences (Theorem 3.10). We see that these Quillen pairs induce Quillen pairs between the overcategories of $\mathcal{S}, \mathcal{D}$ and $\mathcal{C}^0$, and give the conditions under which they are pairs of Quillen equivalences.

We begin with an observation on adjoint pairs. Let $F : \mathcal{A} \rightleftarrows \mathcal{B} : G$ be an adjoint pair. Suppose that $\mathcal{A}$ has pullbacks. Then, the functor

$$F : \mathcal{A}/A \longrightarrow \mathcal{B}/FA$$

has a right adjoint for any $A \in \mathcal{A}$. In fact, the right adjoint $G' : \mathcal{B}/FA \longrightarrow \mathcal{A}/A$ is defined by assigning to an object $q : Z \longrightarrow FA$ the object $G'q : G'Z \longrightarrow A$ which is defined by the pullback diagram in $\mathcal{A}$

$$\begin{array}{ccc} G'Z & \longrightarrow & GZ \\ {\scriptstyle G'q}\downarrow & & \downarrow{\scriptstyle Gq} \\ A & \xrightarrow{i_A} & GFA, \end{array}$$

where i_A is the unit of the adjoint pair (F, G).

From this, we see that the functors

$$| \ |_{\mathcal{D}} : \mathcal{S}/K \longrightarrow \mathcal{D}/|K|_{\mathcal{D}} \ \text{ and } \ \widetilde{\cdot} : \mathcal{D}/X \longrightarrow \mathcal{C}^0/\widetilde{X}$$

have the right adjoints $(S^{\mathcal{D}}\cdot)'$ and R' respectively.

Recall that the overcategory of a model category has a canonical model structure ([**12**, Theorem 7.6.5(2)]) and note that if $F : \mathcal{A} \rightleftarrows \mathcal{B} : G$ is a Quillen pair between model categories $\mathcal{A}$ and $\mathcal{B}$, then $F : \mathcal{A}/A \rightleftarrows \mathcal{B}/FA : G'$ is also a Quillen pair. Then, we have the Quillen pairs

$$| \ |_{\mathcal{D}} : \mathcal{S}/K \rightleftarrows \mathcal{D}/|K|_{\mathcal{D}} : (S^{\mathcal{D}}\cdot)' \ \text{and} \ \widetilde{\cdot} : \mathcal{D}/X \rightleftarrows \mathcal{C}^0/\widetilde{X} : R'.$$

Further we have the following result, which generalizes Theorem 3.10.

PROPOSITION 6.1 ([**15**, Proposition 6.1]). (1) *$| \ |_{\mathcal{D}} : \mathcal{S}/K \rightleftarrows \mathcal{D}/|K|_{\mathcal{D}} : (S^{\mathcal{D}}\cdot)'$ is a pair of Quillen equivalences for any simplicial set K.*

(2) *The Quillen pair*

$$\widetilde{\cdot} : \mathcal{D}/X \rightleftarrows \mathcal{C}^0/\widetilde{X} : R'$$

is a pair of Quillen equivalences if and only if X is in $\mathcal{V}_{\mathcal{D}}$.

Enrichment of overcategories. To further investigate the categories $\mathcal{C}^0/X$ and $\mathcal{D}/X$, we have to consider the enrichment of overcategories of cartesian closed categories.

Let $\mathcal{V}$ be a cartesian closed category with pullbacks. For objects $p : E \longrightarrow X$, $p' : E' \longrightarrow X$ of $\mathcal{V}/X$, define the object $\mathcal{V}/X(E, E')$ of $\mathcal{V}$ by the pullback diagram in $\mathcal{V}$

$$\begin{array}{ccc} \mathcal{V}/X(E,E') & \longrightarrow & \mathcal{V}(E,E') \\ \big\downarrow & & \big\downarrow {\scriptstyle p'_\sharp} \\ * & \longrightarrow & \mathcal{V}(E,X), \end{array}$$

where the lower horizontal arrow is the map from the terminal object $*$ to $\mathcal{V}(E, X)$ corresponding to p. The object $\mathcal{V}/X(X, E)$ is often denoted by $\Gamma(X, E)$. The underlying set of $\mathcal{V}/X(E, E')$, defined to be the set $\mathcal{V}(*, \mathcal{V}/X(E, E'))$, is just the hom-set $\mathcal{V}/X(E, E')$ of the category $\mathcal{V}/X$.

We can see that $\mathcal{M}/X$ is an $\mathcal{M}$-category with hom-object $\mathcal{M}/X(E, E')$, and hence an $\mathcal{S}$-category with function complex $S^{\mathcal{M}}\mathcal{M}/X(E, E')$. Further, the underlying topological space functor $\widetilde{\cdot} : \mathcal{D} \longrightarrow \mathcal{C}^0$ induces the $\mathcal{S}$-functor $\widetilde{\cdot} : \mathcal{D}/X \longrightarrow \mathcal{C}^0/\widetilde{X}$. Hence, we have the natural map between function complexes

$$S^{\mathcal{D}}\mathcal{D}/X(D, E) \longrightarrow S\mathcal{C}^0/\widetilde{X}(\widetilde{D}, \widetilde{E}).$$

$\mathcal{S}$-category structure and model structure on $\mathcal{M}/X$. We can show that $\mathcal{C}^0/X$ is a simplicial model category. Thus, we have the following result by [**12**, Proposition 16.1.3].

LEMMA 6.2. *Let $f : D \longrightarrow X$ and $p : E \longrightarrow X$ be objects of $\mathcal{C}^0/X$. Suppose that D is cofibrant in $\mathcal{C}^0$ and that p is a fibration in $\mathcal{C}^0$. Then, the equality*

$$\mathrm{map}_{\mathcal{C}^0/X}(D, E) = S\mathcal{C}^0/X(D, E)$$

holds. In particular, for a cofibrant object X and a fibration $p : E \longrightarrow X$ in $\mathcal{C}^0$, the equality

$$\mathrm{map}_{\mathcal{C}^0/X}(X, E) = S\Gamma(X, E)$$

holds.

On the other hand, $\mathcal{D}/X$ turns out not to be a simplicial model category. Thus, we must impose much stronger assumptions on X, D, and E to establish a diffeological version of Lemma 6.2. A locally trivial F-bundle is called a $\mathcal{D}$-*numerable* if it admits a $\mathcal{D}$-numerable trivialization open cover.

PROPOSITION 6.3. *Let K be a simplicial set and $p : E \longrightarrow |K|_{\mathcal{D}}$ be a $\mathcal{D}$-numerable F-bundle in $\mathcal{D}$. Then, there exists a homotopy equivalence*

$$\mathrm{map}_{\mathcal{D}/|K|_{\mathcal{D}}}(|K|_{\mathcal{D}}, E) \simeq S^{\mathcal{D}}\Gamma(|K|_{\mathcal{D}}, E).$$

Using these results, we can prove the following smoothing theorem of continuous sections, generalizing Theorem 5.5.

THEOREM 6.4 ([**15**, Theorem 1.8]). *Let $p : E \longrightarrow X$ be a $\mathcal{D}$-numerable F-bundle in $\mathcal{D}$. If X is in $\mathcal{W}_{\mathcal{D}}$ and F is in $\mathcal{V}_{\mathcal{D}}$, then the natural inclusion*

$$S^{\mathcal{D}}\Gamma(X, E) \hookrightarrow S\Gamma(\widetilde{X}, \widetilde{E})$$

is a weak equivalence in $\mathcal{S}$.

7. Dwyer-Kan equivalence between $(\mathsf{P}\mathcal{D}G/X)_{\mathrm{num}}$ and $(\mathsf{P}\mathcal{C}^0\widetilde{G}/\widetilde{X})_{\mathrm{num}}$

In this section, we explain a smoothing result of gauge transformations and unify the smoothing results for principal bundles and gauge transformations using the notion of a Dwyer-Kan equivalence.

For this, we first observe that every category endowed with a faithful functor into $\mathcal{M}$ is enriched over $\mathcal{M}$, and then that the category $\mathsf{P}\mathcal{M}G/X$ of principal bundles has an enriched groupoid structure.

Enrichment of categories embedded into $\mathcal{M}$. For $X \in \mathcal{M}$ and a group G in $\mathcal{M}$, consider the following commutative diagram consisting of faithful functors:

$$\begin{array}{ccccc} & & & \mathcal{M}G & \\ & & \nearrow & & \searrow \\ \mathsf{P}\mathcal{M}G/X & \longrightarrow & \mathcal{M}G/X & & \mathcal{M} \\ & & \searrow & & \nearrow \\ & & & \mathcal{M}/X & \end{array}$$

Since $\mathcal{M}$ is cartesian closed, $\mathcal{M}$ itself is an $\mathcal{M}$-category, and hence the hom-set $\mathcal{M}(A, B)$ has the canonical diffeology/arc-generated topology. Since hom-sets of the other categories have the natural injections into hom-sets of $\mathcal{M}$, we endow them with the initial structures for the injections. Then, all the categories in the above diagram are $\mathcal{M}$-categories and all the functors are $\mathcal{M}$-functors. The $\mathcal{M}$-category structure on $\mathcal{M}/X$ introduced here coincides with that introduced in Section 6.

Since every $\mathcal{M}$-category is also an $\mathcal{S}$-category via the singular functor $S^{\mathcal{M}} : \mathcal{M} \longrightarrow \mathcal{S}$, we can regard the above diagram as a commutative diagram of $\mathcal{S}$-categories.

Enriched groupoid $\mathsf{P}\mathcal{M}G/X$. We introduce the notion of a $\mathcal{V}$-groupoid (i.e., a groupoid enriched over a cartesian closed category $\mathcal{V}$), and then show that the category $\mathsf{P}\mathcal{M}G/X$ is an $\mathcal{M}$-groupoid.

DEFINITION 7.1. Let $\mathcal{V}$ be a cartesian closed category. A $\mathcal{V}$-category $\mathcal{A}$ is called a $\mathcal{V}$-groupoid if for any pair $A, A' \in \mathcal{A}$, there exists a morphism of $\mathcal{V}$

$$\cdot^{-1} : \mathcal{A}(A, A') \longrightarrow \mathcal{A}(A', A)$$

such that the following diagram commutes:

$$\begin{array}{ccccc}
\mathcal{A}(A', A) \times \mathcal{A}(A, A') & \xleftarrow{(\cdot^{-1}, 1_{\mathcal{A}(A,A')})} & \mathcal{A}(A, A') & \xrightarrow{(1_{\mathcal{A}(A,A')}, \cdot^{-1})} & \mathcal{A}(A, A') \times \mathcal{A}(A', A) \\
{\scriptstyle \text{composition}} \downarrow & & \downarrow & & \downarrow {\scriptstyle \text{composition}} \\
\mathcal{A}(A', A') & \xleftarrow[\text{unit}]{} & * & \xrightarrow[\text{unit}]{} & \mathcal{A}(A, A).
\end{array}$$

Define the automorphism group $\mathrm{Aut}_{\mathcal{A}}(A)$ of $A \in \mathcal{A}$ by

$$\mathrm{Aut}_{\mathcal{A}}(A) = \mathcal{A}(A, A).$$

$\mathrm{Aut}_{\mathcal{A}}(A)$ is obviously a group in $\mathcal{V}$.

LEMMA 7.2. *The $\mathcal{M}$-category $\mathsf{P}\mathcal{M}G/X$ is an $\mathcal{M}$-groupoid, and hence an $\mathcal{S}$-groupoid via the singular functor $S^{\mathcal{M}} : \mathcal{M} \longrightarrow \mathcal{S}$.*

Let $\pi : P \longrightarrow X$ be a principal G-bundle in $\mathcal{M}$. Then, the *group* $\mathrm{Gau}_{\mathcal{M}}(P)$ *of gauge transformations* is defined by

$$\mathrm{Gau}_{\mathcal{M}}(P) = \mathrm{Aut}_{\mathsf{P}\mathcal{M}G/X}(P).$$

By the definition, $\mathrm{Gau}_{\mathcal{M}}(P)$ is a group in $\mathcal{M}$, and

$$\begin{aligned}\mathrm{Gau}_{\mathcal{M}}(P) &= \mathsf{P}\mathcal{M}G/X(P, P) \\ &= \mathcal{M}G/X(P, P)\end{aligned}$$

hold. When $\mathsf{P}\mathcal{M}G/X$ is regarded as an $\mathcal{S}$-groupoid, the automorphism group of P in $\mathsf{P}\mathcal{M}G/X$ is just $S^{\mathcal{M}}\,\mathrm{Gau}_{\mathcal{M}}(P)$.

Smoothing result for gauge transformations. We formulate the gauge group $\mathrm{Gau}_{\mathcal{M}}(P)$ as a space of sections and apply the smoothing theorem for sections to establish a smoothing result of gauge transformations.

Let $\pi : P \longrightarrow X$ be a principal G-bundle in $\mathcal{M}$. Then, the (non-principal) G-bundle $P[G, \mathrm{conj}] \longrightarrow X$ is defined by

$$\begin{aligned}P[G, \mathrm{conj}] &= P \times_G (G, \mathrm{conj}) \\ &= P \times G/(ug, h) \sim (u, ghg^{-1}).\end{aligned}$$

The following is an enriched version of a known result (see [**18**, 37.17]).

LEMMA 7.3. *Let $\pi : P \longrightarrow X$ be a principal G-bundle in $\mathcal{M}$.*

(1) *$P[G, \mathrm{conj}] \longrightarrow X$ is a group in $\mathcal{M}/X$. Thus, $\Gamma(X, P[G, \mathrm{conj}])$ is a group in $\mathcal{M}$.*
(2) *$\mathrm{Gau}_{\mathcal{M}}(P)$ is isomorphic to $\Gamma(X, P[G, \mathrm{conj}])$ as a group in $\mathcal{M}$.*

Applying Theorem 6.4, we thus have

PROPOSITION 7.4 ([**15**, Proposition 7.6]). *Let $\pi : P \longrightarrow X$ be a $\mathcal{D}$-numerable principal G-bundle in $\mathcal{D}$. If X is in $\mathcal{W}_{\mathcal{D}}$ and G is in $\mathcal{V}_{\mathcal{D}}$, then the natural inclusion*

$$S^{\mathcal{D}}\,\mathrm{Gau}_{\mathcal{D}}(P) \hookrightarrow S\,\mathrm{Gau}_{\mathcal{C}^0}(\widetilde{P})$$

is a weak equivalence.

Dwyer-Kan equivalence. We can use the notion of a Dwyer-Kan equivalence [**1**, p. 2044] to unify the smoothing results for principal bundles and gauge transformations (Propositions 5.10 and 7.4).

THEOREM 7.5 ([**15**, Theorem 1.9]). *Let X be a diffeological space and G a diffeological group. If X is in $\mathcal{W}_{\mathcal{D}}$ and G is in $\mathcal{V}_{\mathcal{D}}$, then the functor*

$$\widetilde{\cdot} : (\mathsf{P}\mathcal{D}G/X)_{\mathrm{num}} \longrightarrow (\mathsf{P}\mathcal{C}^0\widetilde{G}/\widetilde{X})_{\mathrm{num}}$$

is a Dwyer-Kan equivalence of simplicial groupoids.

8. Applications to infinite-dimensional C^∞-manifolds

Frölicher, Kriegl, and Michor [**18**] established the foundation of infinite-dimensional calculus, which is called convenient calculus and is regarded as a prime candidate for the final theory of infinite-dimensional calculus.

In this section, we recall the notion of a C^∞-manifold (in convenient calculus) from [**18**] and observe that the category C^∞ of C^∞-manifolds can be fully faithfully embedded into the category $\mathcal{D}$. Then, we present smooth homotopical results, explaining the following:

- Most of the important C^∞-manifolds studied in [**18**] are in $\mathcal{W}_{\mathcal{D}}$.
- Every C^∞-manifold is in $\mathcal{V}_{\mathcal{D}}$.

Thus, the smoothing theorems in the preceding sections apply to C^∞-manifolds.

Convenient calculus. We make a brief review of convenient calculus.

Local calculus. Let E be a locally convex vector space. A set-theoretic curve $c : \mathbb{R} \longrightarrow E$ is called *smooth* if all derivatives exist and are continuous - this is a concept without problems. The final topology for the set of smooth curves into E is called the *c^∞-topology* of E, which is finer than the original locally convex topology; for a metrizable space, the c^∞-topology coincides with the original locally convex topology ([**18**, Theorem 4.11(1)]).

A set-theoretic map $f : U \longrightarrow V$ between c^∞-open sets of locally convex vector spaces are defined to be *smooth* or C^∞ if f maps smooth curves in U to smooth curves in V. In the finite-dimensional case, this gives the usual notion of smooth mappings ([**18**, Corollary 3.14]). Smooth maps are obviously continuous for the c^∞-topologies; however, a smooth map need not be continuous for the original locally convex topologies ([**18**, Corollary 2.11]).

The notion of a C^∞-manifold. In convenient calculus, C^∞-manifolds are defined by gluing c^∞-open sets of convenient vector spaces, which are locally convex vector spaces satisfying a weak completeness condition ([**18**, Theorem 2.14]).

Smooth maps between C^∞-manifolds M and N are defined using charts in the obvious manner. We can easily see that a set-theoretic map $f : M \longrightarrow N$ is smooth if and only if f preserves smooth curves.

The underlying topological space $\widetilde{M}$ of a C^∞-manifold M is defined to be the set M endowed with the final topology for the smooth curves ([**18**, 27.4]). In this paper, for simplicity, we require that the underlying topological spaces of C^∞-manifolds are Hausdorff. (In [**15**], we imposed a slightly weaker separation condition on the underlying topological spaces of C^∞-manifolds; see [**15**, Section 2.2].)

C^∞-manifolds as diffeological spaces. The category C^∞ of C^∞-manifolds can be regarded as a full subcategory of the category $\mathcal{D}$ of diffeological spaces. In fact, the fully faithful embedding

$$I : C^\infty \hookrightarrow \mathcal{D}$$

is defined to assign to a C^∞-manifold M the set M endowed with the diffeology

$$D_{IM} = \{C^\infty\text{-parametrizations of } M\}$$

([**15**, Section 2.2]). The diffeological space IM is often denoted simply by M if there is no confusion in context.

Remark 8.1.

(1) We can also observe the following:
- $I : C^\infty \hookrightarrow \mathcal{D}$ preserves finite products. Hence, Lie groups (i.e., groups in C^∞) embed fully faithfully into the category of diffeological groups.
- $I : C^\infty \hookrightarrow \mathcal{D}$ is compatible with the underlying topological space functors. In other words, the following diagram commutes:

$$\begin{array}{ccc} C^\infty & \hookrightarrow & \mathcal{D} \\ & \tilde{}\searrow \quad \swarrow\tilde{} & \\ & \mathcal{C}^0 & \end{array}$$

(see [**15**, Proposition 2.6]).

(2) It was shown by Losik that Fréchet manifolds embed fully faithfully into $\mathcal{D}$. However, if we work with convenient calculus, we can easily see that all C^∞-manifolds embed fully faithfully into $\mathcal{D}$ as above (cf. [**15**, Remark 2.7(2)]).

Smooth homotopy of C^∞-manifolds. As mentioned above, we explain that most of the important C^∞-manifolds studied in [**18**] are in $\mathcal{W}_\mathcal{D}$, and that every C^∞-manifold is in $\mathcal{V}_\mathcal{D}$.

A key to providing a sufficient condition for a C^∞-manifold to be in $\mathcal{W}_\mathcal{D}$ is the following result, which is a diffeological version of a theorem of tom Dieck [**30**, Theorem 4].

Theorem 8.2 ([**15**, Theorem 1.10]). *Let X be a diffeological space and $U = \{U_\alpha\}_{\alpha\in A}$ a $\mathcal{D}$-numerable covering of X by subsets. If U_σ $(:= \bigcap_{\alpha\in\sigma} U_\alpha)$ is in $\mathcal{W}_\mathcal{D}$ for any nonempty finite $\sigma \subset A$, then X is in $\mathcal{W}_\mathcal{D}$.*

We need the following notions on C^∞-manifolds to describe a sufficient condition for a C^∞-manifold to be in $\mathcal{W}_\mathcal{D}$.

Definition 8.3.

(1) A C^∞-manifold M is called *hereditarily C^∞-paracompact* if any open set U of M is C^∞-paracompact.

(2) A C^∞-manifold M is called *semiclassical* if M admits an atlas $\{(U_\alpha, u_\alpha)\}_{\alpha\in A}$ such that $u_\alpha(U_\alpha)$ and $u_\alpha(U_\alpha \cap U_\beta)$ are open in the model vector space E_α with respect to the locally convex topology for any $\alpha, \beta \in A$.

Semiclassicality condition enables us to use local convexity in local arguments. The following remark relates to semiclassicality condition.

REMARK 8.4. The sets $u_\alpha(U_\alpha)$ and $u_\alpha(U_\alpha \cap U_\beta)$ are c^∞-open sets of the model vector space E_α, but they need not be open sets with respect to the original locally convex topology of E_α. Nevertheless, most of the important C^∞-manifolds satisfy semiclassicality condition by the following two facts:

- If E is a Fréchet space or a Silva space, then the c^∞-topology of E is just the original locally convex topology.
- C^∞-manifolds in a classical infinite-dimensional calculus can be regarded as C^∞-manifolds in convenient calculus under the same underlying set and atlas.

The following corollary can be easily derived from Theorem 8.2.

COROLLARY 8.5 ([**15**, Theorem 11.1]). *Every hereditarily C^∞-paracompact semiclassical C^∞-manifold is in $\mathcal{W}_\mathcal{D}$.*

About half a century ago, Palais [**25**], Heisey [**11**], and Milnor [**21**] studied sufficient conditions for an infinite-dimensional topological manifold to have the homotopy type of a CW-complex. The above corollary and the following examples can be regarded as smooth refinements of their results.

EXAMPLE 8.6.

(1) Let M be a paracompact C^∞-manifold whose model vector spaces are
  (i) Hilbert spaces (especially, finite-dimensional vector spaces),
  (ii) nuclear Fréchet spaces, or
  (iii) nuclear Silva spaces.
  Then, M is hereditarily C^∞-paracompact and semiclassical, and hence is in $\mathcal{W}_\mathcal{D}$.
(2) Let M, N be second countable finite-dimensional C^∞-manifolds. Then, the following C^∞-manifolds are hereditarily C^∞-paracompact and semiclassical, and hence are in $\mathcal{W}_\mathcal{D}$:
  - $\mathfrak{C}^\infty(M,N)$
  - $\mathrm{Diff}(M)$, $\mathrm{Emb}(M,N)$
  - $B(M,N) = \mathrm{Emb}(M,N)/\mathrm{Diff}(M)$.

  (see [**18**, Theorems 42.1, 43.1, and 44.1] for these C^∞-manifolds.)

Next, we explain that every C^∞-manifold is in $\mathcal{V}_\mathcal{D}$. We can show the following general result.

THEOREM 8.7 ([**15**, Theorem 1.11]). *Every locally contractible diffeological space is in $\mathcal{V}_\mathcal{D}$.*

In convenient calculus, it cannot be guaranteed that each point of a C^∞-manifold has a neighborhood that is diffeomorphic to a locally convex open set of the model vector space, but every C^∞-manifold is locally contractible as a diffeological space by [**18**, Lemma 4.17]. Thus, the following result is an immediate consequence of Theorem 8.7.

COROLLARY 8.8 ([**15**, Theorem 11.2]). *Every C^∞-manifold is in $\mathcal{V}_\mathcal{D}$.*

Smoothing results for C^∞-manifolds. Using the results in this section along with the smoothing results in the preceding sections, we have the following smoothing results for C^∞-manifolds. Regard the categories $C^\infty, C^\infty/M$, and $\mathsf{P}C^\infty G/M$ as $\mathcal{D}$-categories (and hence $\mathcal{S}$-categories) via the faithful embeddings into $\mathcal{D}$.

THEOREM 8.9 ([**15**, Theorem 1.1]). *Let M and N be C^∞-manifolds. If M is hereditarily C^∞-paracompact and semiclassical, then the natural inclusion*

$$S^{\mathcal{D}}C^\infty(M,N) \hookrightarrow SC^0(\widetilde{M},\widetilde{N})$$

is a weak equivalence of simplicial sets.

THEOREM 8.10 ([**15**, Theorem 1.2]). *Let $p : E \longrightarrow M$ be a smooth fiber bundle of C^∞-manifolds. If M is hereditarily C^∞-paracompact and semiclassical, then the natural inclusion*

$$S^{\mathcal{D}}\Gamma(M,E) \hookrightarrow S\Gamma(\widetilde{M},\widetilde{E})$$

is a weak equivalence of simplicial sets.

THEOREM 8.11 ([**15**, Theorem 1.3]). *Let M be a C^∞-manifold and G a Lie group. If M is hereditarily C^∞-paracompact and semiclassical, then the functor*

$$\tilde{\cdot} : \mathsf{P}C^\infty G/M \longrightarrow \mathsf{P}C^0\widetilde{G}/\widetilde{M}$$

is a Dwyer-Kan equivalence of simplicial groupoids.

REMARK 8.12. The π_0-parts of Theorems 8.9-8.11 give answers to the following problems:

(a) When do the smooth homotopy classes of smooth maps between M and N bijectively correspond to the continuous homotopy classes of continuous maps?
(b) Let $p : E \longrightarrow M$ be a smooth fiber bundle. When do the vertical smooth homotopy classes of smooth sections of E bijectively correspond to the vertical continuous homotopy classes of continuous sections?
(c) Let G be a Lie group. When do the isomorphism classes of smooth principal G-bundles over M bijectively correspond to those of continuous principal G-bundles over M? Let $\pi : P \longrightarrow M$ be a smooth principal G-bundle. When do the isotopy classes of smooth gauge transformations of P bijectively correspond to those of continuous gauge transformations of P?

These problems had been addressed by several authors. The most basic result is the Steenrod approximation theorem [**29**, Section 6.7], which states that if all C^∞-manifolds involved are finite-dimensional, then the correspondences in Problems (a) and (b) are always bijective.

In more general settings, Problems (a), (b), and (c) were addressed in [**19**], [**32**], and [**22, 31**], respectively. Roughly speaking, as answers to these questions, it had been shown that the correspondences in the questions are bijective, provided that M is finite-dimensional. Precisely, Müller and Wockel [**22, 31, 32**] did not work in convenient calculus but in Keller's C^∞_c-theory; moreover, to prove the smoothing result of continuous gauge transformations, Wockel imposed even compactness condition on M along with additional conditions on G [**31**, Proposition 1.20]. However, no essential answer had been known in the case where M is infinite-dimensional since the existing approaches are essentially based on the finite dimensionality (or local compactness) of M (cf. [**32**, Section 1] and [**22**, Introduction]).

We end this section with the following remark on the classes $\mathcal{W}_{\mathcal{D}}$ and $\mathcal{V}_{\mathcal{D}}$.

REMARK 8.13. We have introduced the classes $\mathcal{W}_{\mathcal{D}}$ and $\mathcal{V}_{\mathcal{D}}$ to establish smoothing results for diffeological spaces and C^∞-manifolds. However, these classes also play an important role in other aspects of smooth homotopy theory.

Oh and Tanaka showed that if G is the diffeomorphism group of a compact C^∞-manifold or the group of Liouville automorphisms of a Liouville sector, then $\pi_*^{\mathcal{D}}(G)$ and $\pi_*(\widetilde{G})$ coincide (i.e., such a G is in $\mathcal{V}_{\mathcal{D}}$); they used this fact to construct homotopy-coherent group actions of G [**24**], proving a conjecture of Teleman from the 2014 ICM in the Liouville and monotone settings [**23**].

The classes $\mathcal{W}_{\mathcal{D}}$ and $\mathcal{V}_{\mathcal{D}}$ also appear in an essential way in de Rham theorem for diffeological spaces. More precisely, we establish the de Rham theorem for diffeological spaces using smooth topological conditions such as $\mathcal{D}$-paracompactness and (weak) local contractibility. However, since de Rham cohomology is a smooth homotopy invariant [**13**, 6.88], we would like to provide smooth homotopical conditions under which de Rham theorem holds. Such smooth homotopical conditions are given by using the classes $\mathcal{W}_{\mathcal{D}}$ and $\mathcal{V}_{\mathcal{D}}$ [**17**].

References

[1] J. E. Bergner, *A model category structure on the category of simplicial categories*, Trans. Amer. Math. Soc. **359** (2007), no. 5, 2043–2058, DOI 10.1090/S0002-9947-06-03987-0. MR2276611

[2] J. D. Christensen, G. Sinnamon, and E. Wu, *The D-topology for diffeological spaces*, Pacific J. Math. **272** (2014), no. 1, 87–110, DOI 10.2140/pjm.2014.272.87. MR3270173

[3] J. D. Christensen and E. Wu, *The homotopy theory of diffeological spaces*, New York J. Math. **20** (2014), 1269–1303. MR3312059

[4] J. D. Christensen and E. Wu, *Smooth classifying spaces*, Israel J. Math. **241** (2021), no. 2, 911–954, DOI 10.1007/s11856-021-2120-6. MR4242549

[5] A. Dold, *Partitions of unity in the theory of fibrations*, Ann. of Math. (2) **78** (1963), 223–255, DOI 10.2307/1970341. MR155330

[6] W. G. Dwyer and J. Spaliński, *Homotopy theories and model categories*, Handbook of algebraic topology, North-Holland, Amsterdam, 1995, pp. 73–126, DOI 10.1016/B978-044481779-2/50003-1. MR1361887

[7] A. Frölicher and A. Kriegl, *Linear spaces and differentiation theory*, Pure and Applied Mathematics (New York), John Wiley & Sons, Ltd., Chichester, 1988. A Wiley-Interscience Publication. MR961256

[8] P. G. Goerss and J. F. Jardine, *Simplicial homotopy theory*, Progress in Mathematics, vol. 174, Birkhäuser Verlag, Basel, 1999, DOI 10.1007/978-3-0348-8707-6. MR1711612

[9] T. Haraguchi and K. Shimakawa, *A model structure on the category of diffeological spaces*. Preprint, `arXiv:1311.5668`, 2013.

[10] G. Hector, *Géométrie et topologie des espaces difféologiques* (French, with English summary), Analysis and geometry in foliated manifolds (Santiago de Compostela, 1994), World Sci. Publ., River Edge, NJ, 1995, pp. 55–80. MR1414196

[11] R. E. Heisey, *Manifolds modelled on R^∞ or bounded weak-* topologies*, Trans. Amer. Math. Soc. **206** (1975), 295–312, DOI 10.2307/1997158. MR397768

[12] P. S. Hirschhorn, *Model categories and their localizations*, Mathematical Surveys and Monographs, vol. 99, American Mathematical Society, Providence, RI, 2003, DOI 10.1090/surv/099. MR1944041

[13] P. Iglesias-Zemmour, *Diffeology*, Mathematical Surveys and Monographs, vol. 185, American Mathematical Society, Providence, RI, 2013, DOI 10.1090/surv/185. MR3025051

[14] H. Kihara, *Model category of diffeological spaces*, J. Homotopy Relat. Struct. **14** (2019), no. 1, 51–90, DOI 10.1007/s40062-018-0209-3. MR3913971

[15] H. Kihara, *Smooth Homotopy of Infinite-Dimensional C^∞-Manifolds*, vol. 289, Memoirs of the AMS, Amer. Math. Soc., Providence, RI, 2023, pp. 1–129.

[16] H. Kihara, *Smooth singular complexes and diffeological principal bundles*, to appear in Algebr. Geom. Topol., Preprint, `arXiv:2202.00131`, 2022.

[17] H. Kihara, *De Rham calculus on diffeological spaces*, in preparation.

[18] A. Kriegl and P. W. Michor, *The convenient setting of global analysis*, Mathematical Surveys and Monographs, vol. 53, American Mathematical Society, Providence, RI, 1997, DOI 10.1090/surv/053. MR1471480

[19] A. Kriegl and P. W. Michor, *Smooth and continuous homotopies into convenient manifolds agree*, unpublished preprint, 2002.

[20] J. P. May, *Simplicial objects in algebraic topology*, Chicago Lectures in Mathematics, University of Chicago Press, Chicago, IL, 1992. Reprint of the 1967 original. MR1206474

[21] J. Milnor, *On spaces having the homotopy type of a* CW*-complex*, Trans. Amer. Math. Soc. **90** (1959), 272–280, DOI 10.2307/1993204. MR100267

[22] C. Müller and C. Wockel, *Equivalences of smooth and continuous principal bundles with infinite-dimensional structure group*, Adv. Geom. **9** (2009), no. 4, 605–626, DOI 10.1515/ADVGEOM.2009.032. MR2574141

[23] Y.-G. Oh and H. L. Tanaka, *Continuous and coherent actions on wrapped Fukaya categories*, Preprint, arXiv:1911.00349, 2019.

[24] Y.-G. Oh and H. L. Tanaka, *Smooth constructions of homotopy-coherent actions*, Algebr. Geom. Topol. **22** (2022), no. 3, 1177–1216, DOI 10.2140/agt.2022.22.1177. MR4474782

[25] R. S. Palais, *Homotopy theory of infinite dimensional manifolds*, Topology **5** (1966), 1–16, DOI 10.1016/0040-9383(66)90002-4. MR189028

[26] D. Pavlov, *Projective model structures on diffeological spaces and smooth sets and the smooth Oka principle*, Preprint, arXiv:2210.12845, 2022.

[27] J.-M. Souriau, *Groupes différentiels* (French), Differential geometrical methods in mathematical physics (Proc. Conf., Aix-en-Provence/Salamanca, 1979), Lecture Notes in Math., vol. 836, Springer, Berlin, 1980, pp. 91–128. MR607688

[28] J.-M. Souriau, *Groupes différentiels et physique mathématique. Group Theoretical Methods in Physics*. Springer, Berlin, Heidelberg, 1984. 511-513.

[29] N. Steenrod, *The topology of fibre bundles*, Princeton Landmarks in Mathematics, Princeton University Press, Princeton, NJ, 1999. Reprint of the 1957 edition; Princeton Paperbacks. MR1688579

[30] T. tom Dieck, *Partitions of unity in homotopy theory*, Compositio Math. **23** (1971), 159–167. MR293625

[31] C. Wockel, *Lie group structures on symmetry groups of principal bundles*, J. Funct. Anal. **251** (2007), no. 1, 254–288, DOI 10.1016/j.jfa.2007.05.016. MR2353707

[32] C. Wockel, *A generalization of Steenrod's approximation theorem*, Arch. Math. (Brno) **45** (2009), no. 2, 95–104. MR2591666

Center for Mathematical Sciences, University of Aizu, Tsuruga, Ikki-machi, Aizu-Wakamatsu City, Fukushima, 965-8580, Japan

Current address: Center for Mathematical Sciences, University of Aizu

Email address: kihara@u-aizu.ac.jp

Selected Published Titles in This Series

794 **Jean-Pierre Magnot, Editor,** Recent Advances in Diffeologies and Their Applications, 2024

790 **Peter Crooks and Alexandru I. Suciu, Editors,** Compactifications, Configurations, and Cohomology, 2023

789 **I. S. Krasil′shchik, A. B. Sossinsky, and A. M. Verbovetsky, Editors,** The Diverse World of PDEs, 2023

788 **I. S. Krasil′shchik, A. B. Sossinsky, and A. M. Verbovetsky, Editors,** The Diverse World of PDEs, 2023

787 **Hemen Dutta, Editor,** Mathematical Modelling, 2023

786 **Hemen Dutta, Editor,** Mathematical Modelling, 2023

785 **A. Leroy and S. K. Jain, Editors,** Algebra and Coding Theory, 2023

784 **Dinh-Liem Nguyen, Loc Hoang Nguyen, and Thi-Phong Nguyen, Editors,** Advances in Inverse Problems for Partial Differential Equations, 2023

783 **Alla Detinko, Michael Kapovich, Alex Kontorovich, Peter Sarnak, and Richard Schwartz, Editors,** Computational Aspects of Discrete Subgroups of Lie Groups, 2023

782 **Galina Filipuk, Alberto Lastra, and Sławomir Michalik, Editors,** Recent Trends in Formal and Analytic Solutions of Diff. Equations, 2023

781 **Emanuel Indrei, Diego Marcon, and Levon Nurbekyan, Editors,** Geometric and Functional Inequalities and Recent Topics in Nonlinear PDEs, 2023

780 **Erik Koelink, Stefan Kolb, Nicolai Reshetikhin, and Bart Vlaar, Editors,** Hypergeometry, Integrability and Lie Theory, 2022

779 **Samuele Anni, Valentijn Karemaker, and Elisa Lorenzo García, Editors,** Arithmetic, Geometry, Cryptography, and Coding Theory 2021, 2022

778 **Carlos Galindo, Alejandro Melle Hernández, Julio José Moyano-Fernández, and Wilson A. Zúñiga-Galindo, Editors,** p-Adic Analysis, Arithmetic and Singularities, 2022

777 **Bang-Yen Chen, Nicholas D. Brubaker, Takashi Sakai, Bogdan D. Suceavă, Makiko Sumi Tanaka, Hiroshi Tamaru, and Mihaela B. Vajiac, Editors,** Differential Geometry and Global Analysis, 2022

776 **Aaron Wootton, S. Allen Broughton, and Jennifer Paulhus, Editors,** Automorphisms of Riemann Surfaces, Subgroups of Mapping Class Groups and Related Topics, 2022

775 **Fernando Galaz-García, Cecilia González-Tokman, and Juan Carlos Pardo Millán, Editors,** Mexican Mathematicians in the World, 2021

774 **Randall J. Swift, Alan Krinik, Jennifer M. Switkes, and Jason H. Park, Editors,** Stochastic Processes and Functional Analysis, 2021

773 **Nicholas R. Baeth, Thiago H. Freitas, Graham J. Leuschke, and Victor H. Jorge Pérez, Editors,** Commutative Algebra, 2021

772 **Anatoly M. Vershik, Victor M. Buchstaber, and Andrey V. Malyutin, Editors,** Topology, Geometry, and Dynamics, 2021

771 **Nicolás Andruskiewitsch, Gongxiang Liu, Susan Montgomery, and Yinhuo Zhang, Editors,** Hopf Algebras, Tensor Categories and Related Topics, 2021

770 **Stéphane Ballet, Gaetan Bisson, and Irene Bouw, Editors,** Arithmetic, Geometry, Cryptography and Coding Theory, 2021

769 **Kiyoshi Igusa, Alex Martsinkovsky, and Gordana Todorov, Editors,** Representations of Algebras, Geometry and Physics, 2021

768 **Dražen Adamović, Andrej Dujella, Antun Milas, and Pavle Pandžić, Editors,** Lie Groups, Number Theory, and Vertex Algebras, 2021

For a complete list of titles in this series, visit the AMS Bookstore at **www.ams.org/bookstore/conmseries/**.